Analytical Calorimetry

Analytical Calorimetry

Proceedings of the American Chemical Society Symposium on Analytical Calorimetry, San Francisco, California, April 2-5, 1968

Edited by

Roger S. Porter

Head, Polymer Science and Engineering Program
University of Massachusetts
Amherst, Massachusetts

and

Julian F. Johnson

Department of Chemistry
University of Connecticut
Storrs, Connecticut

PLENUM PRESS · NEW YORK · 1968

ISBN 978-1-4757-0003-9 ISBN 978-1-4757-0001-5 (eBook)
DOI 10.1007/978-1-4757-0001-5

Library of Congress Catalog Card Number 68-8862

Softcover reprint of the hardcover 1st edition 1968
A Division of Plenum Publishing Corporation
227 West 17 Street, New York, N. Y. 10011

Preface

Techniques of thermal measurement for analytical purposes have been in use for many years. Importantly, tremendous improvements have been made in instrumentation in recent years and the increased need for characterization results have led to significant advances and to an increased interest in thermal methods. This volume reflects the recent activity in the field. It contains all the papers from a symposium on Analytical Calorimetry held in San Francisco, California, April 2-5, 1968, at the 155th National Meeting of the American Chemical Society. The symposium was sponsored jointly by the Divisions of Analytical Chemistry and Polymer Chemistry.

Covering advanced areas over a broad range of thermal methods, the contributions include instrumental techniques and a critical review of requirements. Illustrations of analytical calorimetry range from organic to inorganic substances and extend from low molecular weight compounds to high polymers. The collection of papers on organic polymers is particularly comprehensive and current. In general it is hoped that this work will facilitate research in analytical calorimetry. We wish to extend our thanks to all the authors, the participants in the symposium, and to the Perkin-Elmer Corporation for their financial support.

June 20, 1968

Roger S. Porter
Julian F. Johnson

Contents

CALORIMETRIC STUDIES OF THE FUSION OF LINEAR POLYETHYLENE

by

J. F. Jackson and L. Mandelkern

Department of Chemistry and Institute of Molecular Biophysics

Florida State University, Tallahassee, Florida

Recent advances in instrumentation, as exemplified by the Perkin-Elmer Differential Scanning Calorimeter,[1][2] have enabled calorimetric studies to be made on molecular weight fractions of crystalline polymers since only small amounts of sample are required. Recently, the extremely important influence of molecular weight in governing all aspects of the crystallization behavior of polymers and in allowing for a critical interpretation of morphology, including the interfacial structure, has been pointed out.[3][4][5] As a prelude to discussing the fusion properties of linear polyethylene, as observed calorimetrically, and their dependence on molecular weight, morphology, and mode of crystallization it is appropriate to examine certain of the operational aspects of the instrument. This involves the absolute calibration of the instrument itself and an assessment of the effect of polymer mass, heating rate and scale factor on the measured enthalpy of fusion with particular concern for the base line delineation and the detection of the onset of melting. For the calibration, the enthalpy of fusion of indium is taken as 6.8 cal./g. Detailed studies have shown that the area under the fusion curve is in direct proportion to the indium mass and, for the DSC-1B, is inversely proportional to the degree of attenuation. Gray[6] has shown that using the above value of ΔH^* for indium, the corresponding values obtained for tin, lead, zinc, and a variety of organic compounds agree almost exactly with the values quoted in the literature as obtained by conventional calorimetry. This is also quantitatively shown by the work of Yoncoskie[7] in the present symposium.

In examining the behavior of linear polyethylene, samples crystallized in a variety of ways to give a diversity of morphologies[8] have been chosen

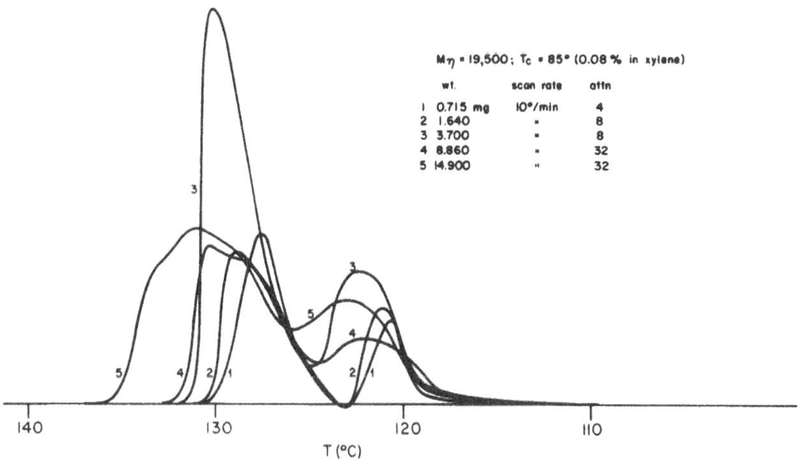

FIG. 1 Fusion curves for molecular weight fraction ($M\eta$ = 19, 500) polyethylene crystals formed from dilute xylene solutions. Mcal. / sec. vs. temperature. Data summarized in Table I.

for study; the major distinction made is between samples crystallized from the pure melt and from dilute solution. As has been previously reported[9] high molecular weight samples, $M\eta > 1 \times 10^5$, crystallized from dilute solution, give broad but very simple fusion curves at all crystallization temperatures. There are no difficulties in base line delineation or in area evaluation in this case. For the lower molecular weights, however, multiple peaks are resolved which are indicative of partial melting and recrystallization;[9][10] in certain extreme cases definite exothermic processes are observed.[9] Examples are given in Fig. 1 of the more complex type of fusion observed with crystals formed from dilute solution. A summary of the analyses of the data is given in Table I. Here T(1), T(p), and T(2) represent the onset of melting, main peak maximum and the end of melting respectively.

Table I
$M\eta = 19,500$; $T_c = 85^\circ C$ from Xylene Solution

Curve #	mass, mg.	attenu-ation	rate °/min.	ΔH^* cal/g	T(1)	T(p)	T(2)
(1)	0.715	4	10	49.2	112.1	128.4	131.0
(2)	1.640	8	10	51.0	109.5	129.8	132.5
	2.065	8	10	49.3	107.0	129.0	133.2
	2.505	8	10	51.8	102.5	129.2	132.5
	3.000	8	10	50.7	102.0	129.5	134.0
(3)	3.700	8	10	50.6	106.0	129.2	132.5
	4.005	8	10	49.6	114.0	129.2	131.9
	4.655	8	10	51.3	110.0	128.0	131.8
(4)	8.860	32	10	50.3	109.0	130.2	133.0
(5)	14.500	32	10	49.6	110.0	133.0	136.1

The different character of the fusion curves with varying experimental conditions can be discerned. Increasing the sample mass or the heating rate yields curves which are easier to analyze. However, despite the apparent complex nature of the curves obtained the data in the Table clearly indicate that there is very little variation in ΔH^* as a function of either sample mass, up to 14.5 mg., or of heating rate. For this sample the average value of ΔH^* is 50.6 ± 0.8 cal./g. For samples showing a more pronounced exothermic peak a proper base line can also be drawn. However, as is indicated above, this difficulty can be alleviated by increasing sample mass or the heating rate. Such a situation exists in a sample kindly supplied to us by Dr. Hamada ($M\eta = 56,000$, $T_c = 83.9^\circ C$) from decalin. The exothermic peak disappears when the heating rate is increased from 5°/min to 10°/min and ΔH^* is found to be 51.1 and 50.8 respectively. These results are to be contrasted with the value of 63.0 that has been reported, for the same sample, without, however, any detailed analysis being provided.[11]

Table II
Bulk Crystallization $M\eta = 5.8 \times 10^3$; $T_c = 124^\circ C$

mass, mg.	attenuation	rate °/min.	ΔH^*cal/g	$T_m(1)$	T_p	$T_m(2)$
1.590	4	5	50.7	113	135.2	138.8
3.655	8	5	51.0	115	135.3	138.5
9.865	16	5	51.1	114	135.4	140.2
17.020	32	5	51.3	109.5	136.7	143.5

Molecular weight fractions crystallized isothermally from the pure melt and then cooled to room temperature yield fusion curves on subsequent heating which usually contain multiple peaks.[12] The areas under these curves can be easily evaluated and very good agreement is obtained among identical samples. A typical set of data for such a crystallization process is given in Table II. The average value of ΔH^* in this case is 51 $\pm$ 0.2 cal/g. There is no appreciable effect of sample mass (1.590 to 17.020 mg.) or heating rate on this latter quantity. Molecular weight fractions covering the range 1×10^4 to 3×10^6 have the same properties and about the same uncertainty, as is indicated in Table II, when crystallized in the manner described above. Superheating with increased heating rate has been previously demonstrated.[12] We can also note from the Table that an apparent superheating is obtained, at a constant heating rate, by increasing the mass of the sample. Thus, in contrast to ΔH^*, reliable absolute melting temperatures cannot be determined from these measurements. An exception to the reliability of the values of ΔH^* has been found in an unfractionated very high molecular weight sample ($M\eta = 7 \times 10^6$) crystallized very rapidly from the melt. In this case the measured ΔH^* vary from 22.0 to 27.0 cal./g. for repetitive observations and under a variety of instrumental conditions. These large deviations can be attributed to the exceptionally broad fusion curves that are observed for this system and the attendant difficulty in drawing a consistent base line.

With this background, studies have been made of the enthalpy of fusion, ΔH^*, for molecular weight fractions of linear polyethylene crystallized under a variety of conditions. In one set of experiments, crystallization was conducted at 130° in the pure melt for a sufficient length of time to insure that no further crystallization occurred at this temperature and the samples were then cooled to room temperature. The measured ΔH^* ranged from 66 cal./g. in the lower molecular weight range to 34 cal./g. for the highest molecular weight sample.[5][13] These results can be explained in terms of the crystallite morphology. Of particular importance is the relationship between the crystallite size in the chain direction and the extended chain length and, as the molecular weight increases, the development of interzonal regions connecting crystallites. In these regions the chain units are in random conformation. There is a similar variation of ΔH^* with molecular weight for the lower crystallization temperatures.[13] However, the level of the enthalpy of fusion decreases with decreasing crystallization temperatures. For example, enthalpies of fusion as low as 24 cal./g. have been directly observed for high molecular weight samples crystallized at low temperatures. Similar results have been reported by Hendus and Illers.[14] It is also found that the degree of crystallinity calculated from the enthalpy data is always less than that deduced from density, infrared,[15] or wide-angle x-ray[16] for identical samples which were cooled to room temperature. This disparity becomes greater as the density is decreased or the molecular weight increased. However, for samples which were never cooled below the elevated crystallization temperature very good agreement is calculated for the degree of crystallinity from calorimetric and dilatometric observations.[13] The aforementioned disparity can be attributed to the interfacial enthalpy contribution of the smaller sized crystallites formed on cooling. The amount of crystallinity developed on cooling becomes more profuse with increasing molecular weight.[17] An analysis of the ΔH^*-density relations indicates an extrapolated value of 69-70 cal./g. for the enthalpy of fusion of the high molecular weight perfect crystal. This value is in excellent accord with that deduced previously from studies of the melting point depression of linear polyethylene with monomeric diluents.[18]

The enthalpy of fusion of crystals formed from dilute xylene solution has also been studied in detail.[13] As has been previously indicated, the fusion is relatively broad, characterized by partial melting and recrystallization and the onset of melting is detected at relatively low temperatures ranging from 100 to 114°. Care must therefore be exercised in evaluating the data to properly account for the contribution from the low temperature region. For the molecular weight range studied, 1.9×10^3 to 1.2×10^6, ΔH^* is independent of chain length and depends only on the crystallization temperature. ΔH^* ranges from 48 cal./g. for crystallization at 60° to 54.5 cal./g. when the crystallization temperature is 90°C.[13] Within the limited range of crystallite sizes available for this system a strictly formal analysis shows that the data are consistent with either the model for a regularly folded interface or for one with an amorphous or non-crystalline overlayer. For the regularly folded interface, the deduced interfacial enthalpy of 10,500 ± 1500 cal/mole of crystalline sequence is almost an order of magnitude greater than that which has been estimated for this model.[19] For a non-regularly structured interface, the corresponding interfacial enthalpy is found to be 4500 ± 100 cal./mole and is consistent with the interfacial free energy deduced from solubility temperature measurements.[20][21] For either case, with our data, the extrapolated value of the enthalpy of fusion for the infinite size crystal, ΔH_u, is found to be 69 ± 1 cal./g. Other data confirm these results for the higher molecular weight fractions.[11] However, values as high as 64 cal./g. are reported for the molecular weights at the lower end of the scale studied.[11] As has been noted in the introductory portion identical samples yield values of 51 cal./g. in our hands, which is in agreement with the analyses just given. Similarly for an unfractionated Marlex-50 crystallized at 90°C, a value of 61 ± 2 cal./g. has been recently reported.[22] The reasons for these differences in the reported values of ΔH_u are not apparent at present. However, it should be noted that the crystallite thickness is also independent of molecular weight. As a consequence, if these higher values of ΔH^* are accepted, the extrapolated value for ΔH_u becomes the order of 85 cal./g. Such an abnormally high value evades any theoretical explanation.

Studies have also been conducted to attempt to elucidate the origin of the multiple peaks commonly observed during the fusion of bulk crystallized experiments. It was found originally[12] that when the samples are rapidly cooled to room temperature after fusion, all the melting on subsequent

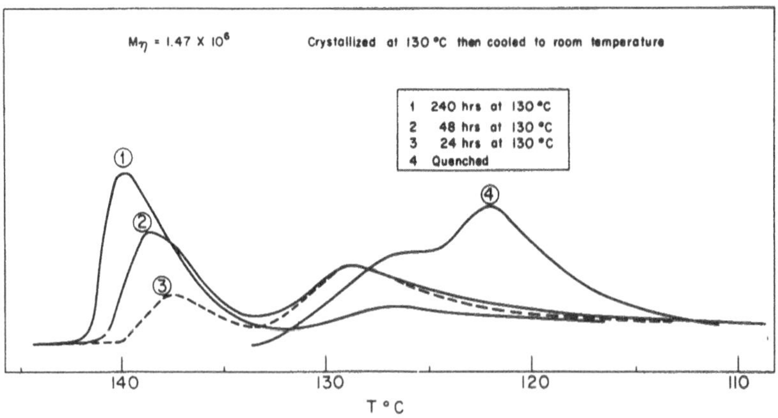

FIG. 2 Fusion curves for molecular weight fraction ($M\eta = 1.47 \times 10^6$) of bulk crystallized polyethylene for various indicated crystallization procedures. Mcal./sec. vs. temperature.

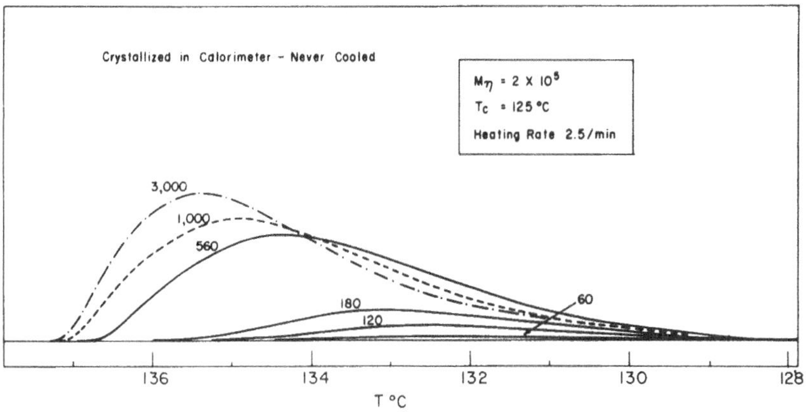

FIG. 3 Fusion curves for molecular weight fraction ($M\eta = 2 \times 10^5$) of bulk crystallized polyethylene for indicated crystallization times at 125°C. Samples never cooled below crystallization temperature. Mcal. /sec. vs. temperature.

heating takes place at a temperature corresponding to the initial low temperature peak. Furthermore, if the original sample is heated to a temperature so that only the first peak disappears and is then cooled to room temperature this peak reappears on subsequent heating. Both of these observations strongly suggest that the low temperature peak is a manifestation of the melting of crystallites formed at low temperatures on cooling after crystallization at an elevated temperature. This suggestion has been confirmed by more detailed experiments, summarized in Fig. 2, on a high molecular weight fraction ($M\eta = 1.4 \times 10^6$).(23) We can observe that the location of the melting peaks and their relative heights are very dependent on the crystallization procedure, particularly the time spent at the elevated temperature prior to cooling. A wide diversity of types of fusion curves is observed by this very simple variation in crystallization procedures. Since extended chain crystals are not formed by this high molecular weight fraction, it becomes clear that multiple peaks cannot, as has been casually proposed,(24)(25)(26) be attributed to major differences in crystallite morphology or to molecular weight segregation during the crystallization process. Rather they appear to be a consequence of the crystallite size distribution that is developed and the dependence of the thermodynamic stability on the size. Virtually similar results are obtained with unfractionated Marlex-50 and in light of the data in Fig. 2 the same explanation can be given for this case also.

Fig. 3 illustrates the results for a molecular weight fraction ($M\eta = 2 \times 10^5$) that was crystallized isothermally in the calorimeter and never cooled. The times of crystallization are indicated on each curve. As would be anticipated from the preceding discussion, it is clear that multiple peaks are not observed in this case. An increased level of crystallinity develops with time and there is a concomitant increase in the melting temperature as has been observed by other techniques.(27)

In consideration of the above results, it becomes a matter of interest to ascertain as to whether any molecular weight segregation occurs in a mixture of two fractions of widely differing molecular weights. Two fractions, $M\eta = 9 \times 10^3$ and 3.7×10^5 were homogeneously mixed in various proportions by rapid crystallization from a dilute solution.(28) The samples were rendered molten and then crystallized from the pure melt at 130.2°C for

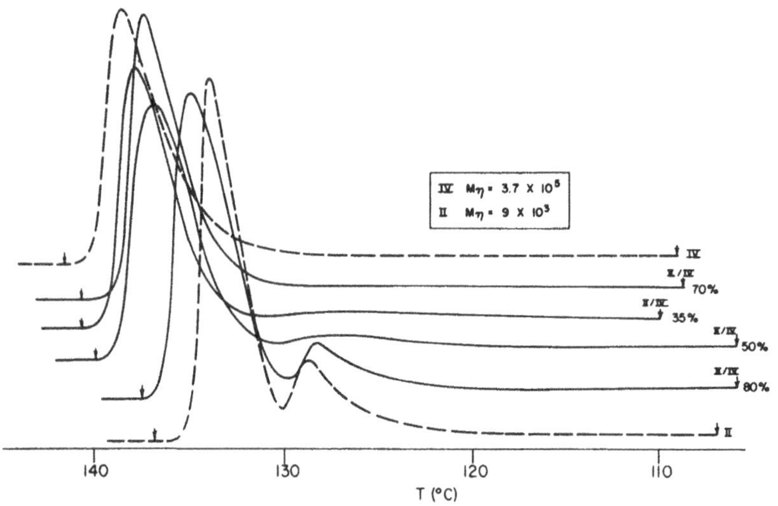

FIG. 4 Fusion curves for fractions and homogeneous mixtures of bulk crystallized polyethylene. Mcal. / sec. vs. temperature.

thirty days. The resulting fusion curves are shown in Fig. 4. It is seen that there are no indications of any multiple peaks or of the melting of two species, except for the small peak in the vicinity of 130°C which is due to the cooling process. As the composition changes from that of one pure fraction to that of the other, the peak shifts to a higher temperature and the fusion range broadens. However, over the complete composition range each mixture shows only a single peak. In contrast when two small films, each of a pure fraction, were placed in the sample pan and then crystallized in the above manner, after melting at 150°C for 20 minutes, separate crystallization does occur as is illustrated in Fig. 5. The dashed curves in this figure refer to the pure fractions and the solid curves to mixtures. It is clear that the sample has still remained heterogeneous as is reflected in the two peaks that are observed. In this special system, 10% by weight

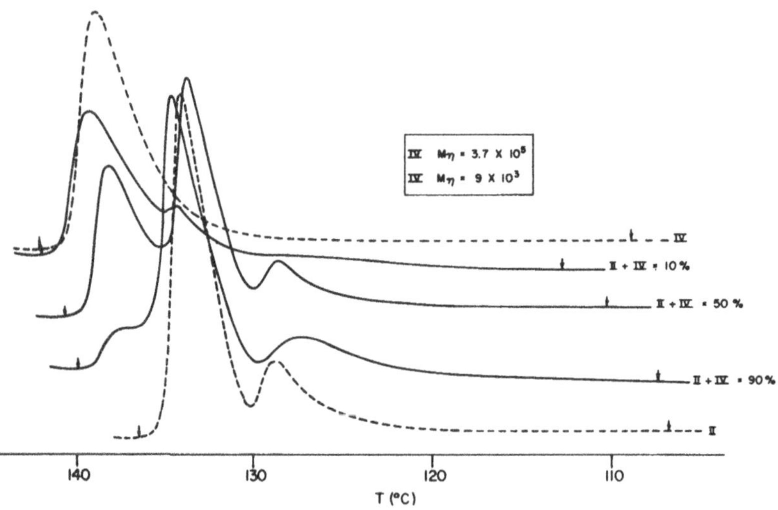

FIG. 5 Fusion curves for heterogeneous and molecular weight fractions of bulk crystallized polyethylene. Mcal. / sec. vs. temperature.

of a component which crystallizes separately can easily be detected. The contrasting features of the two sets of fusion curves in Figs. 4 and 5 for the same materials at the same compositions are self-evident. One can conclude from these results that molecular weight segregation during the crystallization from a homogeneous melt should be a rare event if it ever occurs at all.

This work was supported by Air Force Materials Laboratory for Systems Engineering Group under contract no. AF33(615)-3811.

References

(1) E.S. Watson, M.J. O'Neill, J. Justin and N. Brenner, Anal. Chem. 36, 1233 (1964).
(2) A.P. Gray and K. Casey, J. Polymer Sci. B2, 381 (1964).
(3) L. Mandelkern, J.M. Price, M. Gopalan, and J.G. Fatou, J. Polymer Sci. A2, 4, 385 (1966).
(4) L. Mandelkern, J. Polymer Sci. 15C, 129 (1966).
(5) L. Mandelkern, Poly Sci and Engineering J. 7, 232 (1967).
(6) A.P. Gray, private communication.
(7) R.A. Yoncoskie, Polymer Preprints 9, 923 April 1968 (No. 1).
(8) L. Mandelkern, Crystallization of Polymers, McGraw-Hill Book Co., New York, 1964.
(9) L. Mandelkern and A.L. Allou, Jr., J. Polymer Sci. B4, 447 (1966).
(10) E.W. Fischer and G.F. Schmidt, Angew. Chem. 1, 488 (1962).
(11) F. Hamada, B. Wunderlich, T. Sumeda, S. Hayashi and A. Nakajima, J. Phys. Chem. 72, 178 (1968).
(12) L. Mandelkern, J.G. Fatou, R. Denison and J. Justin, J. Polymer Sci. B3, 803 (1965).
(13) L. Mandelkern, A.L. Allou, Jr. and M. Gopalan, J. Phys. Chem. 72, 309 (1968).
(14) H. Hendus and K.H. Illers, Kunstoffe 57, 193 (1967).
(15) T. Okada and L. Mandelkern, J. Polymer Sci. A2, 5, 239 (1967).
(16) M. R. Gopalan and L. Mandelkern, J. Polymer Sci. B5, 925 (1967).
(17) J.G. Fatou and L. Mandelkern, J. Phys. Chem. 69, 417 (1965).
(18) F.A. Quinn, Jr. and L. Mandelkern, J. Am. Chem. Soc. 80, 3178 (1958); 81, 6533 (1959).
(19) J.D. Hoffman, Soc. Plastic Eng. 4, 315 (1964).
(20) J.B. Jackson, P.J. Flory and R. Chiang, Trans. Fara. Soc. 59, 1906 (1963).
(21) J.F. Jackson and L. Mandelkern, in preparation.
(22) H.E. Bair, T.W. Huseby and R. Salovey, Polymer Preprints 9, 795 (1968), (No. 1).
(23) L. Mandelkern and A.L. Allou, Jr., unpublished results.
(24) F.R. Anderson, J. Appl. Phys. 35, 64 (1964).
(25) P.H. Geil, F.R. Anderson, B. Wunderlich and T. Arakawa, J. Polymer Sci. A2, 3707 (1964).
(26) J.L. Kardos, E. Baer, P.H. Geil and J.L. Koenig, Kolloid-Z 204, 1 (1965).
(27) M. Gopalan and L. Mandelkern, J. Phys. Chem. 71, 3833 (1967).
(28) K. Ohno, Master of Science Thesis, Florida State University, 1968.

MEASUREMENT OF THE HEAT OF FUSION OF CRYSTALLINE POLYMERS BY A CALORIMETRIC MELTING POINT DEPRESSION TECHNIQUE

J. R. Knox
Avisun Corporation
Marcus Hook, Pennsylvania

The accurate measurement of percent crystallinity in polymers by calorimetric techniques depends upon having a reliable value for the heat of fusion. Use of adiabatic calorimeters to obtain heats of fusion, however, requires knowledge of the percent crystallinity, at the temperature from which the enthalpic base line is drawn. This temperature is generally well above the temperature at which independent measurements of the degree of crystallinity are made, and heat of fusion values obtained via this technique are no more accurate than the percent crystallinity values used in the calculation.

The Flory[1] expression of the theory of melting point depression allows the calculation of the heat of fusion independently of the level of crystallinity. The expression depends only on the molar volumes of the polymer repeat unit and the diluent, and on the interaction between polymer and diluent. These parameters are expressed as

$$1/T_m - 1/T_m^o = (R/\Delta H_u)\ (V_u/V_1)\ (v_1 - X_1 v_1^2) \qquad (1)$$

where V_u and V_1 are the molar volumes of the repeat unit and the diluent, and v_1 is the volume fraction diluent. The equilibrium melting point, T_m^o, of the polymer is difficult to determine experimentally, but may be estimated rather accurately by extrapolation to zero of a plot of $1/T_m$ vs. v_1.

The heat of fusion is obtained from the intercept at zero volume fraction diluent of a plot of $(1/T_m - 1/T_m^o)/v_1$ vs. v_1 (or v_1/T_m).

An extension of this theory to the case of random crystalline copolymers allows the calculation of the fraction of the non-crystallizable copolymer units by the amount of the melting point depression when the heat of fusion is known. If these copolymer units are incorporated within the lattice, however, the heat of fusion must be changed. When one considers defects such as stereoinversions in polypropylene or branches in polyethylene, it is possible by measurement of the heat of fusion to establish the presence or absence of these defects in the crystal lattice.

Experimental

The diluent is added by volume with a calibrated micropipet into a volatile sample pan of the Perkin-Elmer DSC-1 Scanning Calorimeter and then weighed. Three to five miligrams ($\pm$0.02 mg.) of the polymer in powder form is weighed on top of the diluent. The pan is then crimped shut and the melting temperature of the polymer is determined at a scanning rate of 5°C./min. The melting temperature is measured below the peak

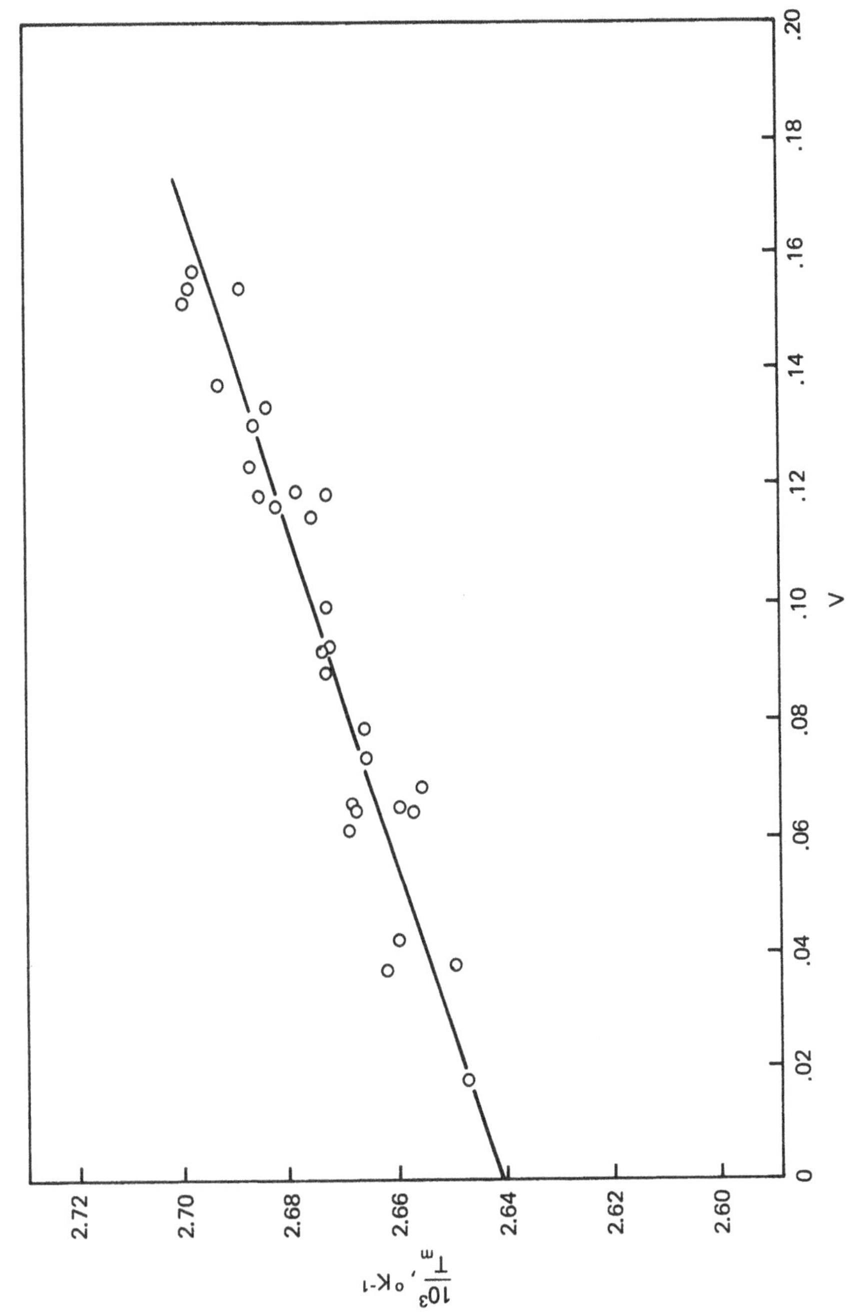

Fig. 1. Relationship of Reciprocal Melting Temperature to Volume Fraction Solvent

temperature by an amount determined by use of the instrument response line obtained from a sample of pure indium.

The polymer volume at the melting point is calculated from dilatometrically measured melt volumes and volume coefficients of expansion. The volume fraction of diluent, v_1, is calculated at the observed melting point. A plot of $1/T_m$ vs. v_1 is made and a best fit line is extrapolated to zero v_1. The value of T_m^o is obtained from the intercept. By analysis of Eq. 1 it is seen that the slope of a plot of $1/T_m$ vs. v is related to the latent heat of fusion by

$$\text{Slope} = \frac{1/T_m - 1/T_m^o}{v_1} \tag{2}$$

$$= \frac{RV_u}{\Delta H_u V_1} \tag{3}$$

To obtain a reliable value for T_m^o and the slope, it is necessary to run each concentration in triplicate. The major problem is in evaporation of the diluent, both in weighing and in the melting run if the pan is improperly crimped. A typical plot of the data is shown in Fig. 1. Least squares regression analysis of the data lead to correlation coefficients between .91 and .97 for all samples tested. The standard error in the slope measurement was about 2% relative. It is estimated that errors in the values of V_u and V_1 lead to values of the latent heat of fusion which are precise within ±5% relative.

The samples of polypropylene which have been examined by this technique were prepared using a variety of catalyst systems. Different catalyst systems were selected on the basis of their ability to polymerize propylene to polymers with different levels of crystalline perfection. The three polyethylene samples contained 3.3, 2.0, and ca. 0.5 methyls/100 methylenes. The diluent used was reagent grade cis-decalin in all cases.

Results and Discussion

Measurements of the heat of fusion on the polypropylene samples yielded the results shown in Table I.

TABLE I

Heat of Fusion of Polypropylene Samples

Sample		ΔH_f, cal/g.
I		42.7
II	Decreasing	38.7
III	Crystalline	35.6
IV	Perfection	34.2
V		33.3
VI		30.7

As the sample numbers increase, the catalyst system becomes less and less sterospecific and the shape of the polymer melting endotherm becomes broader, but the peak temperature is not significantly reduced. The fact that the measured heat of fusion decreases as the crystalline perfection decreases is consistent with reports that chain defects may be incorporated

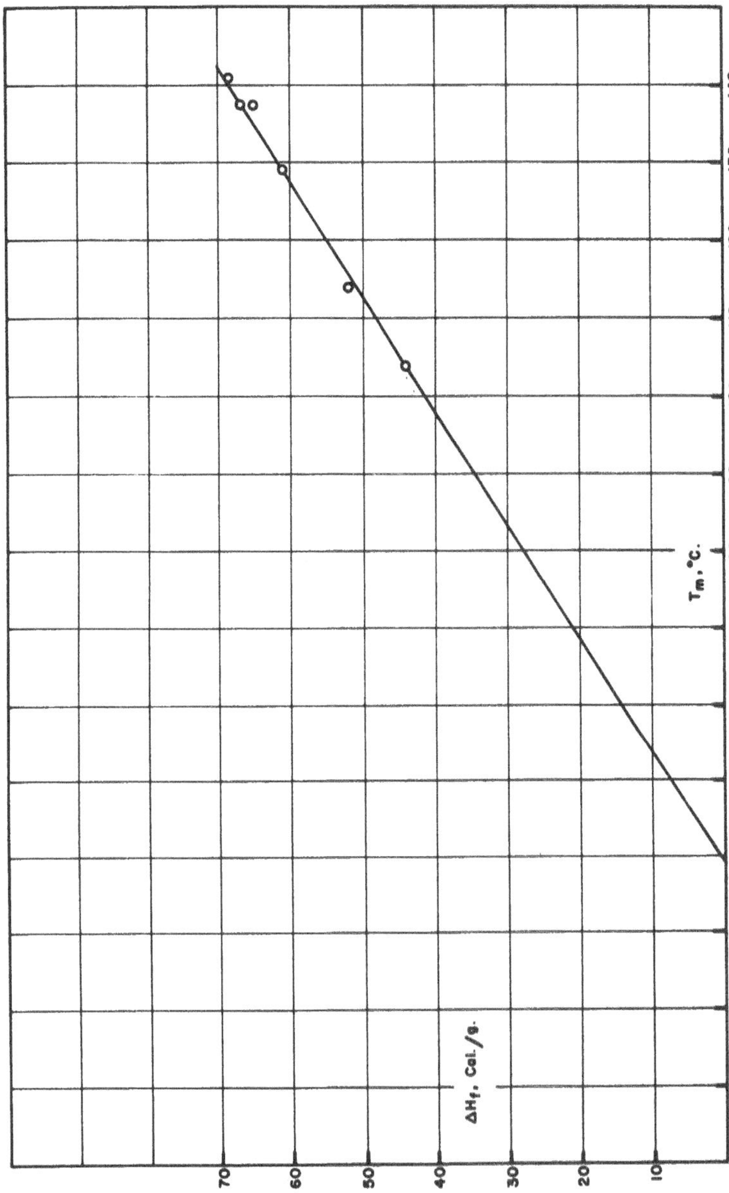

Fig. 2. Relationship of Latent Heat of Fusion of Polyethylene to the Melting Temperature

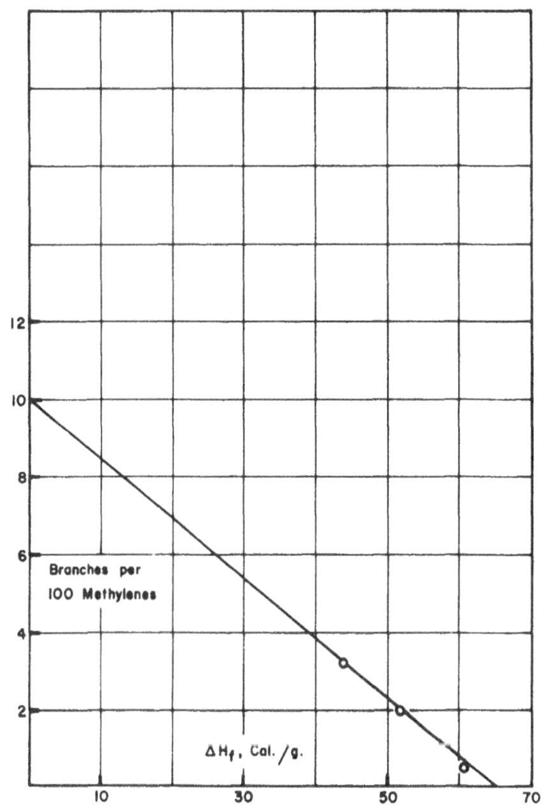

Fig. 3. Relationship of Latent Heat of Fusion of Polyethylene to the Degree of Branching

within the lamellar crystalline lattice. Our preliminary estimates from these samples suggest that an extrapolation of the values to 100% crystalline perfection will lead to a heat of fusion of ca. 46-50 cal/g.

It was considered that a lattice having a three-fold helix chain structure might accomodate a lattice defect more easily than a two-fold helix or planar zig-zag lattice such as polyethylene. Measurements of the heat of fusion of polyethylenes with different degrees of branching led to the results shown in Table II.

TABLE II

Heat of Fusion of Polyethylene

Sample	$CH_3/100CH_2$	ΔH_f, cal/g.
A	ca 0.5*	60.8
B	2.0	51.8
C	3.3	44.1

*Not measured. Calculated from the melting point[2].

The heat of fusion decreases with an increase in the branching concentration, indicating that the branches in polyethylene are probably incorporated at least in part within the crystalline structure. Plots of heat of fusion vs. melting point (Fig. 2) and heat of fusion vs. methyls/100 methylene (Fig. 3) extrapolate to a value of 65-68 cal/g. for the heat of fusion of a sample with no branches and a melting point of 141°C. This is in excellent agreement with the value of 68±2 cal/g. reported[3] on a sample with melting point of 137.5°C. By extrapolation to zero heat of fusion it is predicted that 10 branches /100 methylenes is the level beyond which crystallinity cannot exist. This is equivalent to 20 mole percent of an α-olefin copolymer.

Summary

A scanning calorimeter has been used to measure heats of fusion of crystalline polymers by the melting point depression technique. It is shown that the heat of fusion is decreased by increasing the number of geometric or chemical irregularities in the chain, indicating that these irregularities are at least in part, incorporated within the crystalline regions.

References

(1) L. Mandelkern, R. R. Garrett, and P. J. Flory, J. Am. Chem. Soc., 74, 3949 (1952).

(2) J. R. Knox, J. Polymer Sci. C18, 69 (1967).

(3) F. A. Quinn and L. Mandelkern, J. Am. Chem. Soc., 80, 3178 (1958).

CALORIMETRY OF FUMING NITRIC ACID TREATED POLYETHYLENE

by

G. Meinel, A. Peterlin and K. Sakaoku*

Camille Dreyfus Laboratory

Research Triangle Institute, Research Triangle Park, North Carolina 27709

SYNOPSIS

Polyethylene drawn at different temperatures to a draw ratio between 7 and 25 and polyethylene rolled at room temperature to a draw ratio about 3 was treated with fuming nitric acid. From the two characteristic maxima of the thermograms, crystallite length and the number of unoxidized tie molecules are calculated. In the drawn sample the crystallite length is about 50A smaller than the long period found by small angle X-rays. This is a consequence of the non-crystalline layer between the lamellae which is first eaten away by the acid. The number of tie molecules is practically constant in samples drawn between 14 and 110°C, but increases nearly linearly with draw ratio λ up to $\lambda = 20$ and stays constant at higher draw ratios. The so obtained numbers are compared to the minimum number of tie molecules needed for the explanation of the elastic modulus and the ultimate tensile strength of the same samples before oxidation.

INTRODUCTION

Fuming nitric acid (FNA) preferentially attacks the non-crystalline portions of polyethylene (PE). From the investigation of the etched debris, conclusions can be drawn about the original morphology.[1-6] In the study of the debris, calorimetry proves very valuable giving data about the polydispersity of the molecular weight of the etched sample. This data, however, is obtained only after remelting of the etched sample (second run), because the fractionation of different molecular weights, which were closely mixed together immediately after etching, takes place on recrystallization after the first melting.

The thermal investigation of single crystals and quenched films, which have been treated with FNA for 28 hrs. or longer show a rather narrow melting peak - the first run nearly coinciding with the second run - which indicates a rather narrow molecular weight distribution in the debris in agreement with the results of gel permeation chromatography.[7] In drawn PE the single melting peak of the first run of the etched material separates into two maxima during the second run, indicating the presence of essentially two molecular weights.[6] The ratio of the areas under the two maxima did not change after etching times longer than 26 hrs. The interpretation of these data is based on the assumption that the acid preferentially attacks the folded chains as observed in single crystals and does not attack the strained tie molecules connecting adjacent crystal lamellae. In support of this the smaller molecular weight can indeed be correlated with the thickness of the crystalline core of a single lamella while the larger molecular weight corresponds to a length slightly greater than twice this value, which is interpreted as the thickness of the sandwich formed by two superimposed lamellae and the intervening amorphous layer. The schematic model for the drawn material after etching is shown in Fig. 1a. (The dimensions refer to a sample drawn at 60°C, draw ratio $\lambda = 10$.) Both molecular weights are thoroughly mixed in the etched sample and separate on recrystallization of the melt as shown in Fig. 1b.

The study of the morphology is here further pursued by the investigation of etched materials of rolled PE and of drawn PE as functions of draw ratio and temperature of drawing. The conclusions are compared with the resulting changes in tensile properties and density.

* Present address: Tokyo University of Agriculture and Technology, Koganei-shi, Tokyo, Japan

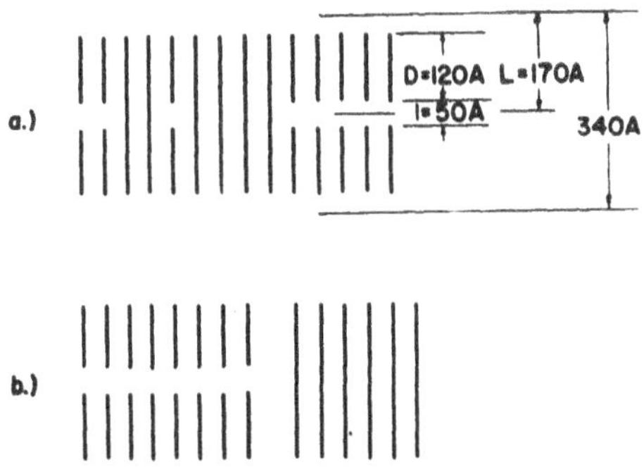

Figure 1

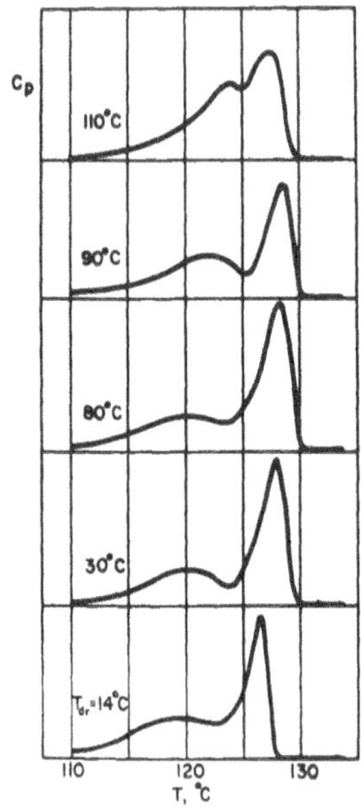

Figure 2

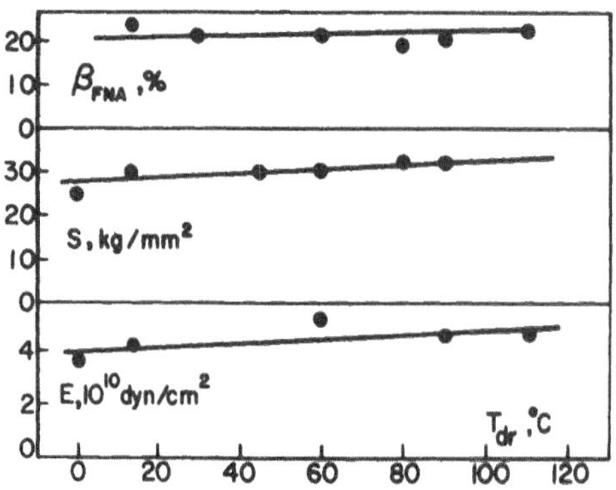

Figure 3

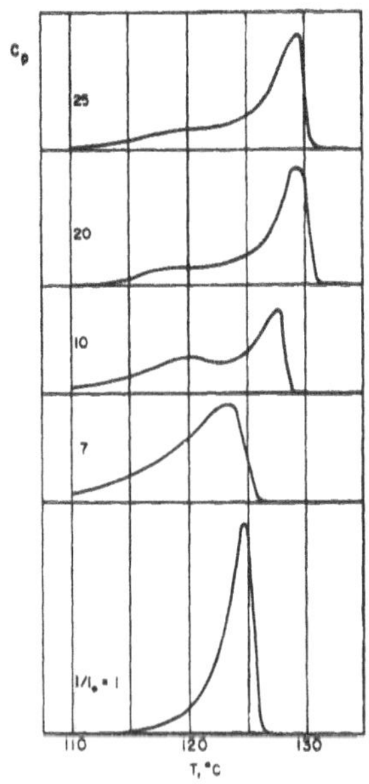

Figure 4

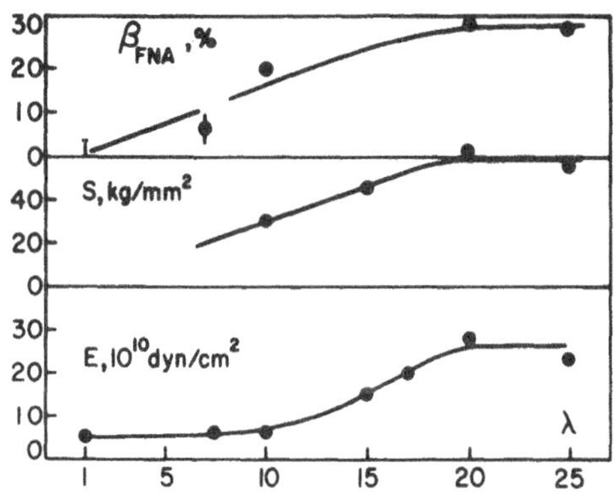

Figure 5

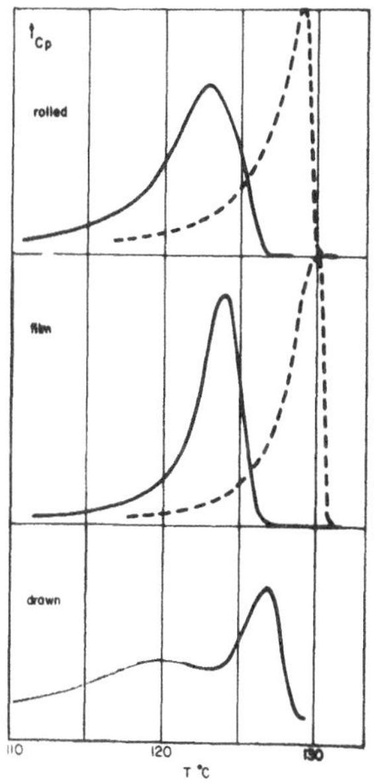

Figure 6

CAPTIONS TO FIGURES

Fig. 1 - a) Schematic diagram of drawn material after FNA treatment.
b) The same material after recrystallization from the melt (T_d = 60°C, λ = 10)

Fig. 2 - Melt curves (2nd run) of drawn samples after the FNA treatment (Draw ratio ≈ 9). Parameter: Draw temperature.

Fig. 3 - Relative number of tie molecules β_{FNA}, elastic modulus E and ultimate strength S as function of the draw temperature. Draw ratio ≈ 9

Fig. 4 - Melt curves (2nd run) of samples drawn at T_d = 60°C after FNA treatment. Parameter: Draw ratio.

Fig. 5 - Relative number of tie molecules β_{FNA}, elastic modulus E and ultimate strength S of PE drawn at T_d = 60°C as function of draw ratio.

Fig. 6 - Melt curves (2nd run) of 2 films of different long period (full line: L = 190A; broken line: L = 290A) before and after rolling (λ = 3) and drawing (λ = 9), respectively.

EXPERIMENTAL

Strips of quenched or slowly cooled films of linear PE, Fortiflex A 60/500 were drawn or rolled. The drawing was performed at a rate of 0.5 cm/min in a thermostated water bath below 90°C and in a nitrogen atmosphere bath above 90°C. Rolling was done in a rolling press at room temperature whereby the sample thickness was reduced to one-third of the original value. All samples were etched in fuming nitric acid at 80°C. After 50 hrs. of etching the material was filtered, washed in distilled water and rinsed with acetone in a Soxhlet extractor.

The thermograms were obtained by a PERKIN ELMER Differential Scanning Calorimeter (DSC 1). The samples of about 3 mg in weight were usually heated at 2.5°C/min. The heating was stopped 10°C above the point where the last crystalline traces vanish. The melted samples were cooled at 2.5°C/min. The subsequent thermogram (second run) obtained from this newly crystallized was taken at 2.5°C/min.

The elastic modulus and the tensile strength were measued with an Instron TM-M tester at room temperature. The test specimen of 2.5 cm gauge length were extended at 0.5 and at 5 cm/min for the measurement of the elastic modulus and the ultimate tensile strength, respectively.

RESULTS AND DISCUSSION

1. Influence of Draw Temperature

Fig. 2 shows the second run thermograms for samples, etched after drawing (draw ratio $\lambda = 9$) at draw temperatures between 14 and 110°C. The thermograms show the two maxima mentioned before. The low temperature maximum shifts from 119°C to 124°C with the draw temperature increasing from 14 to 110°C, the second maximum rises from 126.5°C to about 128.5°C between the draw temperature of 14 and 90°C and decreases again to about 127.5° for drawing at 110°C.

The peak at the lower temperature T_1 can be attributed to a molecular length corresponding to the thickness of the crystallite as mentioned in the introduction. This thickness increases with increasing draw temperature as indicated by the increasing melting temperature T_1 of the low molecular weight material. The exact values L_1 calculated from the melting temperature (see Ref. 6) are compared in Table I to the long periods L_x obtained from small angle X-ray scattering. The two values show a difference, which is mainly due to the preferential digestion of non-crystalline surface portions of the lamellae. The mass fraction of the non-crystalline portions, $(1-\alpha)$ can be calculated from the long periods L_1 and L_x according to

$$(1-\alpha) = \frac{(L_x-L_1)\rho_a}{L_1\,\rho_c+(L_x-L_1)\rho_a} \tag{1}$$

Hereby one assumes that after 50 hrs of acid treatment only non-crystalline material was digested.[6] This assumption is justified by the measurements of the weight loss.[3] With the density of the non-crystalline material[9] $\rho_a = 0.89\ g/cm^3$ the values of crystallinity obtained from equation (1) compare well with those obtained from density for samples drawn between 60° and 110°C. The crystallinity of samples drawn below 60°C cannot be determined from density as a consequence of the void content of the samples and the presence of the monoclinic phase.[10]

The interpretation of the thermograms in terms of the model given before[6] allows, in addition, a quantitative determination of the number b of chains of length L_1 which are oxidized at the surface of each lamella (the folded chains) and the number c of chains of length L_2 which go through two superimposed lamellae and remain unoxidized in the section between the lamellae (strained tie molecules). The area under the T_1 melting peak yields the mass $m_1 = M_1b/N$ and that under the T_2 peak the mass $m_2 = M_2c/N$, where M_1 and M_2 are the molecular weights of the short (L_1) and long (L_2) chains and N is Avogadro's number. The probability of nonfolding β_{FNA} can be expressed as

$$\beta_{FNA} = n_2/(n_1+n_2) = c/(b+2c) = 1/(2+m_1M_2/m_2M_1) \quad (2)$$

where $n_1 = b+c$ is the number of folds at the lamella surfaces and $n_2 = c$ is number of chains going through the interface and entering the next lamella. The ratio m_1/m_2 slightly depends on the cooling rate after the first melting. But the observed variation is small and does not appreicably influence the number β_{FNA}. The so obtained β_{FNA} are plotted in Fig. 3 for samples drawn between 14 and 110°C. The draw ratio $\lambda = 9$ was the same for all samples. The fraction of tie molecules turns out to be nearly independent of the draw temperature over the range investigated.

This data seems to be supported by the dependence of mechanical properties on draw temperature, if one assumes that the folds do not contribute to the longitudinal modulus E or tensile strength S. Such an assumption is corroborated by the fact that in a sandwich model the elastic modulus of an amorphous polymer above the glass transition temperature (T_g is at about -20°C with PE) has a nearly 10^3 times smaller value than the glass or the crystal.[11] In a fracture occurring in the non-crystalline region between two crystals, folded chains are not broken but only tie molecules bridging the gap between the superimposed lamellae. According to the constancy of the fraction of tie molecules obtained for samples drawn between 14 and 110°C, one would expect the elastic modulus as well as the tensile strength to be nearly constant, which is observed indeed (Fig. 3). This fraction of the molecules also seems to not change upon annealing in this temperature range, since the elastic modulus of a drawn sample (T_d-60°C, $\lambda = 10$) stays constant up to an annealing temperature of 115°C.[12]

Assuming that only tie molecules carry the load, one obtains the fraction of tie molecules β_E as the ratio between E and $E_{id} = 2.4\times10^{12}$ dyne/cm^2, where E_{id} is the c-axis elastic modulus of the ideal crystal containing fully extended polyethylene chains of infinite length (13-15). In a similar way one can calculate the number of tie molecules per unit area which are broken from the knowledge of the ultimate tensile strength S_{id} of a C-C bond,[16] which is $S_{id} = 1400$ kg/mm^2. The fraction derived from the elastic modulus as well as from the tensile strength are, however, smaller than those from the nitric acid experiment as shown in Table I. A discussion of this behavior will be given at the end of the next section.

2. Influence of Draw Ratio

The second run thermograms of the etched material with different draw ratios are shown in Fig. 4 (draw temperature 60°C). The melting point of the low molecular weight material stays practically constant, while that of the high molecular weight material increases with increasing draw ratio. This trend can be explained from the behavior of a binary mixture of different concentrations as discussed previously.[6]

The area of the low temperature maximum decreases first with increasing draw ratio as can be seen by comparing the sample of draw ratio 10 with that of draw ratio 20 in Fig. 4. It indicates, according to results discussed above, that with higher draw ratios a smaller portion of the chains folds back and a higher portion of the chains transverses the amorphous region between the lamellae. Yet there seems to be a limit to this decrease, since a sample with the highest draw ratio ($\lambda = 25$) shows a thermogram similar to that of draw ratio 20.

This behavior of the changing number of tie molecules is paralleled by that of the elastic modulus and the tensile strength (Fig. 5), showing an increase up to a draw ratio of 20 and a constant value with higher draw

TABLE I

T_d	T_1	L_1	L_X[1]	α[2]	ρ	α_ρ[3]	β_{FNA}	E	β_E[4]	S	β_S[5]
14°C	119°C	112A	165A	70%	-		23%	4.2×10^{10} dyne/cm^2	2%	30 kg/mm^2	2%
30	120	120	170	73	-		21	-	-	-	-
60	120	120	170	69	0.956g/cm^3	63	21	5.0	2	30	2
80	120	120	190	65	0.959	66	19	-	-	32	2
90	122	137	200	71	0.959	66	20	4.5	2	32	2
110	124	149	230	68	0.957	64	22	4.5	2	-	-

1 data partly taken from (8).
2 calculated according to equation (1) with $1/\rho_a = 1.124$ cm^3/g (9).
3 calculated from density, $1/\rho = 1.124$ cm^3/g.
4 β_E calculated from E/E_{id}; $E_{id} = 2.4 \times 10^{12}$ dyne/cm^2 (11)
5 β_S calculated from S/S_{id}; $S_{id} = 1400$ kg/mm^2 (14)

TABLE II

λ	β_{FNA}	E	β_E	S	β_S
7	5%	6.10^{10} dyne/cm^2	3%	-	-
10	20	6	3	30 kg/mm^2	2%
20	30	27	11	60	4
25	30	23	10	60	4

ratios. Since both elastic modulus and tensile strength are proportional to the number of tie molecules, as stated before, the mechanical data seem to support the above results at least qualitatively.

The quantitative evaluation of the data is shown in Table I, where the probability for the nonfolding molecule, β_{FNA}, β_E and β_S calculated from equation (2) and from the ratio E/E_{id} and S/S_{id}, respectively, are given as a function of the draw ratio. The quantitative agreement between the different values is rather poor. Especially the values derived from the tensile strength are too low. This could be due to an error in the tensile strength of the C-C bond. The more probable reason, however, seems to result from the inhomogeneity of deformation in which a great many chains break before the final fracture of the sample. This was shown in polyethylene terephthalate, polyethylene, nylon 6 and natural silk by Zhurkov and Tomashevsky[17] and in nylon 66 by Campbell and Peterlin[18] who measured the number and formation rate of radicals and the lifetime of samples during application of tensile stress. Therefore, the value β_s, obtained from the ultimate tensile strength, is expected to be markedly lower than β_{FNA} or β_E.

The fraction β_E derived from the elastic modulus is at least of similar magnitude as β_{FNA} obtained from the thermograms of the etched samples. The quantitative disagreement is not too serious so long as there are uncertainties concerning the exact value of the elastic modulus of the extended PE-chain and the inhomogeneity of the structure of the drawn sample which may be seen on electron micrographs of surface replicas.[19] That which matters is the fact that the number of tie molecules obtained from the thermograms of the nitric acid treated samples is not in disagreement with the data derived from mechanical properties: β increases with the draw ratio, reaches a saturation value and is independent of temperature of drawing.

3. Influence of Film Preparation on Drawing and Rolling

During the second heating of a drawn, etched sample two maxima are observed. In the rolled, etched sample, however, the thermogram is nearly the same as that before rolling. This effect is demonstrated in Fig. 6, where the melt curves of the second heating run of two etched PE films of different long periods (L_1 = 190A quenched sample, L_2 = 290A slowly cooled), are compared to the second runs of the etched, drawn and rolled materials, obtained from the two films.

The second run thermograms of the etched, drawn samples are independent of the long period of the original film. Such an independence has already been observed with the long period of drawn PE.[8] The corresponding thermograms of the rolled samples however, do not differ appreciably from those of the etched film. The maximum is shifted about 1°C to lower temperature and the widths of the melting peak is slightly increased compared to the film. The broadening of the melting peak compared to that of the film indicates a higher polydispersity of the rolled sample. This could be due to a wider distribution of molecular weight resulting from either the broadening of the original crystal thickness distribution or by the generation of an additional thickness. Small angle X-rays investigations seem to support the latter assumption, since with a draw ratio higher than 3 in the rolled samples one observes an additional reflection, corresponding to a smaller long period. The investigated sample with the draw ratio 3 is just in the transition region from the practically unchanged original long period to the new long period.

Therefore, from the difference of the thermograms one cannot draw the conclusion, that the two types of plastic deformation are basically different. The main difference is probably a consequence of the large disparity of the draw ratios: less than 3 in rolling, more than 7 in drawing. In our rolled sample the original lamellae and to some extent also the spherulites were only more or less plastically deformed but not yet broken into small blocks of folded chains as it is the case during drawing. Consequently, there was practically no chance for rearrangement of chains and folds leading to a new crystal thickness.

CONCLUSIONS

The fraction of tie molecules is independent of temperature of drawing but increases with the draw ratio λ. The increase is first nearly linear with λ and levels off beyond λ = 20. Concurrently, the longitudinal elastic modulus and ultimate tensile strength increase in nearly the same manner as the number of tie molecules. This supports the assumption that tie molecules interconnecting the stacked lamellae, are an important, maybe even the main contributor to the longitudinal mechanical strength of drawn samples and in particular of fibers.

The rolled samples, however, yield a second run melting curve which is only slightly broader but otherwise very similar to that of the original film. Specifically the thickness of the lamellae differs very little from the value before rolling. This is in striking contrast with observations on drawn samples and is probably a consequence of the small draw ratio λ of about 3 compared to λ = 9 in the drawing experiment. At such a low draw ratio the original lamellae seem not yet destroyed so that the folded chain blocks remain nearly unchanged.

ACKNOWLEDGMENT

The authors gratefully acknowledge the financial support of this investigation by the Camille and Henry Dreyfus Foundation. They thank Mr. Harry Sugg for assistance in the experimental work.

REFERENCES

1. R. P. Palmer and A. J. Cobbold, Macromol. Chem. 74, 174 (1964).
2. A. Keller and S. Sawada, Macromol. Chem. 74, 190 (1964).
3. A. Peterlin and G. Meinel, J. Polymer Sci., B3, 1059 (1965).
   A. Peterlin, G. Meinel and H. G. Olf, J. Polymer Sci., B4, 399 (1966).
4. F. H. Winslow, M. Y. Hellmann, W. Matreyek and R. Salovey, J. Polymer Sci., B5, 89 (1967).
5. K. H. Illers and H. Hendus, Koll.-Z., 218, 56 (1967).
6. G. Meinel and A. Peterlin, J. Polymer Sci., B5, 197 (1967).
   J. Polymer Sci., A-2 (in press).
7. Unpublished data.
8. R. Corneliussen and A. Peterlin, Makromol. Chem. 105, 193 (1967).
9. P. R. Swan, J. Polymer Sci., 42, 525 (1960).
10. R. Corneliussen, private communication.
11. M. Takayanagi, Mem. Fac. Eng. Kyushu University 23, 8 (1963).
12. G. Meinel and A. Peterlin, J. Polymer Sci., B5, 613 (1967).
13. I. Sakurada, Y. Nukushina and T. Ito, J. Polymer Sci., 57, 651 (1962).
14. L. R. G. Treloar, Polymer 1, 95 (1960).
15. T. Shimanouchi, M. Asahina and S. Enomoto, J. Polymer Sci., 59, 93 (1962).
16. H. Mark, Cellulose and Cellulose Derivatives (ed. E. Ott), Interscience, New York; 1943.
17. S. N. Zhurkov and E. E. Tomashevsky, Physical Basis of Yield and Fracture, Edited by A. C. Stickland, Oxford, Sept. 1966, Institute of Physics and Phys. Soc. Conference Series No. 1, p. 200.
18. D. Campbell and A. Peterlin, J. Polymer Sci., B (in press).
19. A. Peterlin and K. Sakaoku, J. Appl. Phys. 38, 4152 (1967), Kolloid-Z. 212, 51 (1966); K. Sakaoku and A. Peterlin, J. Macromol. Sci. (Phys.) B1, 103 (1967).

THE THERMODYNAMIC DISSOLUTION TEMPERATURE OF SUSPENSIONS OF POLYETHYLENE SINGLE CRYSTALS BY DTA

By
H. M. Schleinitz
Plastics Department
E. I. du Pont de Nemours & Co., Wilmington, Delaware

INTRODUCTION

Although numerous investigations have been made of the crystallization of polymers from solution (1-3) and of the melting or annealing of single crystals or agglomerate structures in the absence of solvent, very few results have been reported on the effects of heating crystals in solvents. It is the purpose of this paper to discuss the utility of differential thermal analysis of crystal suspensions and to show how the technique can be used to determine the thermodynamic dissolution temperature.

EXPERIMENTAL

The polymer used throughout this investigation was Rigidex 50, a non-fractionated sample of linear polyethylene with a number average molecular weight of 12,300, a weight average molecular weight of 104,000, and fewer than 1 side branch per 1000 backbone carbon atoms.

The solvent was 99% p-xylene, chosen as a moderate solvent for polyethylene.

Differential thermal analysis was carried out in 1 cm. diameter Dural cells in a large heating block of the same material. Thermocouples of fine iron and constantan wire were axially located in the cells.

Calibration of the apparatus demonstrated that heating rates were linear within 0.2%. Melting points in the range of interest could be determined to ± 0.3°C., and heats of fusion to ± 7%.

THEORY

The thermodynamic melting point of solvent free polymer, $T_m°$, is the temperature at which perfect, infinitely thick crystals are in equilibrium with pure molten polymer. It cannot be measured by direct experiment, but Hoffman and Weeks (4) showed that if the melting temperature of the most stable crystals, T_m, is plotted versus the crystallization temperature, T_x, extrapolation to the point where the two are equal yields $T_m°$. The distribution of crystal thickness is Gaussian with a maximum at

$$\ell_p^* = 4\sigma_e T_m°/\Delta h_f (T_m° - T_x) \quad (1)$$

where σ_e is the end surface free energy and Δh_f is the enthalpy of fusion per unit volume of crystal. The thickness of the highest melting crystals is $\zeta \ell_p^*$, where ζ increases from unity as the means of detecting the last trace of crystallinity becomes more sensitive. Rearrangement of equation (1) in terms of ζ yields

$$T_m = T_m^\circ\,(1 - 1/(2\zeta)) + T_x/(2\zeta) \qquad (2)$$

The thermodynamic dissolution temperature for a given polymer-solvent pair, T_s°, is the temperature at which an infinitely thick polymer crystal is in equilibrium with pure solvent. Since, in principle, DTA can be used to determine the dissolution temperature of a given crystal suspension, T_s, the extrapolation procedure can be used to obtain T_s°.

RESULTS

It is well known that the conditions of crystallization effect the morphology of crystalline polymers, and that these morphological differences are reflected by DTA. The phenomenon of recrystallization, annealing, when crystalline polymers are heated above the crystallization temperature is also known to complicate the thermal analysis of these materials. Consequently, it is necessary to prepare samples for DTA in a manner that ensures analogous crystal forms at all conditions and to eliminate or account for recrystallization during the run. Furthermore, it is necessary to prepare stable suspensions to avoid the possibility that settling will change the concentration of the active species.

Crystal suspensions of 0.375, 0.75, 1.0 and 1.5% w P.E. in p-xylene were prepared by slow cooling of 0.1% w solutions followed by filtration and resuspension of the crystals in fresh solvent. Suspensions of the lozenge-shaped crystals with a concentration of 0.75% w or greater did not settle noticeably.

The thermograms for the suspensions are shown in Figure 1, and the data in Table I.

TABLE I

DTA of Single Crystal Suspensions

	Conc. Wt. %	T_s °C	A x 10^3 $(mV)^2$	$D_{H_{70°C}}$ x 10^3 $cm^2/sec.$	$D_{H_{110°C}}$ x 10^3 $cm^2/sec.$
1	0.0	—	—	3.1	4.9
2	0.375	89.2	0.4	1.18	3.76
3	0.75	91.2	4.1	0.93	2.98
4	1.00	91.0	5.6	0.92	2.69
5	1.50	91.2	10.4	0.93	3.10

DTA of the pure solvent indicates a thermal diffusivity far higher than the literature value, but the thermal diffusivity of the visually stable suspensions is nearly that expected for pure solvent. The higher experimental values for solvent and solutions are probably due to heat transfer by convection. The suspensions form a tenuous agglomerate of polymer crystals that effectively prevents convection so they can be treated as homogeneous solids. Therefore, DTA can be applied to suspensions of single crystals at concentrations from 0.75% w to the limit of miscibility of polymer with solvent at the dissolution temperature.

To study recrystallization during DTA suspensions were prepared by isothermal crystallization of 0.1% w solutions at 70, 80, and 85°C., followed by slow cooling to room

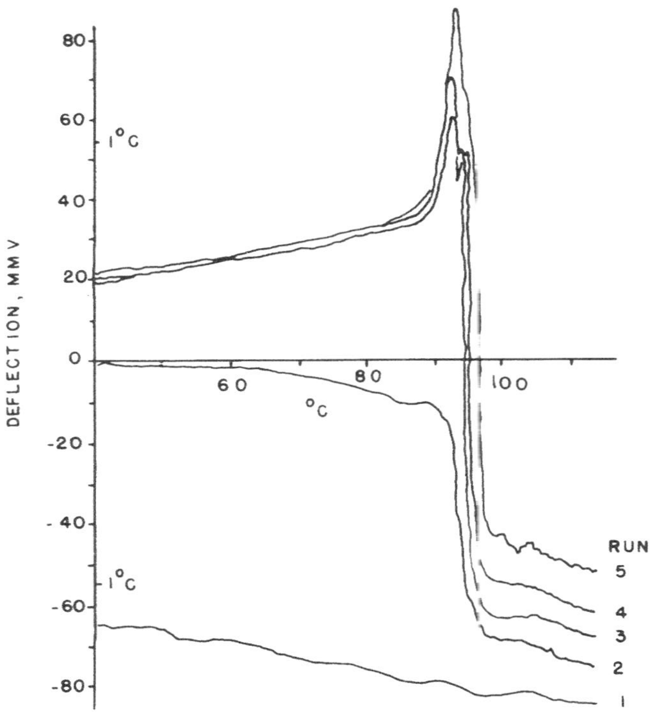

Figure 1

Thermograms of suspensions of pclyethylene single crystals at various concentrations. See Table I. Figures reprinted by permission of the editors of Polymer.

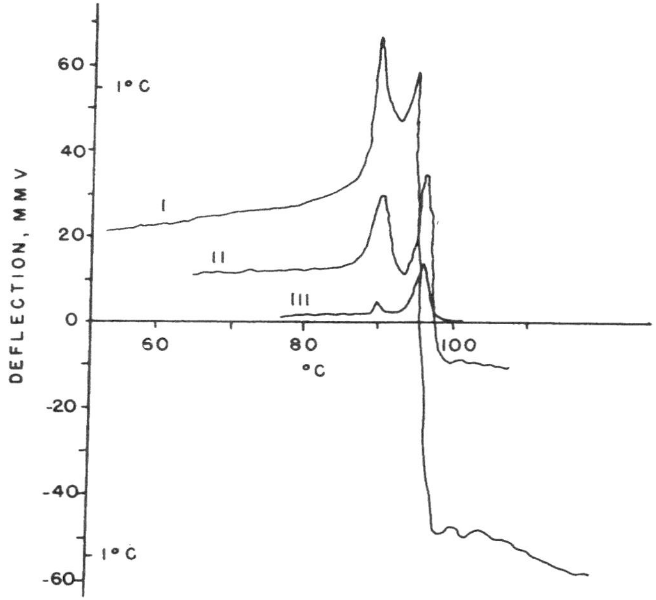

Figure 2

Thermograms of suspensions of polyethylene single crystals prepared at 70°C. at 1.5% wt. concentration. I, 2.35°C./min.; II, 0.94°C./min.; III, 0.235°C./min.

temperature and concentration by filtration to 1.5% w. Typical results are shown in Figure 2 for the crystals prepared at 70°C. Although the total peak area was proportional to the rate of heating, the relative area of the second peak increased at lower heating rates, indicating a greater amount of recrystallization.

For crystals prepared at 80°C., the second peak was not significant at a heating rate of 2.35°C./min. and the second peak was observed only at 0.235°C./min. for the crystals prepared at 85°C.

Recrystallization during DTA appears to involve the reprecipitation of dissolved polymer at a temperature between the original crystallization temperature and the dissolution temperature of the original crystals. Crystals which are not filtered at the crystallization temperature exhibit a thinner border formed during the course of cooling which redissolves at lower temperatures than the original crystals. This increases the concentration of polymer in solution at any temperature and promotes recrystallization.

The effect can be seen in the thermograms of Figure 3. The crystals used to obtain Trace II were not filtered at the crystallization temperature and had a thinner border. The peak corresponding to the recrystallized polymer was in this case completely eliminated by isolation of the crystals at the crystallization temperature. Therefore, it is necessary to separate the crystals from the mother liquor at the crystallization temperature to inhibit recrystallization during DTA.

It is desirable to extend observations to the highest possible temperatures. However, isothermal crystallization of 0.1% w solutions of polyethylene in p-xylene produced no crystals even in 1000 minutes at 90°C. so samples representative of higher temperatures were prepared by annealing. A suspension prepared at 70°C. was annealed at 90° for varying lengths of time, and the crystals were separated at the annealing temperature.

DTA of the original crystals resulted in two peaks, due to recrystallization during the run. For suspensions annealed for increasing periods of time, the relative area of the second peak increased until the first peak disappeared entirely.

Electron microscopy corroborated the results of DTA. In the earliest stages of annealing, the edges of the crystals dissolved, Figure 4. Later, a thickened border formed, Figure 5, and finally virtually all of the crystal with the original thickness dissolved, Figure 6, leaving only a "picture frame" crystal with thickness appropriate to the annealing temperature. The second peak, therefore, is characteristic of crystals prepared at the temperature of annealing.

For the determination of the thermodynamic dissolution temperature suspensions characteristic of 70, 80 and 85°C. were prepared by isothermal crystallization of 0.1% w solutions; and those characteristic of 90, 92, and 94°C., by annealing suspensions prepared at 70, 80, and 85°C., respectively. In each case the crystals were separated at

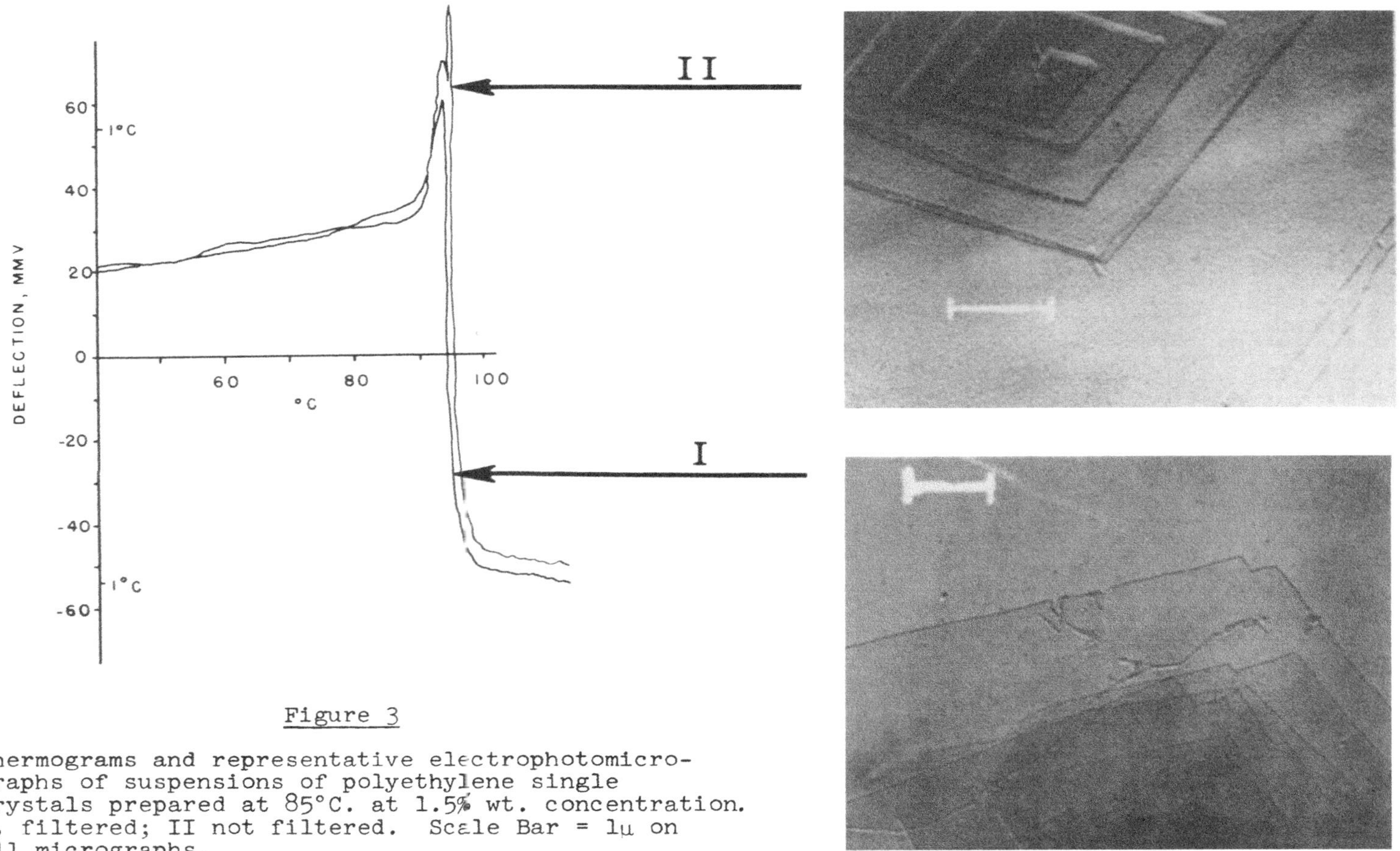

Figure 3

Thermograms and representative electrophotomicrographs of suspensions of polyethylene single crystals prepared at 85°C. at 1.5% wt. concentration. I, filtered; II not filtered. Scale Bar = 1μ on all micrographs.

the temperature of preparation and resuspended in fresh solvent to a nominal concentration of 1.5%w.

DTA was run on the suspensions at 2.35°C./min. The peak temperature was taken as the dissolution temperature, T_s, in accordance with a visual experiment. However, a small number of more nearly perfect crystals might exist beyond the temperature when most disappear. Therefore, a second measure of the dissolution temperature, T_s', was taken as the temperature at which the differential temperature falls to the level of the baseline extended.

The results are reported in Table II, with equations for the best straight lines obtained by least squares analysis. The equation for T_s neglects the data for 70°C.

TABLE II

Determination of $T_s°$, $T_s°'$

T_x or T_a °C.	T_s °C.	T_s' °C.	$T_s°$ °C.	ζ
70	89.2	—		
80	90.6	—		
85	92.6	93.1		
90	94.2	95.5		
92	95.0	95.6		
94	95.9	96.7		
Best Line:	$T_s = 60.9 + 0.372\ T_x$		96.8	1.35
:	$T_s' = 60.4 + 0.368\ T_x$		98.3	1.30

because the point lies well above the line defined by the other points, presumably because crystallization occurred above 70° while the solution cooled to that temperature. The equation for T_s' excludes the data for both 70° and 80° because recrystallization occurred during DTA, making it impossible to use the second measure of dissolution temperature.

DISCUSSION

Reference to Table I shows that in the range of concentration in which a stable suspension is formed, the observed dissolution temperature of identical single crystals is virtually independent of the concentration of the suspension. Within experimental error, the melting point of crystals in 1.5% w. suspension can be taken as the melting point at extreme dilution.

The elimination of recrystallization during DTA gives reasonable assurance that the measured melting points are accurate. The hypothesis that the annealed crystals are representative of the annealing temperature is corroborated not only by electron microscopy, but also by the fact that the melting points lie on the same straight line when T_s is plotted versus T_x or T_a. DTA is a useful tool to study the solution behavior of polymer crystals, and it can be used quantitatively to determine the extent of recrystallization.

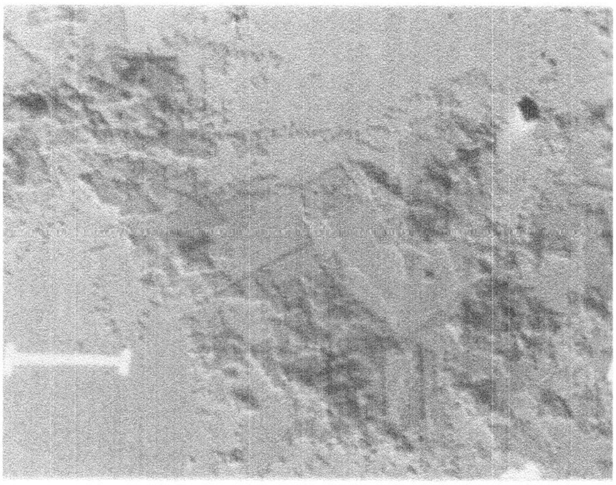

Figure 4

Crystals prepared at 70°C. and annealed at 90°C. for 6 min.

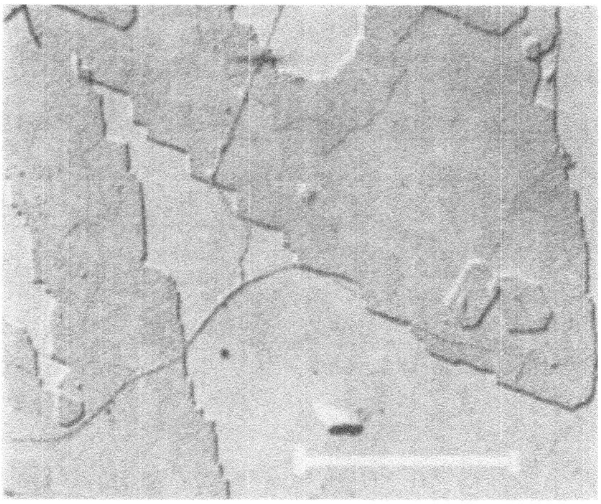

Figure 5

Crystals prepared at 70°C. and annealed at 90°C. for 40 min.

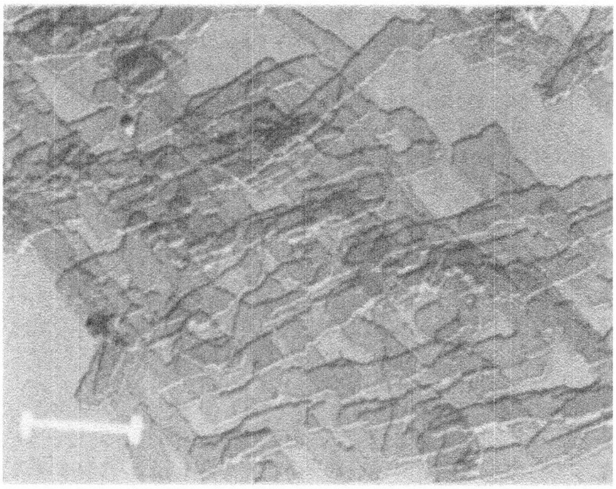

Figure 6

Crystals prepared at 70°C. and annealed at 90°C. for 1000 min.

The results for the thermodynamic dissolution temperature agree favorably with those of Coran and Anagnostopoulous (5) and of Peterlin and Meinel (6). They differ from Holland's (7) results, possibly due to solvent or molecular weight differences or possibly because the formation of a thickened border stabilizes the original crystal. The use of DTA has the advantage that the dissolution behavior of a particular species can expediently be observed without serious interference from other crystal entities.

It would be of interest to see whether this technique could be used to identify crystallizable components of polymer blends and to obtain a measure of their concentration.

ACKNOWLEDGMENT

The author wishes to thank Mr. J. Bell, Cavendish Laboratories, University of Cambridge for valuable assistance with electron microscopy and Dr. D. A. Blackadder, Department of Chemical Engineering, University of Cambridge, for many useful suggestions and discussions during the course of the work.

REFERENCES

1. P. H. Geil, Polymer Single Crystals, Interscience: London, 1963.
2. L. Mandelkern, Crystallization of Polymers, McGraw-Hill: New York, 1964.
3. P. H. Lindenmeyer, J. Polymer Sci., C, 1, 5, (1963).
4. J. D. Hoffman and J. J. Weeks, J. Res. Nat. Bur. Stand., A, 66, 13 (1962)
5. A. Y. Coran and C. E. Anagnostopoulous, J. Polymer Sci., 57, 13 (1962).
6. A. Peterlin and G. Meinel, J. Appl. Phys., 35, 3221 (1964).
7. V. F. Holland, J. Appl. Phys., 35, 59 (1964).

THE EQUILIBRIUM MELTING TEMPERATURE AND SURFACE FREE ENERGY OF POLYETHYLENE SINGLE CRYSTALS

by

H. E. Bair, T. W. Huseby and R. Salovey
Bell Telephone Laboratories, Incorporated
Murray Hill, New Jersey

Chain folded polyethylene single crystals grown from dilute solutions at undercoolings of 20-40°C are metastable. It is well known that under certain conditions the crystals begin to thicken rapidly at elevated temperatures.[1,2,3] In fusion studies of polyethylene in our laboratory, the observed melting temperature of solution grown crystals (where the crystallization temperature is 77°C or lower) has been found to increase by nearly 10°C as a result of thickening while the crystals were being heated at 10°C/min. Thus, it was impossible to correlate the melting results with structural data which had been obtained on the original crystals prior to heating.

Fast heating rates on the order of 100°C/min. are not sufficient to entirely suppress lamellar thickening during fusion experiments.[4] Moreover, the accurate determination of melting temperatures becomes more difficult with increasing heating rates. Thus we attempted to circumvent this experimental difficulty by irradiating the polyethylene crystal aggregates in vacuum with 1 MeV electrons from a van de Graaff generator. Subsequent melting of irradiated films at 10°C/min. in a differential calorimeter revealed that incorporation of a few crosslinks between polymer chains reduced the rate of lamellar thickening.[4] It was found that an irradiation dose of 26 Mrads suppressed lamellar thickening of crystals grown isothermally at 85°C in xylene so effectively that nearly 90% of the crystals melted sharply within a 3°C temperature interval.

We decided to utilize this technique of linking lamellae together in an attempt to measure accurately the melting temperatures of solution grown crystals of various thicknesses. From the information obtained in these experiments we intend to estimate an equilibrium melting temperature, independent of crystal size effects, and to calculate the chain folded surface free energy of the crystal.

EXPERIMENTAL

Materials

Solutions of polyethylene (Marlex 6000, type 50) in boiling xylene were poured into a tenfold excess of xylene, thermostatically controlled at a crystallization temperature of 70°C, 77°C, 85°C, or 90°C to produce a final concentration of 0.04 g/100 ml. Crystals were grown at 95°C by a self-nucleation technique.[5] After several days, the suspensions were filtered at the growth temperature and crystal aggregates dried in a vacuum. These separate preparations had different morphologies as revealed by electron and optical microscopy and low angle x-ray scattering (Table 1).

Melting Points

Five melting compounds were used to obtain a temperature correction curve between 70° and 156°C at a heating rate of 10°C/min. for the platinum resistance thermometers in the Perkin-Elmer differential scanning calorimeter. The melting temperatures, T_m, of polyethylene samples were determined as described in the Appendix.

TABLE 1

The Melting Points and Apparent Heats of Fusion as a Function of Dose Rate For Polyethylene Single Crystals of Different Morphologies

Sample	Dose, Mrads	T_m, °C	Q_f, cal/g.	Crystal Thickness, Å	Remarks
1 (T_c=70°C)	0	129.4±1.6	50±2	104±5	Dendritic, many-layered crystals; several microns across
	51	123.3±1.5	50±2		
	100	120.8±1.3	47±2		
	300	115.7±1.3	41±2		
2 (T_c=77°C)	0	129.8±1.7	52±2	111±5	Smooth edged, multiple-layered crystals; several microns across
	26	125.6±2.0	49±2		
	51	128.8±1.5	50±2		
	100	121.3±1.1	51±2		
3 (T_c=85°C)	0	130.3±1.9	56±2	126±6	Monolayered, truncated platelets; about 10 microns across
		(124.9±3.5)[a]	-		
	26	125.3±1.2	54±2		
	50	125.2±1.0	57±2		
	100	124.1±1.0	54±2		
	300	121.1±1.0	48±2		
4 (T_c=90°C)	0	131.6±1.4	61±2	145±6	Single platelets; approximately 10 microns across
		(127.8±1.8)[a]	-		
	25	127.4±1.1	58±2		
	51	128.3±1.1	59±2		
	102	127.5±1.0	57±2		
5 (T_c=95°C)	0	131.0±1.1	56±2	178±15	Small crystalline overgrowths
	25	131.1±0.9	57±2		
6 (T_c=85, annealed 4000 minutes 125.5±1.0°C)	0	133.8±1.1	-	215±20	
7 (T_c=90, annealed 570 hrs. at 128±1°C)	0	136.9±1.5	65±2	278±30	

[a]Melting temperature of unthickened and unirradiated crystals from multiple peaked thermograms (Fig. 1 and 2)

RESULTS AND DISCUSSION

The morphologies of the single crystal aggregates ranged from dendritic stacks of crystals to large monolayered platelets each having a characteristic thickness dependent on the crystallization temperature, T_c. The dimensions of each group of single crystals are tabulated in Table 1. In the results which follow, we would like to emphasize that it is the crystal thickness between fold surfaces which determines the melting temperature, T_m, of each sample.

Multiple peaked melting patterns were observed in each case for crystals grown at 85° and 90°C. (Solid lines in Fig. 1 & 2.) The multiple peaks are due to lamellar reorganization coincident with melting of the crystals.[4] With the proper conditions or irradiation dose and heating rate, lamellar thickening can be successfully suppressed. Then, instead of multiple peaks, a single sharp endotherm should result. It is our contention that a melting point determined in this manner represents melting of the original lamellae whose thickness has not increased during heating.

The melting of crystals grown at 85°C is represented by the solid black line in Fig. 1. This thermogram exhibits three maxima between 121° and 131°C in the apparent heat capacity. However, after the sample is irradiated to a dose of 26 Mrads a single sharp melting peak occurs in the same temperature region as the two lower temperature maxima of the unirradiated crystals (open circles Fig. 1). The observed melting temperature of the irradiated crystals was 125.3°C. If the two lower temperature maxima in the thermogram of unirradiated crystals are due to melting of unthickened crystals and treated as a single peak, a melting point nearly identical to the irradiated crystals (26 Mrads) is observed (Table 1). When crystals are subjected to doses of 50 and 100 Mrads there may be a slight lowering of the observed melting point due to irradiation damage on the order of 1°C. However, the melting temperature of crystals irradiated to 300 Mrads drops significantly to 121.1°C. It appears that polyethylene single crystals subjected to irradiation doses exceeding 100 Mrads show significant changes in such thermal properties as observed melting temperature and apparent heat of fusion, ΔQ_f. Below 100 Mrads ΔQ_f for these crystals varies within a few percent of 56 cal/g, but at 300 Mrads it decreases by 14% to 48 cal/g. Figure 3 shows the lowering of ΔQ_f which occurs beyond a dose of 100 Mrads for single crystals prepared at 70° and 85°C.

It should be noted that the fraction of unirradiated 85°C crystals which thicken during heating to melt at temperatures above 125°C may be estimated from the ratio of the heat absorbed above 125°C to the total heat absorbed during fusion. In this case about 50% of the lamellae thickened to melt at temperatures between 125 and 132°C. Analogously, about 13% of the crystals prepared at 90°C thicken on heating at 10°C/min. and melt between 127° and 132°C (Fig. 2). Following a 25 Mrad dose a single peaked endotherm results with a melting temperature of 127.4°C (Fig. 2). The peak almost coincides with the lower temperature peak in the melting of the unirradiated crystals. Since this lower temperature peak (solid black line in Fig. 2) between 125° and 129°C has been associated with melting of 87% of the unirradiated single crystals which do not thicken during the experiment, we may conclude that irradiation of 26 Mrads has not affected significantly the observed melting temperature. In addition, irradiation of these crystals with doses up to 100 Mrads causes no further lowering of the melting temperature (Table 1).

In order to obtain crystals from solution at temperatures above 90°C, a self-nucleation technique was employed.[5] Crystals grown in this manner at 95°C give x-ray long periods of about 180Å. These crystals melted at 131.0°C without any sign of reorganization (Fig. 4). If the unirradiated crystals melted without thickening, then irradiation of these crystals to 25 Mrads should indicate any possible deleterious effect of low doses of

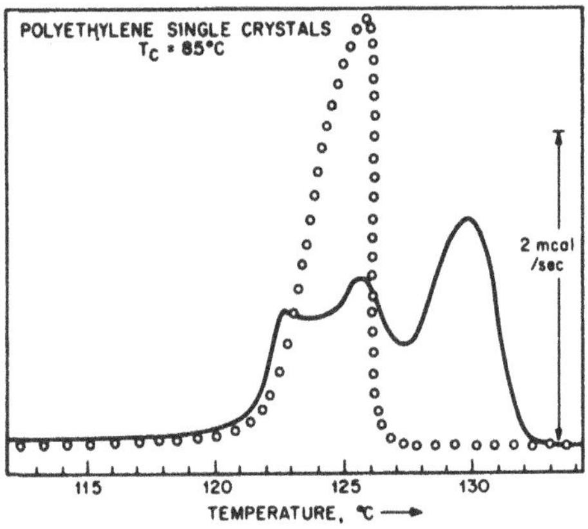

Fig. 1. Comparative melting behavior of unirradiated and irradiated 85°C crystals at 10°C/min. (~800 μg)

_ Unirradiated crystals

O Irradiated crystals, 25 Mrads

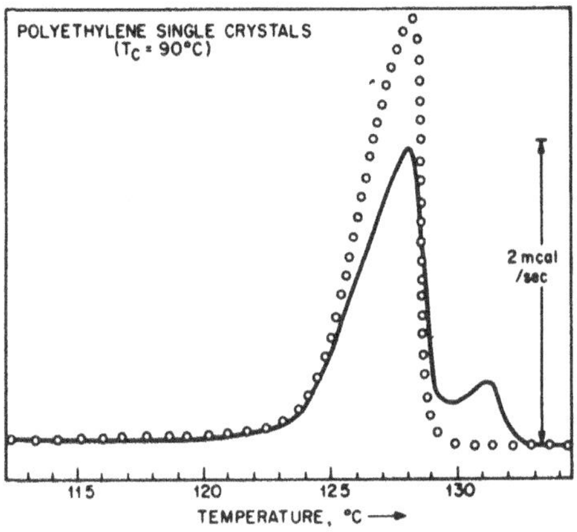

Fig. 2. Comparative melting behavior of unirradiated and irradiated 90°C crystals at 10°C/min. (~600 μg)

_ Unirradiated crystals

O Irradiated crystals, 25 Mrads

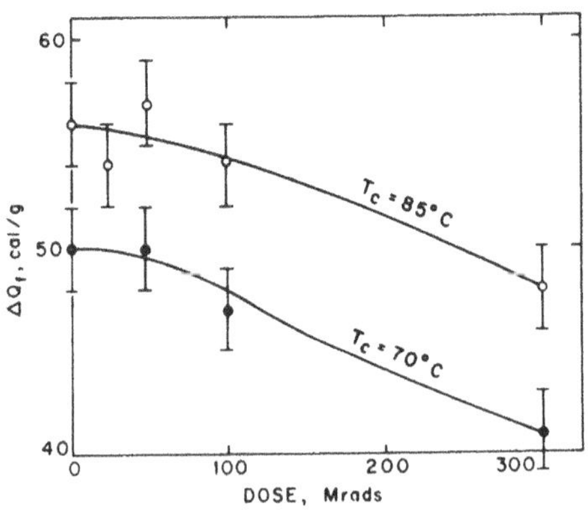

Fig. 3. Apparent heat of fusion for crystals prepared at 70°C and 85°C as a function of dose.

irradiation upon the crystals. Figure 4 shows that the irradiated and unirradiated crystals have the same melting point and apparent heat of fusion. Thus, it is concluded that irradiation of 100 Mrads or less (to which all crystals in this study have been exposed) has not affected significantly the determination of melting points.

We found that samples of unirradiated single crystals prepared at 70° and at 77°C melted in a single endothermic peak.[4] This thermal behavior seems to be in disagreement with the multiple peaked melting behavior of crystals grown at 85°C and higher. However, the former crystals should be less stable than those crystallized at 85°C and above, and consequently should thicken upon heating at 10°C/min. The absence of multiple peaks suggests that reorganization must occur simultaneously with melting in such a manner that the heat liberated by the thickening crystals is balanced by the heat being absorbed by fusing crystals. Under these conditions the baseline of the thermogram would be smooth in the temperature interval where melting and thickening are occurring. With the completion of reorganization the thermogram would display only the melting of the thickened crystals.

The comparative melting behavior of the crystals collected at 77°C with various irradiation doses is shown in Fig. 5. Since the crystals irradiated to 100 Mrads melt between 115° and 121°C, it is apparent that melting of the unthickened crystals and the reorganization of those unirradiated crystals proceeded without thermal detection. Only the melting of thickened crystals is observed between 124° and 130°C. If the melting temperature depression of the sample irradiated to 100 Mrads was caused only by the radiation-induced suppression of lamellar thickening and was not due to crystals damaged by irradiation it is concluded that nearly the total mass of crystalline material increases in fold period upon fusion at 10°C/min. The fact that the apparent heat of fusion for the 77°C crystals is unchanged with doses up to 100 Mrads strengthens this conclusion.

If we now construct a plot of the mass fraction of unirradiated crystals which thicken during heating at 10°C/min. to melt above the melting temperature of the unreorganized crystals against T_c, we find that the amount of crystals which thicken during fusion decreases monotonically with increasing crystallization temperature. Figure 6 shows how rapidly the fraction of thickening crystals falls off as the lamellar fold period increases by 70% from 104Å at T_c = 70°C to 178Å at T_c = 95°C. Therefore, in addition to halting reorganization by crosslinking polymer chains together, thickening may also be suppressed at a constant heating rate by using crystals of sufficient initial thickness that they transform from the solid to liquid state without reorganization. This may be accomplished by growing crystals in solution at higher temperatures or annealing crystals near their melting point. In this regard, unirradiated single crystals (T_c = 85° and 90°C) which had yielded multiple peaks when heated at 10°C/min. were annealed under the conditions shown in Table 1 and increased in thickness from 126 and 145Å to 215 and 280Å, respectively. These annealed samples melted at 10°C/min. without any sign of reorganization during fusion.

Using Hoffman and Lauritzen's thermodynamic equation $[T_m = T_m^o(1-2\sigma_e/\ H_f\ell)]$ for the melting point lowering due to fold surfaces,[6] the equilibrium melting temperature, T_m^o, and surface free energy, σ_e, for polyethylene single crystals can be evaluated by plotting the observed melting points against reciprocal lamellar thickness, $(1/\ell)$, where ΔH_f = heat of fusion of polyethylene.[7] In Fig. 7, at $1/\ell$ equal to zero a melting temperature of 145.8±1.0°C is indicated, and from the slope of the line a surface free energy of 89±5 erg/cm^2 was calculated.

Flory and Vrij,[8] and later Broadhurst,[9] found the equilibrium melting temperature for polyethylene to be 145±1.0°C from an extrapolation of the melting points of n-paraffins to large chain lengths ($n \to \infty$). Hoffman[10]

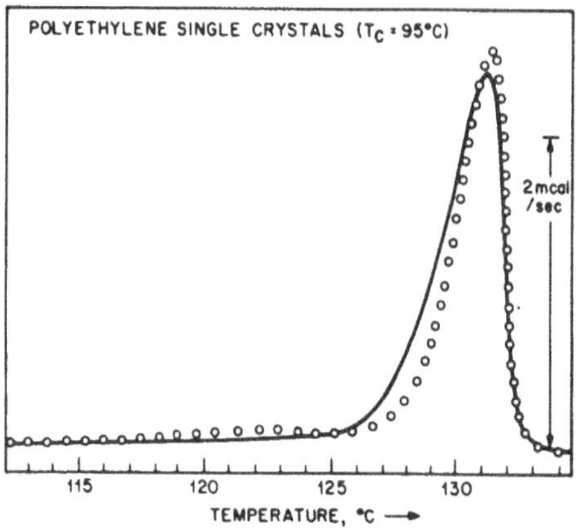

Fig. 4. Comparative melting behavior of unirradiated and irradiated 95°C crystals at 10°C/min. (~900 μg)
_ Unirradiated crystals
O Irradiated crystls, 25 Mrads

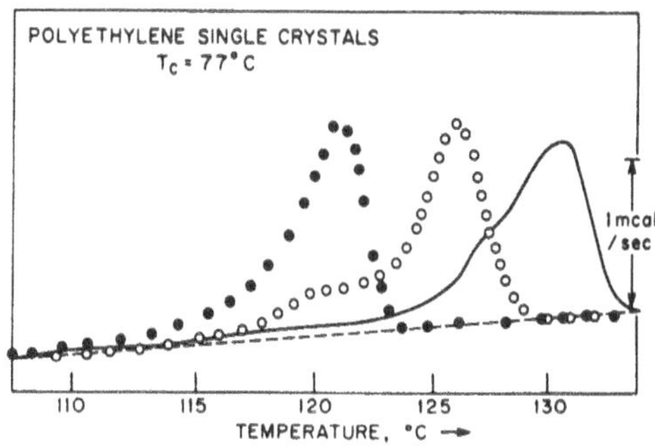

Fig. 5. Comparative melting behavior of unirradiated and irradiated 77°C crystals at 10°C/min. (~750 μg)
_ Unirradiated crystals
O Irradiated crystals, 25 Mrads
● Irradiated crystals, 100 Mrads

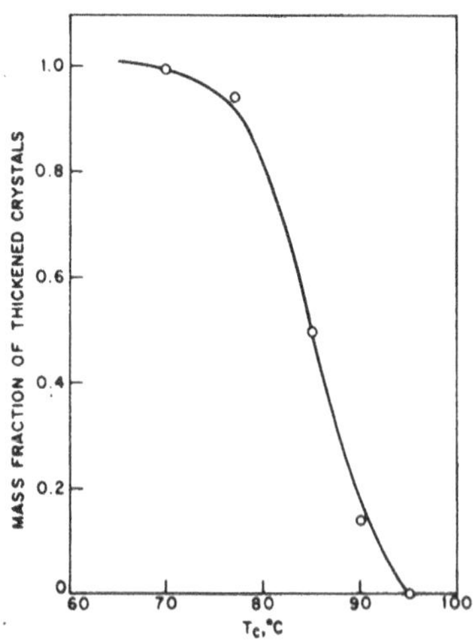

Fig. 6. Mass fraction of unirradiated crystals which thicken during heating at 10°C/min to melt above the melting temperature of the unreorganized crystals as a function of crystallization temperature.

has analyzed the experimental melting results of Brown and Eby[11] on bulk crystallized polyethylene of various long periods and found $T_m^o = 143 \pm 0.5^oC$ and $\sigma_e = 57 \pm 4$ erg/cm^2. However, since Brown and Eby determined melting points by heating samples at 2°C/min. on an optical hot stage, it is probable that thin lamellae (< 250A) increased in thickness during heating to give anomalously high melting temperatures. Since their observed melting point for lamellae which were about 300A thick agree closely (~1°C) with our findings for crystals of similar dimensions, it is possible the estimated surface free energy and equilibrium melting temperature for the bulk and dilute solution grown crystals would be the same if all measurements were performed in the absence of lamellar reorganization.

Illers and Hendus[12] have combined their results for the melting points of melt and solution crystallized polyethylene with selected data of other investigators. A linear relationship was observed between melting points and reciprocal lamellar thickness. From this analysis a value of 141±0.5°C for T_m^o and 79.5±2.5 erg/cm^2 for σ_e were calculated.

If the surface free energy for all crystals in this study is the same, then from the fit of the data to the line in Fig. 7 it is concluded that a dose of 100 Mrads was sufficient to suppress lamellar reorganization at a heating rate of 10°C/min. for the crystals grown at 77°C and possibly for those grown at 70°C.

Measurement of the heat absorbed by the fusing crystals in each experiment revealed that the apparent heat of fusion increases from 50 to 61 cal/g as the crystallization temperature is raised from 70° to 90°C.[12] A graph of ΔQ_f versus undercooling, ΔT, where $\Delta T = T_d^o - T_c$* for each unirradiated single crystal sample appears to be linear (Fig. 8). Extrapolation of this data to $\Delta T = 0$ yields a heat of fusion, ΔH_f, of 68±2 cal/g. This result agrees closely with the value of 69±2 cal/g obtained by Mandelkern and Quinn[7] from melting point depression experiments. Figure 8 also indicates a 100% crystalline sample could be obtained from solution at zero undercooling. Unfortunately, the amount of time required for the nucleation and growth of single crystals from dilute solution at temperatures approaching the dissolution temperature is so inordinate that such experiments are impractical. It might be expected that the use of prenucleated crystals would make it possible to prepare highly crystalline samples at small undercoolings. However, the regular increase in ΔQ_f for single crystals grown at progressively smaller undercoolings in dilute solutions does not seem to hold, at least, for seeded crystals produced in concentrated solution at 95°C. This result is consistent with the more diffuse low angle x-ray scattering pattern and smaller lateral dimensions of these prenucleated crystals than those crystallized directly at 85° and 90°C in dilute solution.

CONCLUSIONS

1) Lamellar thickening of polyethylene single crystals can be suppressed at a constant heating rate by low doses of irradiation. The apparent heat of fusion and melting temperature were not affected significantly at radiation doses up to 100 Mrads. However, at 300 Mrads, ΔQ_f was lowered by about 15% and the melting point by roughly 4°C. The melting of crosslinked crystals in the absence of lamellar reorganization occurs sharply with nearly 90% of the single crystals melting in less than a 3°C temperature interval.

---

*The dissolution temperature, T_d^o, of polyethylene single crystals in dilute xylene is 113.7°C as calculated from a T_m versus $1/\ell$ plot on crystals in suspension.[14]

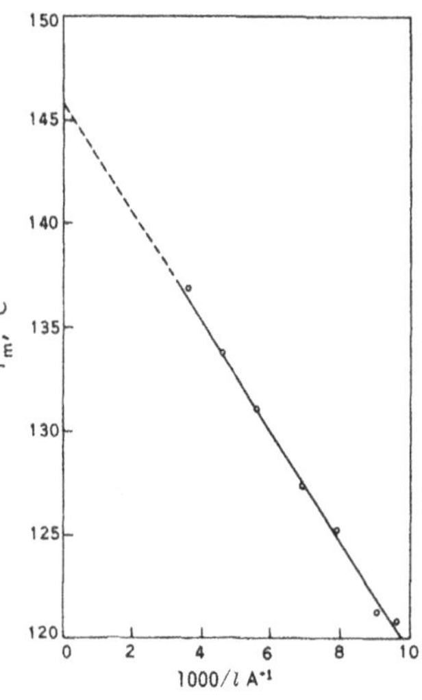

Fig. 7. Melting temperatures of polyethylene single crystals as a function of reciprocal lamellar thickness.

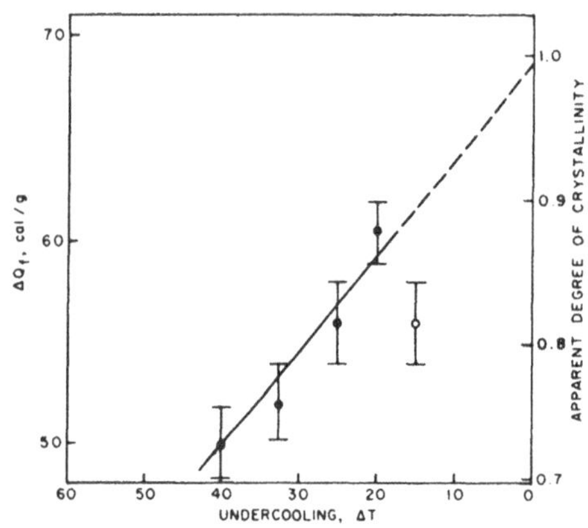

Fig. 8. Apparent heat of fusion for polyethylene single crystals as a function of undercooling.
O Seeded crystals
● Unseeded crystals

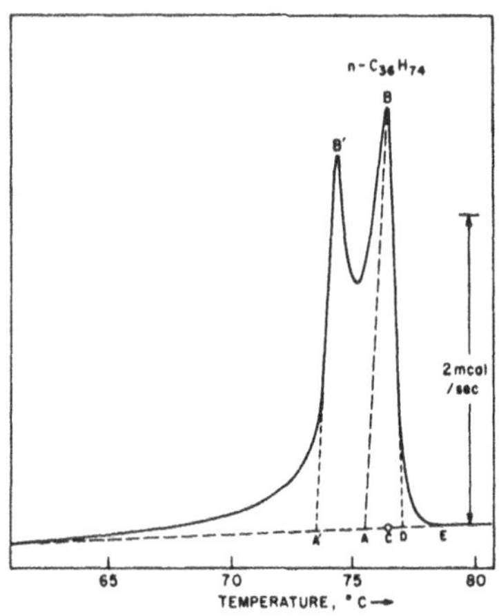

Fig. 9. Free rotation and melting transitions of a paraffin, n-$C_{36}H_{74}$.

2) Without lamellar thickening observed melting temperatures of polyethylene single crystals of various sizes were found to obey the Hoffman and Lauritzen equation $T_m = T_m^o(1-2\sigma_e/\Delta H_f \ell)$. From this relationship an equilibrium melting temperature of 145.8±1.0°C and surface free energy of 89±5 erg/cm^{-2} were estimated.

3) At a constant heating rate the fraction of thickening crystals was found to decrease monotonically with increasing fold period.

4) The apparent heat of fusion increases as polyethylene single crystals are prepared in dilute solution at progressively smaller undercoolings. It is estimated that a 100% crystalline sample could be obtained from dilute solution when the crystallization temperature is equal to the dissolution temperature.

REFERENCES

1. Fisher, E. W. and Schmidt, G. F., Angew. Chem., 74, 551 (1962).

2. Statton, W. O. and Geil, P. H., J. Appl. Poly. Sci., 3, 357 (1960).

3. Takayanagi, M. and Nagatoski, F., Mem. Fac. Engng. Kyushu, 24, 33 (1965).

4. Bair, H. E., Salovey, R. and Huseby, T. W., Polymer, 8, 9 (1967).

5. Blundell, D. J., Keller, A. and Kovacs, A. J., J. Polymer Sci. Part B, 4, 481 (1966).

6. Hoffman, J. D. and Lauritzen, J. I., Jr., J. Res. Natl. Bur. Std., 65A, 297 (1961).

7. Quinn, F. A., Jr. and Mandelkern, L., J. Am. Chem. Soc., 80, 3178 (1958); Mandelkern, L., Rubber Chem. Tech., 32, 1403 (1959).

8. Flory, P. J. and Vrij, A., J. Am. Chem. Soc., 85, 3548 (1963).

9. Broadhurst, M. G., J. Res. Natl. Bur. Std., 70A, 481 (1966).

10. Hoffman, J. D., S.P.E. Trans., 4, 315 (1964).

11. Brown, R. G. and Eby, R. K., J. Appl. Phys., 35, 1156 (1964).

12. Illers, K. A. and Hendus, A., to be published.

13. Fisher, E. W. and Hinrichsen, G., Polymer, 7, 195 (1966).

14. Huseby, T. W. and Bair, H. E., to be published.

15. Perkin-Elmer Corporation, Thermal Analysis Newsletter, 5, Norwalk, Connecticut.

16. Broadhurst, M. G., J. Res. Natl. Bur. Std., 66A, 241 (1962).

APPENDIX

A thermal lag develops in the differential scanning calorimeter between a sample undergoing fusion (and thereby maintaining a constant temperature) and the calorimeter vessel whose temperature is increasing at a programmed rate. An explanation of how to correct for this thermal lag is available from the Perkin-Elmer Corporation.[15] The multiple peaked thermogram in Fig. 9 indicates the magnitude of the correction for the melting of a paraffin, $\underline{n}$-$C_{36}H_{74}$. The lower temperature peak results from a crystal transformation from orthorhombic to hexagonal[16] which allows free rotation about the main chain axis. The high temperature peak is due to melting of the paraffin.

The correct transition temperatures do not occur at peak points B and B′. Points A and A′ result from the correction for thermal lag at a heating rate of 10°C/min. These points yield corrected transition temperatures of 75.4°C and 73.5°C which are in good agreement with literature values of 76.0°C and 73.9°C.[16]

The area under the curve BE (BCE) in Fig. 9 normally represents the energy required by the sample to catch up to the programmed temperature of the calorimeter after termination of melting. The area under this curve is normally the same for samples of similar heat capacity and mass. However, the area under this portion of the curve for irradiated polyethylene single crystals is approximately twice that of the paraffin crystals. Since similar amounts of material were melted in each instance, it is concluded that the polymer crystals are melting over a wider temperature interval, AD, rather than over AC, typical for paraffins. The fraction of crystals which melt beyond the peak temperature is small and decreases rapidly with increasing temperature. We have chosen point C which is the projection of the peak B onto the baseline, AD, as the melting temperature, T_m. In these studies the melting temperature of polyethylene single crystals was 0.8±0.1°C above point A.

The magnitude of the temperature interval, CD, for the polyethylene single crystals is indicated by the temperature limits of T_m in Table 1 ($AC \leq 0.9$°C). Analysis of the melting temperatures indicates that with the exception of the crystals grown at 70°C, the width of the melting peak decreases from about 2°C to 1°C as lamellar reorganization is suppressed.

MELTING BEHAVIOUR OF LINEAR POLYETHYLENE FRACTIONS

by

R.W. Ford, J.D. Ilavsky and R.A. Scott
Research and Development Laboratory
Dow Chemical of Canada, Limited
Sarnia Ontario Canada

Introduction

The advent of commercial differential thermal analyzers of high resolution requiring only milligram quantities of samples has provided the polymer chemist with a powerful tool for studying the melting behaviour of semi-crystalline polymers. In the present case, a Perkin Elmer DSC-1 has been used to study the relationship between melting characteristics and molecular weight of two slightly different linear polyethylenes.

The equilibrium melting point and heat of fusion for Marlex 50 have been determined both calorimetrically[1] and dilatometrically[2]. Wunderlich and Dole[1] report the melting point to be 134.8 ± 0.2°C and the heat of melting to follow the relationship:

$$\Delta H_f = 54.6 + 0.1716t - 6.36 \times 10^{-4}t^2 \text{ cal/g}$$

where t is temperature in °C. Thus at the melting point, ΔH_f = 66.2 cal/g for linear polyethylene.

Recently, Mandelkern et al[3] reported a study of the heats of melting of linear polyethylene fractions in which they used a crystallization technique which assured a close approach to equilibrium conditions. In this case, the heat of melting for the low molecular weight fractions was in close agreement with that quoted above. Higher molecular weight fractions, however, showed significantly lower values, indicating the reduced ability of these fractions to crystallize.

The present work on molecular weight fractions makes use of faster cooling rates than used by previous workers, and demonstrates the ability of differential scanning calorimetry to detect subtle differences in polymer behaviour which are not observed on the whole polymer.

Experimental

The two polyethylenes, designated Sample D-4 and K-1, are similar to some samples previously studied in this laboratory[4,5]. Both are linear Ziegler-type homopolymers of ethylene of 0.960g/cc density and 0.7 dg/min melt index. They were produced, however, in different plants, and with somewhat different catalyst systems.

Fractionations were carried out using the method of Francis, Cooke and Elliot[6] as modified in this laboratory[7]. Molecular weights were determined by inherent viscosity as previously described[8].

Measurements of melting points and heats of melting were carried out using a Perkin Elmer DSC-1 calibrated with an indium standard. Samples weighing about 5 mg were contained in covered aluminum sample pans. Areas of the exotherms were measured with a planimeter. Heats of melting and melting points were determined on both annealed samples and samples which were subjected to various prior thermal histories. The annealing procedure was to heat the sample to 170°C, cool at 10°C/min to 10°C below the melting point, hold for 10 min., then cool to 75°C at 10°C/min. Melting points were run at 5°C/min.

Results and Discussion

The relationship between melting point and molecular weight for the fractions from samples D-4 and K-1 is shown in Fig. 1. Within the limits of precision of the method, the two samples are identical in this regard. The sharp drop in melting point at molecular weights below 20,000 is probably due to the increasing number of chain ends, and possibly branch points, which act as impurities by depressing the melting temperature. All data points in Fig. 1 were obtained on samples which had been cooled at 5°C/min through their freezing point and then subsequently melted, also at 5°C/min.

The effect of prior cooling rate on the melting point of the fractions from sample K-1 is shown in Fig. 2. Increasing the cooling rate from 5° to 40°C/min results in a drop in melting point of 1 to 1.5°C for each fraction. It should be noted that fraction nine of sample K-1 with a molecular weight of 470,000 gave a melting peak with a distinct shoulder 2° to 2.5°C higher than the main peak. The temperatures of both the peak and the shoulder are plotted in Fig. 1 and 2. No such phenomenon was observed with sample D-4.

Also shown in Fig. 2 is the melting point of the whole polymer. At the lowest cooling rate, the melting point is 131.8°C compared with the equilibrium value of 134.8°C as determined by Wunderlich and Dole[1] for Marlex 50. Simple linear extrapolation of the data suggests that a cooling rate of no more than 1°C/hr would be required in order to approach the equilibrium value. It is interesting to note that both Wunderlich and Dole[1], and Quinn and Mandelkern[2] allowed their samples to cool over a period of several days, suggesting a cooling rate in the range of a few degrees per hour.

Conversely, the curve of Fig. 2 for the whole polymer provides a means of estimating cooling rate in a part fabricated from this polymer. A similar proposal has been made for blown film in a previous publication[9].

Despite the similarity of their melting points, the two samples show significant differences in the relationship between heat of fusion and molecular weight for annealed specimens. As shown in Fig. 3 the fractions of sample D-4 show a gradual decrease in heat of melting with molecular weight whereas the fractions from sample K-1 give a maximum at a molecular weight of about 10,000 followed by a more marked decrease at higher molecular weight. Moreover the heat of melting of all fractions of sample K-1 fall below the comparable fractions from sample D-4. The decrease in the heat of melting at the higher molecular weight is easily explained on the basis of decreased crystallinity due to decreasing mobility of the longer polymer molecules. A different explanation is required, however, for the low value for the lowest molecular weight fraction of sample K-1. Possibly this fraction is more

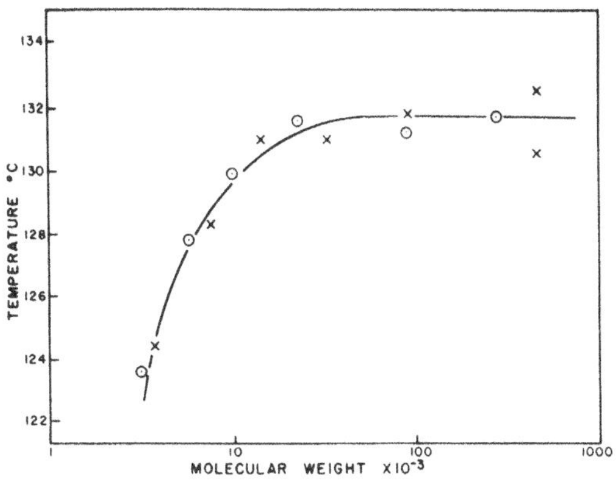

Figure 1

Variation of melting temperature with molecular weight.
⊙ Sample D-4; X Sample K-1

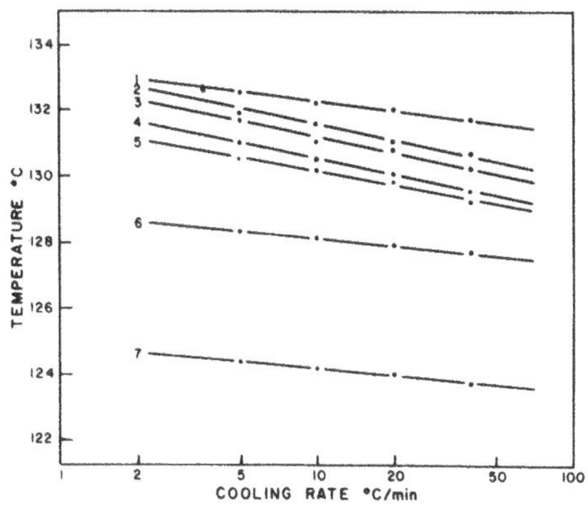

Figure 2

Variation of melting temperature with previous cooling rate for Sample K-1.

1. Fraction IX Shoulder, M.W. = 470,000
2. Whole polymer, M.W. = 80,000
3. Fraction VII, M.W. = 91,000
4. Fraction III, M.W. = 14,500
5. Fraction IX main Peak, M.W. = 470,000
6. Fraction II, M.W. = 7,600
7. Fraction I, M.W. = 3,700

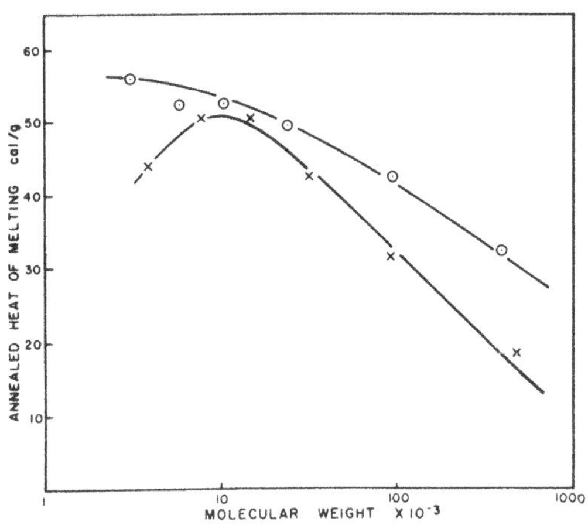

Figure 3

Variation of annealed heat of melting with molecular weight.
⊙ Sample D-4; X Sample K-1

highly branched than the bulk of the molecules, thus reducing crystallinity. Unfortunately the fraction was too small for an I.R. scan which might have confirmed this hypothesis.

It is interesting to compare the present results with those of Mandelkern et al[3]. In their work, the fractions with molecular weights below 10,000 gave a ΔH_f of 70 cal/g after an isothermal crystallization at 130°C, followed by cooling to room temperature over a period of 24 hours. In our work the samples were crystallized for 10 minutes at 10° below the melting point, or in other words as low as 114° for the lowest molecular weight fractions. The samples were then cooled at 10°C/min to 70°C before being remelted. This procedure produced a heat of melting in the lowest molecular weight fraction of sample D-4 of only 58 cal/g. Obviously, crystallization conditions can have a marked effect on the crystallinity even at quite low molecular weights.

Despite the differences in heats of melting for the fractions, the values for the whole polymers are virtually identical, 43.4 cal/g for D-4 and 44.0 cal/g for sample K-1. Apparently the crystallization of the latter is more dependant on the presence of the lower molecular weight species than is the case for the other sample.

In summary, the use of a Perkin Elmer DSC-1 has permitted the study of the melting behaviour of linear polyethylene fractions as a function of molecular weight. Moreover, some differences have been found in the behaviour of fractions from two samples of the same gross characteristics, illustrating the ability of the method to detect subtle differences in polymer properties.

References

1. B. Wunderlich and M. Dole, J. Polymer Sci., 24, 201 (1957).
2. F.A. Quinn, Jr., and L. Mandelkern, J.Amer.Chem.Soc., 80, 3178 (1958).
3. L. Mandelkern, J.G. Fatou, R. Denison and J. Justin. J. Polymer Sci. B3, 803 (1965).
4. R.W. Ford, R.A. Scott and R.J.B. Wilson, J.Appl.Polymer Sci., in press.
5. I.J. Bastien, R.W. Ford and H.D. Mak, J. Polymer Sci. B4, 147 (1966).
6. P.S. Francis, R.C. Cooke, Jr., and J.H. Elliot, J. Polymer Sci., 31, 453, (1958).
7. R.W. Ford and J.D. Ilavsky, in preparation.
8. B.J. Cottam, J. Appl. Polymer Sci., 9, 1853 (1965).
9. R.W. Ford and R.A. Scott, J. Appl. Polymer Sci., 11, 2325 (1967).

MEASUREMENT OF THE CRYSTALLIZATION RATES OF POLYMERS BY DIFFERENTIAL SCANNING CALORIMETRY

J. R. Knox
Avisun Corporation
Marcus Hook, Pennsylvania

Measurements of the crystallization rates of polymers are of great importance both from a practical and a basic viewpoint. Among the techniques most commonly used to date have been volume dilatometry, microscopy, and infrared spectrometry.

All of these techniques have some limitations when data on percent crystallinity vs. time are desired. In dilatometry the large sample required precludes low crystallization temperatures due to slow thermal equilibration. Microscopic techniques give only a spherulite growth rate or a relative crystallinity when light extinction techniques are used. Temperature sensitivity of the infra-red absorption bands makes measurement of anything but relative crystallization rates exceedingly complicated.

With calorimetric techniques, however, percent crystallinities may be calculated as a function of time at a wide range of temperatures provided the heat of fusion of the polymer is known. Sample sizes for the DSC-1 are small, (about 5 mg is sufficient) allowing rapid thermal equilibration, and the sample temperature can be changed rapidly. The combination of these two capabilities allows crystallization rates to be measured at lower temperatures than on a dilatometer, where sample sizes are on the order of 1 g.

The procedure for measuring crystalliation rates using the DSC-1 is to prepare a film about 5 mils thick of the polymer to be examined. A sample is punched out and crimped into the standard holder, making certain that the bottom of the pan is kept flat. After maintaining the sample at the desired fusion temperature for the appropriate length of time, the temperature indicator dial is cranked down by hand to the crystallization temperature. Zero time is taken as the time when the control light comes on, indicating thermal equilibrium. A time-base record (Fig. 1) is then obtained until crystallization is complete. (Although secondary crystallization does occur, it is extremely slow, and any heat liberated is too small to be detected.)

The end of crystallization is easily determined as the time at which the pen returns to the same value, within experimental error that it had at the start of crystallization. The area under the crystallization curve is then integrated in one minute intervals and a percent crystallinity calculation made for each increment according to the usual equation:

$$\%\ \text{Cryst} = \frac{\text{area (in}^2\text{) X range (cal/sec.) X 100}}{\text{wt. sample(g) X } \Delta H_f\text{(cal/g) X cal. fact.(in}^2\text{/sec.)}}$$

A summation with time may then be made and an integral crystallization vs. log time curve plotted.

The instrument calibration factor was obtained by running samples of indium, benzoic acid, and p-toluic acid of high

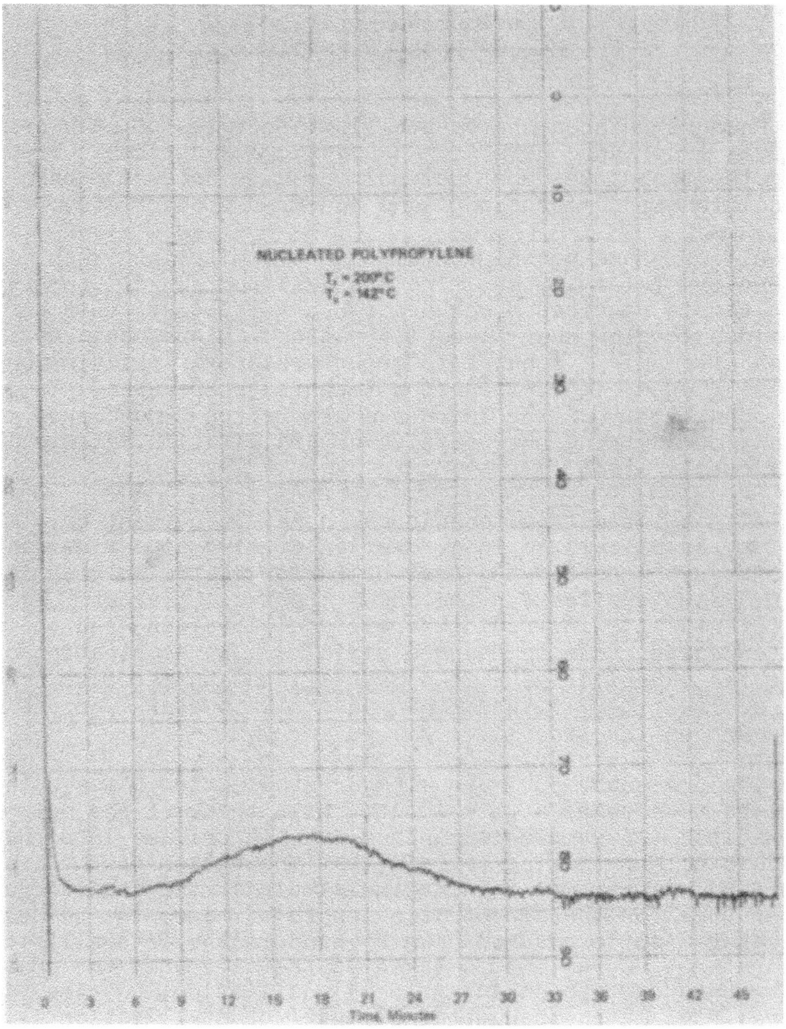

Figure 1.

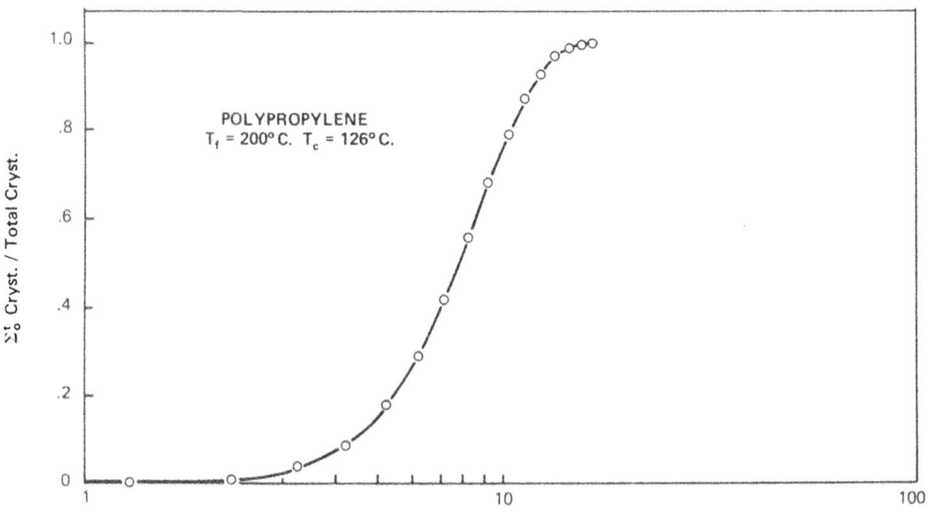

Figure 2.

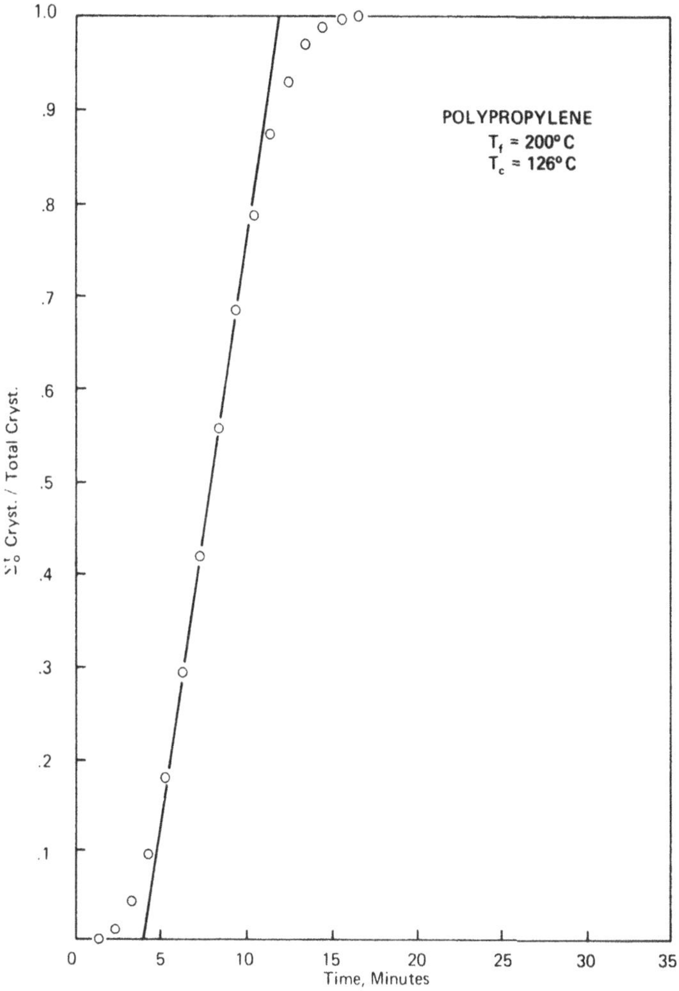

Figure 3.

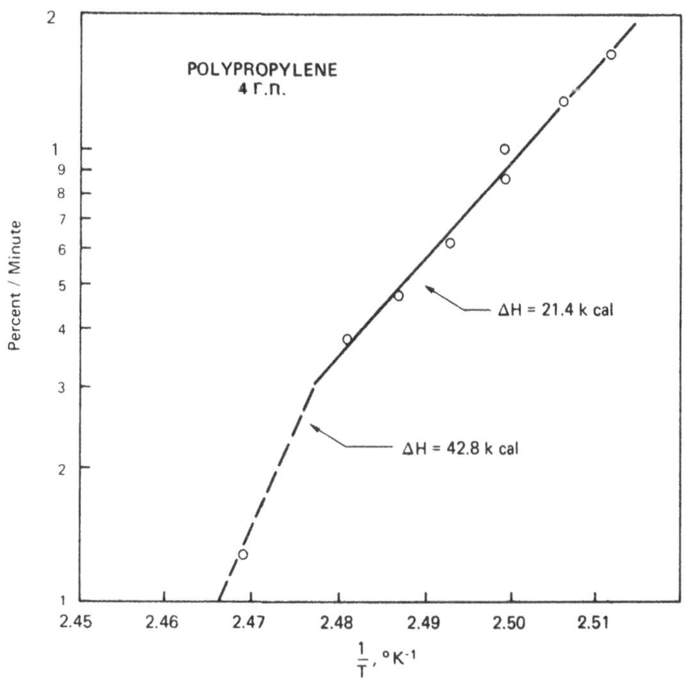

Figure 4.

purity and known heat of fusion. These samples have melting points in the range of crystallization temperatures used, and an average calibration factor of 0.147 in^2/sec. was obtained from the data on all of the standards. The heat of fusion of the polypropylene sample used in the experiments was measured separately by a melting point depression technique and was found to be 35.9 cal/g.

As the crystallization temperature increases, the percent crystallinity which is developed at that temperature decreases. For this reason it was decided to compare rates of crystallinity development on a reduced crystallinity basis. The reduced crystallinity was calculated as the amount of crystallinity developed at any time divided by the total crystallinity attained at that temperature. Plots of reduced crystallinity vs. log time, such as Fig. 2, show that the maximum slope of these curves is independent of crystallization temperature. Exact superposition of the curves by a shifting along the log time axis is not possible, however, because the S-shaped nature of the crystallization curve is much more pronounced at lower temperatures, primarily because of the shorter times involved.

Plots of the reduced crystallinity vs. linear time were made and the rates of crystallization were calculated at each temperature by measuring the linear portion of the curve. The initial and final portions of each crystallization deviated from linear behavior with time, as shown in Fig. 3, but in all cases the rates were linear for over 60% of the total crystallization. The rates obtained are shown in Table I.

TABLE I

Temp., °C.	Rate, %/min.
125	.167
126	.130
127	.102
128	.063
129	.048
130	.038
131	.021
132	.013

The logarithm of the rate was plotted vs. the reciprocal of the absolute temperature in Fig. 4, and it was observed that there were two linear portions of the plot. The data obtained at 130°C. and below all fell on a line of lower slope than the data obtained above 130°C. Application of the Arrhenius equation to this plot leads to values of 21.4 kcal and 42.8 kcal for the activation energies below and above 130°C. This change in temperature dependence of the rate indicates either that different mechanisms for nucleation and/or crystallization are operative above and below 130°C., or that a different crystal form is produced in the two temperature ranges.

Crystallization rate measurements were made on the same polypropylene nucleated with a heterogeneous nucleating agent. This sample was crystallized over the range of 137° to 143°C. The observed rates are shown in Table II and Fig. 5.

Treatment of the rate data on the nucleated sample yielded an activation energy for the crystallization process of 13.8 kcal. Microscopic examination of the samples showed no differences in the nature of the spherulite formed, but the spherulite size was much smaller.

TABLE II

RATE OF CRYSTALLIZATION OF NUCLEATED POLYPROPYLENE, 4 F.R.

Temp., °C.	Rate, %/min.
137	.122
138	.083
139	.077
140	.056
142	.043
143	.037

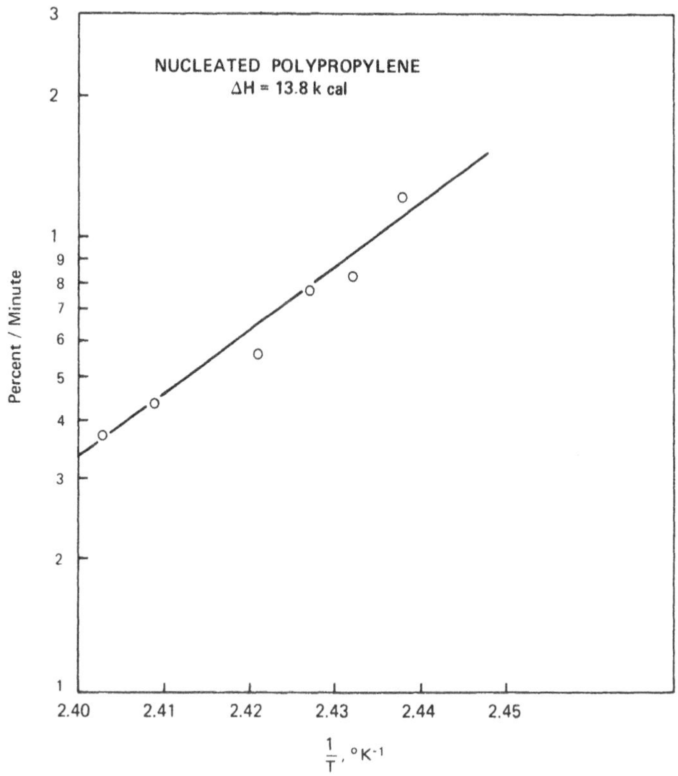

Figure 5.

A tentative explanation of the three activation energies is that the value of 13.8 kcal is for a completely heterogeneously nucleated polypropylene. The value of 21.4 kcal may be assumed to be due to a mixed system--some heterogeneous and some homogeneous nucleation. The value of 42.8 kcal is probably due to a predominantly homogeneously nucleated system although heterogeneous nucleation is undoubtedly present to some degree.

Summary

It is shown that a differential scanning calorimeter can be used to measure rates of total crystallinity development by measurements of the heat liberated from the crystallization process as a function of time. Initial results obtained on polypropylene samples suggest that the nature of the nucleation process changes at about 130°C.

ANALYSIS OF POLYSTYRENE SAMPLES IN THE LIQUID PHASE BY DIFFERENTIAL SCANNING CALORIMETRY

by

James A. Currie and Malcolm Dole
Materials Research Center
Northwestern University, Evanston, Illinois 60201

Introduction

Recently[1,2] there were published two separate studies of the specific heats of amorphous (henceforth called atactic) and isotactic polystyrene, as measured by adiabatic calorimeters. Whereas in the main the results of the two groups agreed rather well, discrepancies did exist amounting to as much as 7% at high temperatures in the liquid range. In order to determine whether the observed differences were the result of the different polystyrene samples used or of experimental errors, we decided to make a careful comparison of the two polystyrenes using a Perkin-Elmer differential scanning calorimeter. Rather to our surprise, we found that differences in the specific heat particularly between the atactic polystyrene samples did exist in the liquid range above T_g and that only after rather long annealing at 200°C did the specific heat of the two atactic samples approach each other.

EXPERIMENTAL

Polystyrene Samples Studied

Abu-Isa et al[1] originally measured the specific heats of a Dow atactic polystyrene called Styron 690. This is a polystyrene of rather broad molecular weight; the molecular weight as calculated from the equation $[\eta] = 1.10\times10^{-4} M^{0.725}$ was 2.4×10^5. In its commercial preparation, the polystyrene was extruded into a water bath and hence rather rapidly cooled. It consisted of transparent rods about 2mm in diameter and 3mm long. Karasz et al[2] on the other hand used an N.B.S. standard sample of atactic polystyrene No. 706 also of broad molecular weight distribution whose M_w was 2.58×10^5 and M_n 1.36×10^5. It consisted of pellets 6mm in diameter. Karasz et al stated that drying at 80° for 24 hr. caused negligible weight loss. We found, however, that the N.B.S. standard polystyrene No. 706 did lose about 0.9% in weight on being heated to 200°C for 10 hrs. In contrast, the Dow Styron 690 contained about 0.4% of volatile matter.

Inasmuch as the molecular weights of the two atactic samples were nearly the same, a nuclear magnetic resonance comparison of the two was carried out in order to see if the samples differed significantly in their tacticity. N.m.r. spectra were taken at 25 and 100°C in 10 and 15% solution using CCl_4 as the solvent. No difference in the spectra of the two samples could be detected. The densities of the atactic samples were also almost identical, differing by not more than 0.004g cm^{-3}.

In preparing the samples for the specific heat measurements, they were molded in a spring loaded press to a size to fit the aluminum calorimeter pans and usually of a thickness of about 0.04 inch. To anneal the samples, they were heated in the calorimeter at 200°C up to 14 hrs. and cooled at the slowest possible rate, i.e., about 0.5°C/min.

Calorimetry Techniques

A Perkin-Elmer differential scanning calorimeter, type DSC-1, was considerably modified and improved for these measurements. The major changes, which brought the instrument more in line with the DSC-1B calorimeter, can be listed as follows:

1. The following electrical sections of the calorimeter were either replaced entirely or largely by new components: (a) the T average amplifier; (b) the ΔT amplifier; (c) the recorder range switch; (d) the readout and program circuit;

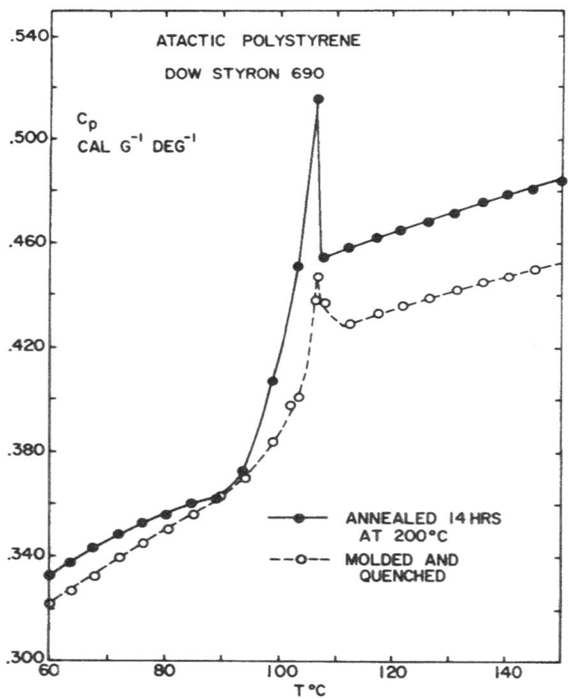

Fig. 1. Comparison of specific heats of Dow atactic polystyrene as received, lower curve, and after a 14 hr. annealing at 200°C, upper curve.

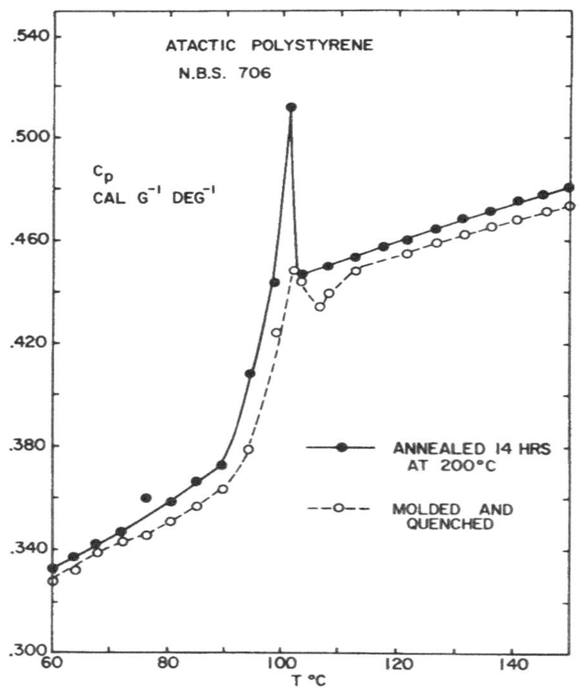

Fig. 2. Same as Fig. 1 except for the N.B.S. atactic polystyrene.

(e) the range-span-selector circuit; (f) the outputs to the recorder; (g) the scan speed switch; (h) the scaler circuit; (i) and the power supply; the single turn potentiometer was replaced by a 10 turn precision potentiometer for more sensitive control of the zero setting of the recorder pen.

2. The sample holder assembly of the DSC-1 model was replaced by that of the DSC-1B.

3. Similarly to the DSC-1B a base plate heater and switch were added to drive off or to prevent the formation of condensed water on low temperature runs.

4. The average temperature control potentiometer was replaced by one of improved accuracy and range to accommodate other changes in the same circuit.

5. A switch for changing the temperature ranges and trimpot controls to adjust the temperature range and span were added to the rear panel of the control cabinet.

6. A Texas Instrument Servo-riter recorder was used and along with it a series of plug-in cards was constructed to enable a quick and accurate change of the recorder scale to be made. Ranges covered were zero to 1.0, 1.3, 1.5, 2.0, 2.5, 3.0, 4.0, 5.0 and 10 mv.

The following experimental techniques and precautions were followed:

1. The recorder, control cabinet and analyzer section were mounted on sponge rubber, Berry shock mounts and Styrofoam, respectively, to reduce vibrations.

2. The analyzer section was housed in a glove box to reduce temperature fluctuations and to enable the system to be operated in a dry nitrogen atmosphere to avoid the condensation of moisture during low temperature experiments.

3. In the analyzer the diffuser plate and shield surrounding the sample holder cups were removed and replaced by a thin sheet of mica at a somewhat lower level. The mica sheet prevented any small object from falling into the electrical circuits below. This procedure reduced the baseline curvature considerably.

4. Two matched sample holder covers were made of polished gold sheet to replace the usual aluminum dome covers. The gold covers made more reproducible heat exchanges by radiation, and enabled a better baseline to be obtained.

5. The socket and pins below the sample holder assembly were frequently cleaned with methylene chloride to reduce signal noise.

6. The Al_2O_3 reference samples were adjusted in size so as to match the heat capacity of the unknown samples as closely as possible.

7. All of the aluminum sample pans and lids were matched in weight to the reference pan and lid to the nearest 0.02 mg.

8. The millivolt range of the recorder was chosen to give the largest deflection possible, compatible with the other instrument settings.

9. The slope was adjusted so that the isothermal chart lines before and after the 30° heating interval were as nearly equal as possible, usually within 2%. A second baseline was determined after recording the deflections

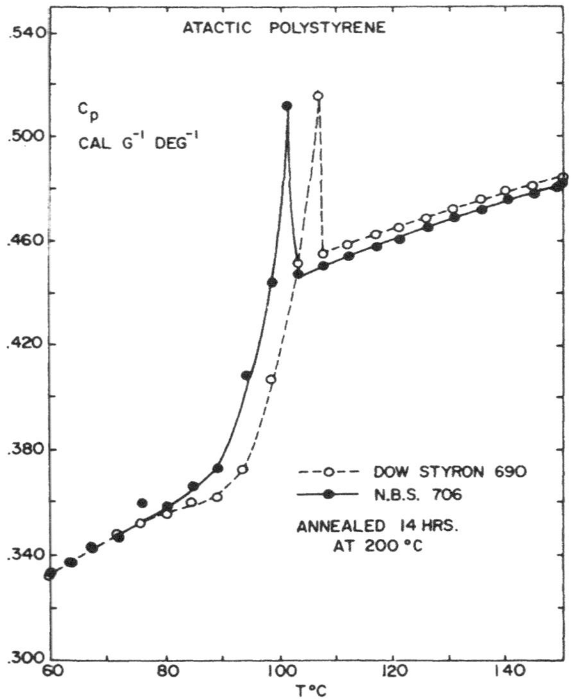

Fig. 3. Comparison of specific heats of annealed Dow and N.B.S. atactic polystyrenes from 60 to 150°C.

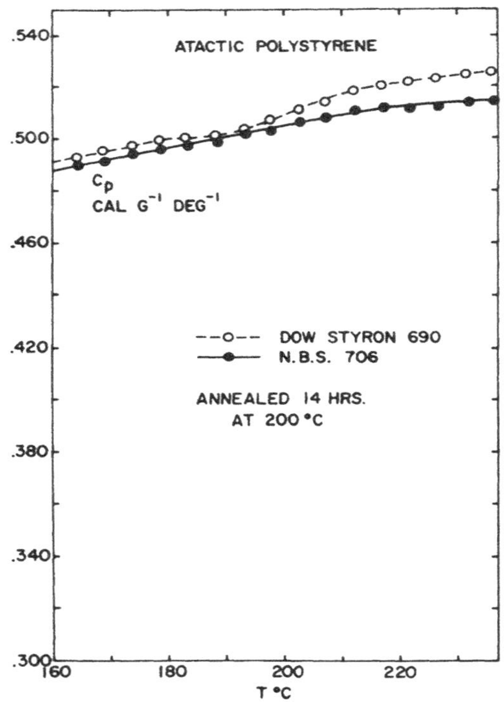

Fig. 4. Same as Fig. 3 except over the temperature range 160 to 240°C.

for the unknown and reference samples. If the second baseline was not superimposable on the first, the experiment was rejected. The heating rate chosen was 5°C/min.

Specific heats of the samples were calculated using the CDC 3400 computer and a 4 point regression equation for the temperature variation of the specific heat of aluminum oxide. (We are indebted to Dr. Edward Barrall of the Chevron Research Co. for the latter equation.)

Except at the lowest temperature range, the temperature scale was calibrated to be accurate to ± 0.3°C. The standard deviation over a temperature range of the specific heat of the Al_2O_3 reference samples was less than 1%. In the temperature ranges where the specific heat of the polymer samples varied linearly with temperature, the standard deviation of data from a straight line was typically less than 1%. Inasmuch as specific heat differences which we wished to study were 5% or higher, our modified DSC-1 calorimeter gave sufficiently accurate results for the purposes of this study.

RESULTS

Atactic Polystyrene

The experimental data for the atactic polystyrene are illustrated in Figs. 1-4. Figure 1 demonstrates that about a 7% difference between the specific heats of the same atactic material with different thermal histories can exist above the glass transition temperature. During the time scale of the experiments done here, the difference persisted without any indication of a reduction in the difference. The lower curve of Fig. 1 is that of the Dow Styron 690 as received, but briefly molded to fit the calorimeter pan and then quenched. Evidently, the long polymer chains in the material as received and molded must have been under considerable mechanical strains which prevented the specific heat above T_g from rising to the value shown by the completely relaxed (annealed) sample. The N.B.S. 706 atactic polystyrene did not show this effect, Fig. 2. After annealing for 14 hrs. at 200°C, the two types of atactic polystyrene had identical specific heats above T_g within the experimental limits of error, Fig. 3. However, above the annealing temperature the specific heats again diverged, Fig. 4. The difference in the data between Abu-Isa et al[1] and Karasz et al[2] is thus seen to be a real difference in the specific heats of the different samples used.

If the Styron 690 was quenched after annealing for 14 hrs. at 200°C, then the specific heat in the liquid range dropped from 0.484 to 0.471 cal $g^{-1} deg^{-1}$ at 150°C as compared to 0.458 for the material as received, or 0.460 as measured by Abu-Isa et al[1]. The physical appearance of the Styron 690 also changed on annealing; the material tended to take on the shape of a sphere. In contrast, the shape of the N.B.S. 706 sample did not change on annealing.

From the above described results, we conclude that the relaxation time of the long polystyrene chains at temperatures between 100 and 200°C is so long that mechanically strained samples did not have time to relax within the period of our measurements.

Isotactic Polystyrene

A striking difference was seen between the Monsanto and Dow samples of the quenched isotactic polystyrene used in this research, Figs. 5 and 6. As shown in Fig. 5, the Monsanto sample of quenched isotactic polystyrene did not crystallize on heating above T_g during our specific heat measurements whereas the Dow quenched polystyrene did, Fig. 6. The DSC results shown in these figures are the average of two runs each; hence, the non-crystallization in one case and the crystallization in the other is a reproducible fact. In Fig. 7, the data of Abu-Isa et al[1] and of Karasz et

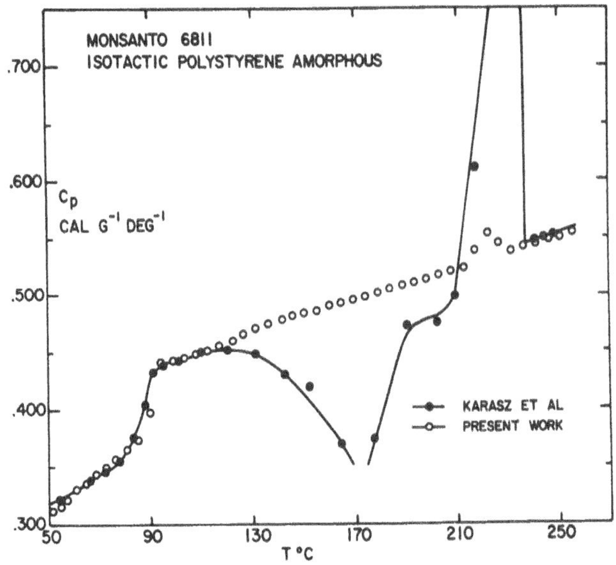

Fig. 5. Specific heats of Monsanto quenched isotactic polystyrene as measured by the DSC, open circles, and by the adiabatic calorimeter of Karasz et al. (2), closed circles.

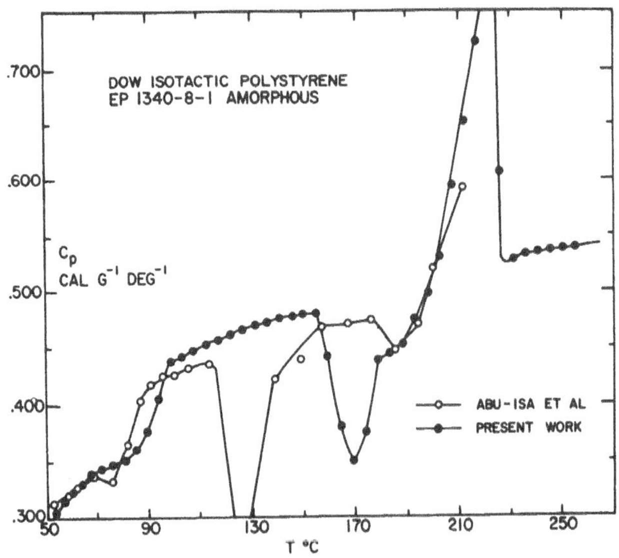

Fig. 6. Specific heats of Dow quenched isotactic polystyrene as measured by the DSC, closed circles, and by the adiabatic calorimeter of Abu-Isa (1), open circles.

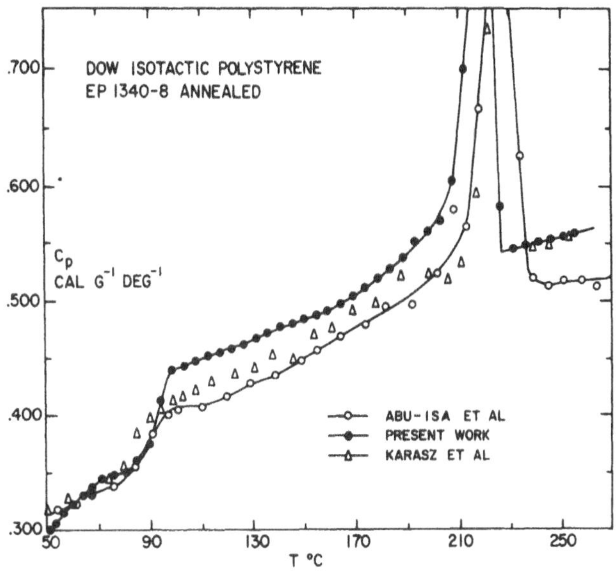

Fig. 7. Comparison of the specific heats of Dow annealed isotactic polystyrene as measured by the DSC, closed circles; and by the adiabatic calorimeters of Abu-Isa et al. (1), open circles, and of Karasz et al. (2), triangles.

al[2] are compared with those of the present research. In this case the data of Abu-Isa et al appear definitely to be too low in the temperature range above the melting point. In this particular experiment, the experimental errors were larger than normal because of the failure to maintain a good vacuum in the region between the adiabatic jacket and the calorimeter.

The data of Fig.6 demonstrate that the intermittent heating of the adiabatic calorimetric method promotes the probability of crystallization. In the case of the Dow material, Fig. 6, the crystallization occurred in the adiabatic calorimeter at a temperature 40° below the crystallization in the DSC experiments. However, in the case of the Monsanto polystyrene, Fig. 5, the sample did not crystallize at any temperature below the melting region when heated at 5°C/min in the DSC experiment despite the fact that considerable crystallization did take place in the adiabatic calorimetric experiments of Karasz et al[2] which were heated through this temperature range over a 24 hr. period.

ACKNOWLEDGEMENTS

This research was supported by the Advanced Research Projects Agency of the Department of Defense through the Northwestern University Materials Research Center. The DSC calorimeter was originally purchased with funds from the U. S. Atomic Energy Commission for another research (results of this research will be published later).

REFERENCES

1. I. Abu-Isa and Malcolm Dole, J. Phys. Chem. 69, 2668 (1965).

2. F. E. Karasz, H. E. Bair and J. M. O'Reilly, J. Phys. Chem. 69, 2657 (1965).

TRANSITION STUDIES OF POLY(2,6-DIMETHYL 1-,4-PHENYLENE ETHER)

F. E. Karasz[1], J. M. O'Reilly[2], H. E. Bair[3] and R. A. Kluge
General Electric Research and Development Center
Schenectady, New York 12301

Introduction:

A study has been made of the transition properties of poly(2,6-dimethyl 1-,4-phenylene ether) (PPO® polymer) using the wide-temperature range adiabatic calorimeter previously described[1] to measure the heat capacity and thereby examine transition behavior of the polymer from 80 to 570°K. In addition, differential scanning calorimetry was used to measure the melting point depressions of mixtures of the polymer with 1-chloronaphthalene, and from these data the heat and entropy of fusion were calculated.

Experimental:

A semi-crystalline sample of PPO polymer in finely powdered form was used for the adiabatic calorimetric measurements. The sample ($\overline{M}w$ = 40,000) was thoroughly dried before use.

Heat capacities were measured in approximately 10°K increments, reduced to 2 or 3°K in transitional regions, with an effective heating rate in these regions of the order of 1 - 2°K/hr.

For the melting-point depression experiments, mixtures of the same PPO sample and 1-chloronaphthalene were prepared by dissolving the polymer in excess diluent and evaporating the latter until approximately the desired compositions were found by chlorine analysis. The melting-points of 5 - 15 mg. samples of these mixtures were determined from DSC traces. T_m was taken as the temperature at which the trace through the endothermic peak returned to the baseline. For the pure polymer, this procedure gave a T_m value about 10° higher than that found by adiabatic calorimetry, but since the ΔH_f analysis essentially involves only differences in T_m's, not absolute values, this shift (provided that it was constant) is negligible in terms of ΔH_f.

Results and Discussion:

The C_p vs. T curve for the polymer is smooth and virtually linear between 100 and 400°K with no indication of any rapid activation of a C_p-contributing torsional or vibrational mode, such as might be expected, possibly, from the observation of a small low-temperature peak in dynamic mechanical loss measurements by Heijboer[2]. In this respect, poly(2,6-dimethyl 1-,4-phenylene ether) resembles other polymers that have been examined by the two techniques[3].

The semi-crystalline polymer did not display a rise in C_p at T_g (480°K, see below) but instead underwent further crystallization starting at a temperature somewhat below T_g. This was observed as an evolution of heat and a consequent drop in calculated values of C_p (Fig. 1). While reorganization of this kind is common and indeed to be anticipated in unannealed polymer samples, it must be considered anomalous for this to start below the observed T_g. It may be that because of the very low entropy of fusion (see below) crystallization is facilitated in regions of poly(2,6-dimethyl 1-,4-phenylene ether), making it kinetically possible even below T_g but certainly this phenomenon needs further study. However, it should also be noted that in absolute terms, the increase in crystallinity noted was still small, less than 5%.

Present Addresses: [1]Goessmann Laboratory, Univ. of Mass., Amherst, Mass., [2]Xerox Research Lab., Webster, New York, [3]Bell Telephone Labs., Murray Hill, New Jersey.
®Registered trademark of the General Electric Company.

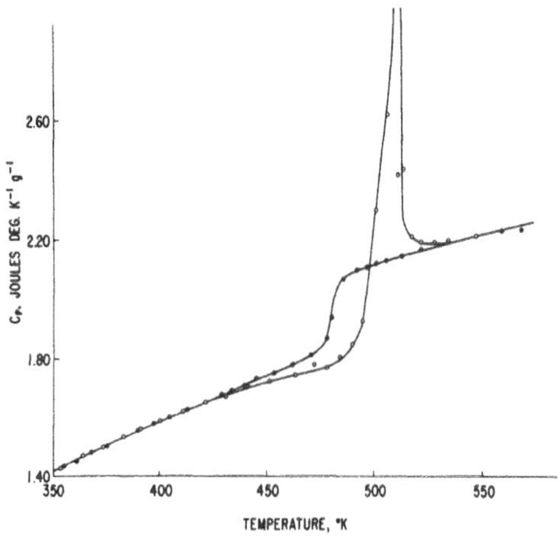

Fig. 1. Heat capacity of poly(2,6-dimethyl-1,4-phenylene ether). Open circles: crystalline polymer. Filled circles: amorphous polymer.

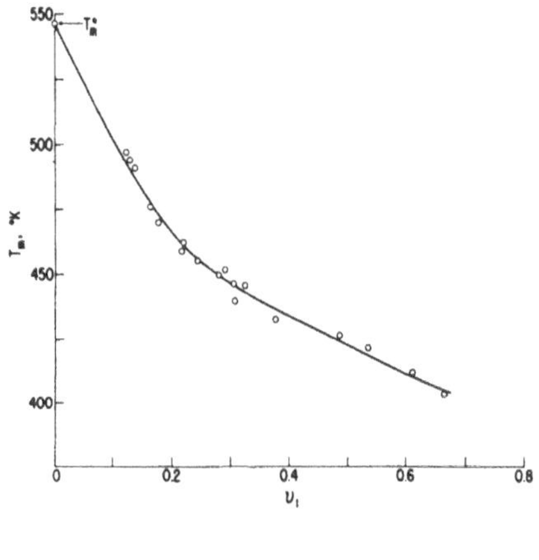

Fig. 2. Melting-points of poly(2,6-dimethyl-1,4-phenylene ether)-1-chloronaphthalene mixtures.

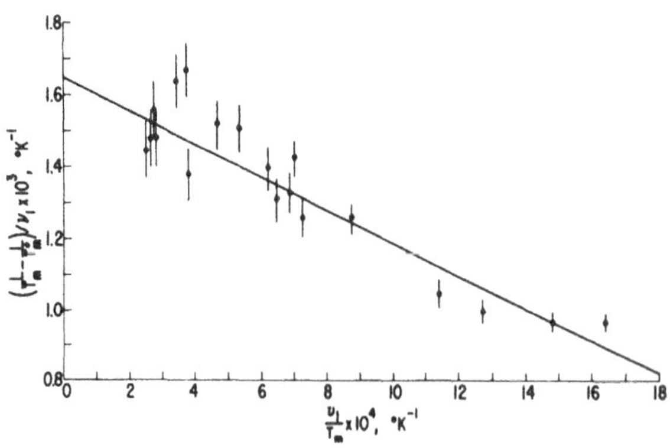

Fig. 3. Data of Fig. 2 plotted as discussed in text.

The melting-peak had a maximum at about 510°K but also displayed a 'tail' on the high temperature side of the endotherm which indicates that further reorganization to more thermodynamically perfect crystalline regions is possible and also suggests that the T_m recorded (535°K) may still be some degrees below the equilibrium melting-point. The area under the peak corresponded to a heat of fusion of 15 joules/g.

The amorphous polymer was produced by cooling the melt in situ. A rise in C_p characteristic of a glass-transition was noted at 480°K, and above this temperature the C_p vs. T curve merged smoothly into the curve for the molten polymer, with no trace of a melting endotherm.

Fig. 2 shows the melting point of mixtures of poly(2,6-dimethyl 1-,4-phenylene ether) and 1-chloronaphthalene up to a diluent volume fraction v_1 of 0.65. These results were analyzed in the customary manner[4], in which $(1/T_m-1/T_m^o)/v_1$ is plotted against v_1/T_m. In these expressions T_m^o and T_m are the melting-points of the pure polymer and of a polymer-diluent mixture containing a volume fraction v_1 of diluent, respectively. The intercept of such a plot is proportional to ΔH_f^{-1} while the slope is related to the temperature independent polymer-diluent interaction parameter,B. Fig. 3 indicates the data fell reasonably close to a straight line and yielded a value of about 900 cals/mole repeating unit for ΔH_f(7.5 cals/g) and about 3.5 cals/ml for the Flory interaction parameter B, both subject to about ± 20% uncertainty. The present value for the ΔH_f may be low; the only other available result, from measurements of precipitation temperatures of the polymer from toluene, yielded about 1400 cals/mole.[5] In any case, it is clear that the heat and entropy of fusion of this polymer are unusually low and currently there is no explanation for this observation in structural terms. Dilute solution measurements can be interpreted as indicating a rather flexible molecule[6]. The other unusual aspect of the transition properties, as has been previously noted[7], is the high value of T_g/T_m. A thermodynamic analysis has been presented elsewhere[8], in which it has been shown that a high T_g/T_m and low ΔS_f can be reconciled in terms of current glass transition theory, if it is assumed that the conformational entropy for poly(2,6-dimethyl 1-,4 phenylene ether) is about the same as for other glass-forming systems.

Acknowledgement:

This work was supported in part by the Directorate of Chemical Sciences, Air Force Office of Scientific Research, through Grant AFOSR 68-1434.

References:

1. F. E. Karasz and J. M. O'Reilly, Rev. Sci. Inst. 37, 255 (1966).
2. J. Heijboer, J. Polymer Sci. C16 (in press).
3. J. M. O'Reilly and F. E. Karasz, J. Polymer Sci. C14, 49 (1966).
4. L. Mandelkern, Crystallization of Polymers, McGraw-Hill, Inc., New York, New York, 1964, pp. 38 ff.
5. A. R. Shultz, private communication.
6. J. M. Barrales-Rienda and D. C. Pepper, J. Polymer Sci. B4, 939 (1966).
7. F. E. Karasz and J. M. O'Reilly, J. Polymer Sci. B3, 561 (1965).
8. F. E. Karasz, H. E. Bair and J. M. O'Reilly, J. Polymer Sci. Part A2 (in press).

MELTING POINT RELATIONSHIPS FOR POLY(ETHYLENE TEREPHTHALATE) SYSTEMS

by

P. E. Slade* and T. A. Orofino
Chemstrand Research Center, Inc.
Durham, North Carolina

A. INTRODUCTION

This investigation was undertaken in an effort to establish a value for the heat of fusion of poly(ethylene terephthalate) using differential thermal analysis (DTA) as the tool for measuring the melting point depression of the polymer due to added solvent.

B. THEORY

The condition for equilibrium between a crystalline polymer and a solution comprised, in whole or in part, of the polymer melt at the melting point T_m of the mixture is equality of the (partial) molar free energies of the polymeric constituent common to both phases. In accordance with the usual simplifying assumption that the molar heat of fusion of the polymer segments, ΔH_u, is independent of temperature, this relationship for polymers (and low molecular weight, crystalline solutes as well) takes the form

$$(1) \quad 1/T_m - 1/T_m^o = -(R/\Delta H_u)\ln a_2$$

Here, T_m^o is the melting point for pure solute, a_2 is the activity of the solute in the co-existing liquid phase and R is the gas constant.

For low molecular weight solutes, the approximation $a_2 \simeq$ mole fraction N_2 transforms Eq. (1) to the classical expression for depression of the melting temperature of the major component (component 2).

$$(2) \quad 1/T_m - 1/T_m^o \simeq RN_1/\Delta H_u; \quad \Delta T_m \simeq R(T_m^o)^2 N_1/\Delta H_u$$

The expression applies equally to polymer/solvent mixtures, if the mole fraction of solvent N_1 is sufficiently low. For solutions of polymers containing higher proportions of diluent, insertion into Eq. (1) of the more appropriate Flory-Huggins expression for ln a_2 yields the equation (1)

$$(3) \quad 1/T_m - 1/T_m^o = (R/\Delta H_u)(V_u/V_1)(v_1 - \chi_1 v_1^2)$$

in which v_1 is volume fraction of solvent, V_u and V_1 are the molar volumes of polymer segments and solvent, and χ_1 is a polymer/solvent interaction parameter. Eq. (3) forms the basis for analyses of polymer/diluent melting, in which mixtures the

* Present Address: Technical Center, Monsanto Co., Pensacola, Fla.

polymer is the major constituent (v_1 is small).

For the systems of present interest, modification of Eq. (3) is required because the polymer units present in the crystalline phase (ethylene terephthalate units) are copolymerized with a small proportion of diethylene glycol (DEG). If the latter units are considered to be excluded from the polymer crystal, i.e., if they function as a diluent, this means that $v_1 \neq 0$ even in the total absence of added solvent. The appropriate revision may be achieved as follows: If N_2 (mole fraction of ethylene terephthalate units present in a copolymer/diluent mixture) is near unity, we may write in accordance with Eq. (2).

$$(4) \quad 1/T_m - 1/T_m^o \simeq (R/\Delta H_u)(N_1 + N_x)$$

where N_x is the mole fraction of DEG units. If n_i denotes moles of constituent i present, then

$$(5) \quad N_1 = n_1/(n_1 + n_2 + n_x)$$

$$N_x = n_x/(n_1 + n_2 + n_x) = p\,(1 - N_1)/(p + 1)$$

where $p = n_x/n_2$ is considered fixed and given. (It is the mole proportion of DEG units in the polymer.) Substituting into Eq. (4) we have

$$(4a) \quad 1/T_m - 1/T_m^o \simeq (R/\Delta H_u)(p+1)^{-1}(N_1 + p)$$

When the amount of added diluent is zero ($N_1 = 0$), we have

$$(6) \quad 1/T_m - 1/T_m^o = 1/T_m^* - 1/T_m^o = pR/\Delta H_u(p+1)$$

where T_m^* is the melting point for the pure copolymer. Rearrangement of the preceeding equations yields the desired limiting relation for the melting point depression, relative to the copolymer, due to solvent addition. Thus

$$(7) \quad 1/T_m - 1/T_m^* = [R/\Delta H_u(p+1)](V_u/V_1)v_1 + O(v_1^2)$$

Here v_1 refers to the volume fraction of added solvent only. The parameter V_u is the number average volume per mole of all units present in the chain. If p is known, Eq. (7) suggests that the heat of fusion of the polymer ΔH_u may be obtained from the limiting slope of a $1/T_m$ versus v_1 plot, in accordance with the usual procedure involving Eq. (3) for homopolymers.

C. ANALYSIS OF DTA THERMOGRAMS

In the measurement of melting temperatures by DTA for the polymer/diluent systems examined here, the question arises as to what compositions in the melt the observed melting peaks refer. It must be recognized, of course, that a polymer/diluent mixture heated from, say room temperature, can display a continuum of melting points corresponding to the changing composition of the liquid phase. The last melting corresponds to equilibrium between a minute amount of crystalline phase and a polymer solution essentially of the composition originally made up. Usually, it is only the latter solvent content which is known and one would hope to record the melting point with which it corresponds. There is some support for believing that this is the case in the DTA experiment.

Suppose in a given experiment φ_1 and φ_2 are the actual volumes of diluent and molten polymer in the liquid phase of the sample at some temperature T_m and v_1 and v_2 are the corresponding volume fractions. Since solvent is excluded from the crystalline phase, φ_1 is also the total solvent volume (a constant). The value of v_1 however, depends both on the amount of solvent originally added to the sample and on the volume of the crystalline phase. If we assume that at various temperatures T_m traced by the DTA profile melting equilibrium is continuously maintained then each point on the T axis corresponds to some definite composition of the liquid phase. We then have

$$(8) \quad v_1 = \varphi_1/\varphi_1+\varphi_2); \; dv_1 = -[\varphi_1/(\varphi_1+\varphi_2)^2]d\varphi_2 = -v_1^2 \, d\varphi_2/\varphi_1$$

According to Eq. (3)

$$(9) \quad 1/T_m - 1/T_m^o \simeq (R/\Delta H_u)(V_u/V_1)v_1; \; -dT_m/T_m^2 = (R/\Delta H_u)(V_u/V_1)dv_1$$

Combining Eqs. (8) and (9), we have

$$(10) \quad d\varphi_2/dT_m = [R\varphi_1 V_u (T_m^o)^2/V_1 \Delta H_u](T_m^o - T_m)^{-2}$$

The quantity in square bracket is constant. Thus, $d\varphi_2/dT_m$ attains its maximum value for T_m values close to T_m^o (i.e., at the higher T_m values). This shows that the <u>amount</u> of polymer which melts per degree rise in program temperature is maximum near the end of the melting range and we might thus feel more confident that the DTA peak (maximum heat absorption by the sample) is close to the "last melting" for which v_1 is the v_1 weighed into the polymer. We have implicity made this assumption in the analysis of data following.

D. EXPERIMENTAL

If poly(ethylene terephthalate) is mixed with some compatible diluent, such as dibutyl phthalate or tricresyl phosphate, the melting point is lowered according to Eq. (7) presented in an earlier section. In this study samples of Goodyear VFR 347 polyester, containing 3.4 mole percent diethylene glycol, (DEG), were melted at about 280°C under a nitrogen atmosphere and various amounts of diluent were added. After cooling the well stirred mixture the samples were crushed and analyzed for diluent content.

The analysis for diluent in the polymer employed standard NMR techniques. In this analysis signals due to both aromatic and aliphatic protons were used in the calculation to minimize the effect of the DEG present in the polymer.

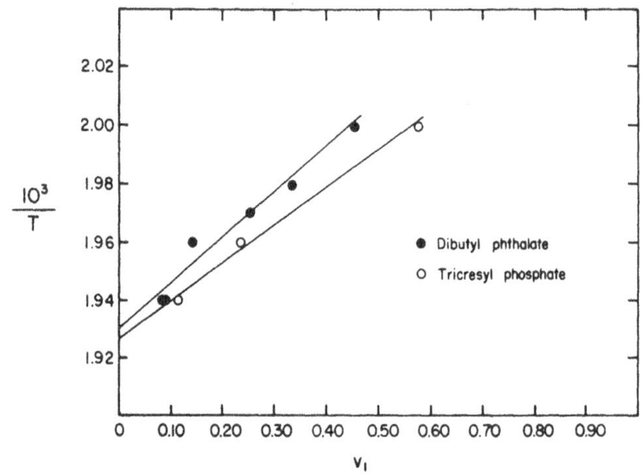

Fig. 1. Melting Temperature vs. Volume Fraction Diluent

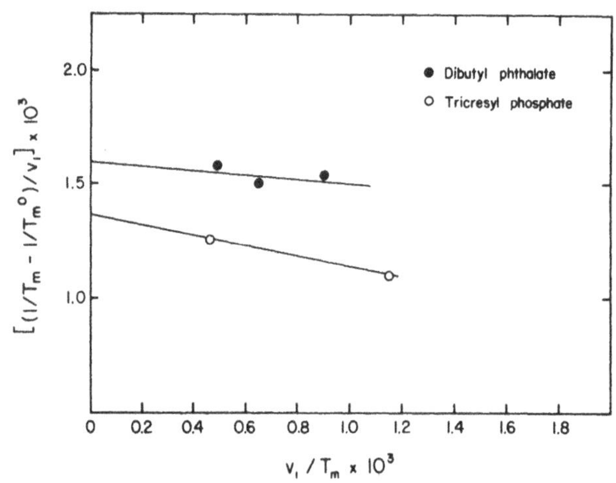

Fig. 2. Plot for Calculating B, the Energy of Interaction Constant

Melting points were measured by DTA utilizing a du Pont Model 900 instrument. The heating rate was 20°C/min. and a nitrogen atmosphere was employed throughout the analysis.

E. RESULTS

A plot of T_m^{-1} vs. volume fraction of diluent is shown in Figure 1. The intercept of the lines, which were drawn using a least squares calculation, is the reciprocal of the melting temperature of the pure PET - DEG copolymer and the limiting slopes are related to the heat of fusion of pure crystalline PET. The parameter p is equal to the mole fraction of repeat units containing DEG divided by the mole fraction of repeat units with ethylene glycol. The DEG content of the polymer is 3.4 mole percent diethyleneglycol terephthalate and thus p is calculated to be 0.04. Substituting into Eq. (7) we may calculate the heat of fusion as follows:

1) Dibutyl phthalate

$$\text{slope} = [R/(p+1)\Delta H_u](V_u/V_1) = 1.57 \times 10^{-4}$$

Here, V_u has been assigned the value for the molar volume of PET and V_1 is the molar volume of dibutyl phthalate. In these calculations the density of PET was taken as 1.34 g/cc, the density of the amorphous polymer. The result for the heat of fusion is ΔH_u = 6560, cal/mole repeat units or

$$\Delta H_f = \frac{6560}{192} = 34.2 \text{ cal/gm.}$$

The latter result is, of course, independent of the definition of what one considers to be the polymer repeat unit. The intercept of the line for the dibutyl phthalate system corresponds to an extrapolated melting point of 245°C. Experimentally this is indeed the measured melting point of this polymer.

2) Tricresyl phosphate

By the method of part 1) above, we find ΔH_u = 6654 cal/mole repeat units; ΔH_f = 34.6 cal/g. The values for ΔH_f are seen to compare favorably between the two polymer/solvent systems, although the slopes differ by about 20%, reflecting the difference in values for molar volume of solvent.

F. ESTIMATES OF POLYMER/DILUENT INTERACTION

For the purpose of estimating the extent of polymer-solvent interactions in the systems considered, we may write in accordance with Eq. (3) an extended version of Eq. (7) as follows:

$$(11) \quad 1/T_m - 1/T_m^* = [R/\Delta H_f(p+1)](V_u/V_1)[v_1 - \chi_1 v_1^2]$$

Here, χ_1 is an interaction parameter between copolymer and diluent. As far as the value of χ_1 is concerned, the approximation introduced in Eq. (11) amounts to neglect of the small proportion of DEG units contributing to the diluent content of the liquid phase. In the calculation of ΔH_f in section B, χ_1 has been taken as zero. In fact, however, this may not be the case.

Flory (6) suggests a method for calculating the interaction in which, in terms of Eq. (11), a plot of $[(1/T_m - 1/T_m^*)/v_1] \times 10^3$ against $v_1/T_m \times 10^3$ is made. From the intercept of such a plot ΔH_f may be calculated. An interaction parameter $B = \chi_1 RT_m/V_1$ may be computed from the ratio of slope to intercept.

By using the data obtained from this investigation such a plot was made and is shown by Figure 2. Values of v_1/T_m of less than 0.4×10^{-3} were not used in drawing the lines since the concentration of diluent is too low to give a significant melting point depression for use in these calculations.

1) Dibutyl phthalate

The extrapolated intercept for dibutyl phthalate/diluted polymer is 1.6×10^{-3}; the result for the heat of fusion is then $\Delta H_f = 33.6$ cal/g. This is in good agreement with the value calculated in section D.

From the slope we calculate $B = 0.5$, or, $\chi_1 = 0.13$ at T_m^o. This positive χ value indicates that dibutyl phthalate is a relatively poor solvent for polyester.

2) Tricresyl phosphate

The same calculation may be made for polyester diluted with tricrysyl phosphate. The extrapolated intercept for this system is 1.35×10^{-3}. Thus,

$$\Delta H_f = 33.6 \text{ cal/gm}$$

This is the same value obtained with dibutyl phthalate and again agrees with the ΔH_f from section B.

From the slope of this line B is found to be 0.6; thus, $\chi_1 = 0.18$ at T_m^o. Tricresyl phosphate is then a fairly poor solvent also. It is evident from the foregoing that no appreciable error in the assignment of ΔH_f is introduced through neglect of the χ_1 term in Eq. (13) for either system.

G. CONCLUSIONS

The application of conventional melting point depression theory to polyester systems has, we feel, led to quite satisfactory results.

This study has provided a reasonable estimate for the heat of fusion of polyester, 34.4 cal/g, which compares very well with the value of 31.8 cal/gm reported by Smith and Dole (2). In their experiments diluents were used but no experimental details were reported. Kirshenbaum (3) calculated a ΔH_f of 30.7 cal/gm from entropy values while Edgar and Ellery (4) have suggested a ΔH_f of 11.5 cal/gm as calculated from the depression of the melting point by copolymerization. The latter technique probably gives the least correct value (3,5) of those described.

The ratio of calorimetrically measured latent heat for a given sample to ΔH_f provides a useful index of crystallinity. This method, it should be noted, is not restricted to pure polyester, provided only (as seems likely) that other co-monomers do not form part of the crystalline phase.

The extension of the methods described here to other fiber-forming polymers should be straightforward and would provide additional useful relations for polymer characterization.

REFERENCES

1. Flory, P. J., Principles of Polymer Chemistry, Cornell University Press, Ithaca, New York, 1953.

2. Smith, C. W. and M. Dole, J. Polymer Sci. 20 37 (1956).

3. Kirshenbaum, I., J. Polymer Sci. A3, 1869 (1965).

4. Edgar, O. B. and E. Ellery, J. Chem. Soc. 1952, 2633.

5. Mandelkern, Leo, Crystallization of Polymers, McGraw-Hill, New York, 1964.

6. Flory, P. J., Leo Mandelkern and H. K. Hall, J. Am. Chem. Soc. 78, 2532 (1951).

THERMAL ANALYSES OF POLYMERS. I. POLYCARBONATE AND POLYETHYLENE TEREPHTHALATE

by
Gerald W. Miller
Mobay Chemical Co.
Pittsburgh, Pa. 15205

INTRODUCTION

Polyethylene terephthalate (PET) and polycarbonate (PC) are two individual aromatic polyesters which have been used for rather different commercial applications. The former is used abundantly in film and fiber applications because of its high strength and ability to crystallize, whereas the latter has found use primarily in applications requiring excellent clarity, high heat distortion, and high impact strength.

A number of investigations (1-7) have been carried out on the thermal behavior of PET, and the work of Illers and Breuer (8) is probably the most illustrative of the relaxations occurring in PET with some data on the behavior of PC. The observed relaxation processes for various terephthalate esters have been reported by Farrow, et.al. (4), showing the effects of chain length of the diol on the dynamic mechanical behavior. A number of authors (9-12) have treated the apparent morphological changes accompanying thermal and solvent crystallization. While the data by Illers and Breuer (8) indicate that only small changes are observed in the mechanical loss data of PET at low temperatures, the dynamic mechanical behavior of PC has been reported by Nielsen (13) and Schnell (14), indicating significant changes in PC loss peaks upon crystallization.

This report illustrates the correspondence of changes in heat capacity, expansion, penetration and strain (tensile modulus) behavior with temperature for both of these polymers as well as reporting some new relaxations at specific temperatures.

EXPERIMENTAL

Materials

The polymer specimens used in this study were either solution cast or extruded without any significant degree of orientation from a ribbon die. The materials used to study the general thermal properties were injection grade MERLON® polycarbonate and Mylar* polyethylene terephthalate. The polycarbonate was crystallized by immersion in CCl_4 several months before drying and obtaining its expansion characteristics.

Methods

1. Thermomechanical Analyses

Thermomechanical analyses were carried out in three basic modes: tension, compression and expansion. A DuPont 940 Thermomechanical Analyzer, an instrument equipped with a probe attached to a linear variable differential transformer (LVDT), amplifier and X-Y recorder and a temperature programmer,

*DuPont Polyethylene Terephthalate

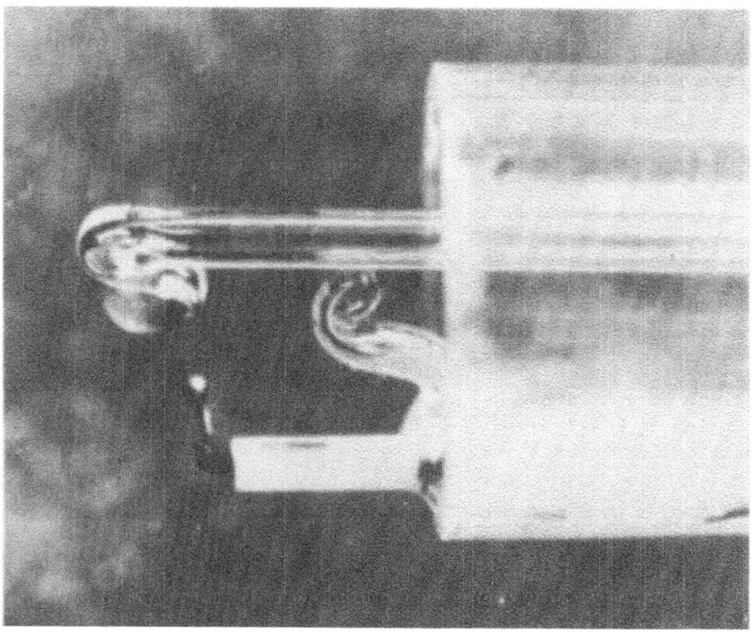

Fig. 1. Quartz probe modification of DuPont 940 TMA for penetration.

Fig. 2. Quartz hook assembly for DuPont 940 TMA for tensile strain measurements.

was modified to use a quartz probe (Figure 1) for compression and expansion and a quartz hook assembly (Figure 2) for strain measurements.

a. Compression

The spherically tipped quartz probe was attached to the core of the LVDT on the 940 TMA, eliminating the stainless steel connecting rod holding the chuck for the normal penetration mode. A thermocouple was attached to a silver pan which rested on the quartz stage assembly (Figure 1). The sample was placed in the pan, cooled with liquid nitrogen, loaded with a desired number of grams, and the temperature programmed from -196°C. to the melt temperature at 5°C./min.

b. Expansion

The expansion of PET and PC samples was measured on a 0.125" thick piece, using a flat-tipped quartz probe. The polymer sample rested on the quartz stage, and the quartz probe was set so that there was no travel of the probe using the maximum sensitivity of the instrument (2.75×10^{-4} in./in. chart). The temperature change was monitored by a chromel-alumel thermocouple which was spread flat on the quartz stage floor next to the sample. The temperature was programmed from -196°C. at a heating rate of 5°C./min.

c. Tension

The tensile measurements versus temperature at constant stress were carried out on solution-cast or extruded films. The film was inserted into the hooks (Figure 2) after punching two holes about 0.250" apart in the film. The hooks were set so that the film was not in tension. The film was then cooled to -196°C., the hooks reset so that no tension was exerted on the sample, and the sample loaded. The temperature was then programmed from -196°C. at 10°C./min.

2. Differential Thermal Analyses (DTA)

These measurements were carried out on the DuPont 900 DTA using a heating rate of 15°C./min. This heating rate tended to sharpen the transitions observed at lower heating rates.

3. Thermogravimetric Analyses (TGA)

These studies were carried out on a DuPont 950 TGA at a heating rate of 15°C./min. The isothermal weight loss comparison between PC and PET was run at 400°C.

RESULTS AND DISCUSSION

A. Thermogravimetric Analyses

While it has been reported that the melting range for polyethylene terephthalate is higher than that for polycarbonate (2), PC is

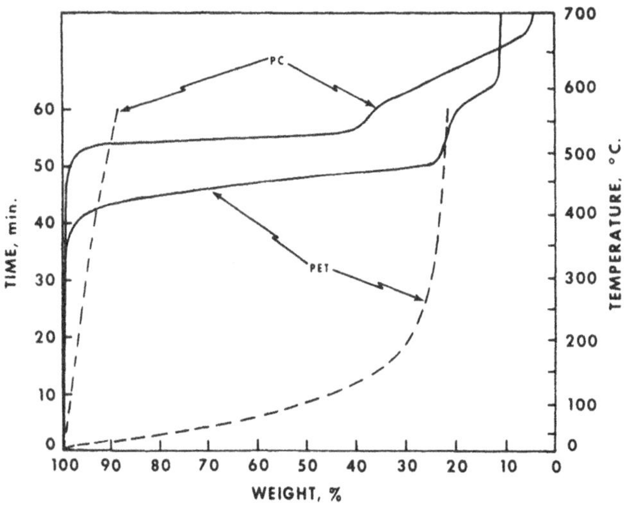

Fig. 3. TGA (___) and isothermal TGA at 400°C, (----) for polycarbonate and polyethylene terephthalate.

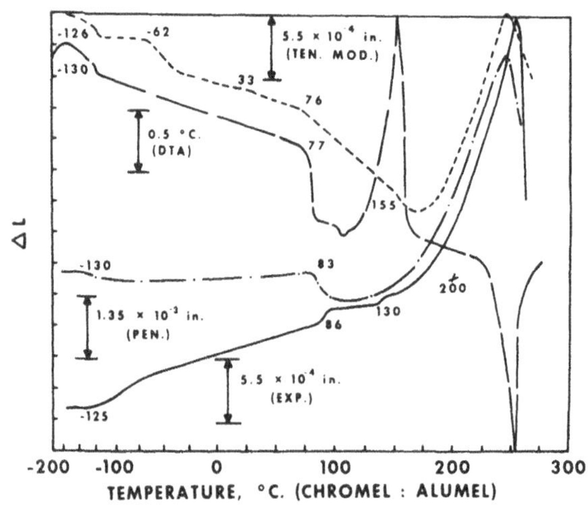

Fig. 4. Linear expansion (___), penetration (-.-), tensile modulus (---) and DTA (--) vs. temperature for polycarbonate.

thermally more stable than PET. This can readily be shown in Figure 3 from typical TGA data, showing the weight loss with temperature. The relative stabilities of each polymer become even more apparent when their weight loss is measured isothermally at 400°C. MERLON polycarbonate decreases in weight by 12% in one hour at 400°C., whereas PET loses 88% of its weight under the same conditions. In fact, this 88% loss is achieved in 40 minutes. Hence, their thermal degradative stabilities are very different.

B. TMA and DTA Correlations

1. Modifications for the DuPont 940 Thermomechanical Analyzer

The comparison among the different thermal techniques was made possible by the design of two new probe systems for the DuPont 940 TMA. While expansion measurements are very tedious and measured over limited temperature ranges, the 940 with a quartz probe (Figure 1) permits the automatic recording of both expansion and penetration behavior over the temperature range -196°C. to 650°C. The substitution of the quartz probe for the stainless steel connecting rod apparently eliminates possible heat transfer to the linear variable differential transformer, because the metallic tip on the present 940 probe behaves similarly to quartz, i.e., very low coefficient of expansion.

The use of the quartz hook assembly (Figure 2) is a totally new adaptation for automatic thermomechanical analysis, because this modification permits the use of the 940 TMA for measuring very small changes in strain. The sample can be investigated under conditions for pure Hookean behavior. The film or fiber can be subjected to a number of stress levels to measure either pure Hookean behavior or predominantly viscoelastic effects. Since polymers exhibit four stages of stress-strain relationships, i.e., rigid, viscoelastic, elastomeric and viscous flow, the amount of tolerable stress for the sample to behave as a purely elastic body over a broad temperature range can be calculated from stress-strain measurements at a temperature where the polymer exhibits elastomeric behavior, or in some cases, the temperature range prior to the heat distortion temperature. This latter phenomenon is true for many semi-crystalline polymers.

If the polymer sample is loaded so that no creep occurs during the length of the measurement, the curve of strain versus temperature represents the change in tensile modulus with temperature. However, if the loading causes creep within the sample, a plot of tensile modulus can be made by correcting for the induced creep. The third alternative is to measure only relative changes in modulus and their relations to the temperatures at which these changes occur; this is the choice we have made in this study. Hence, there is no information in this study from the calculation of any absolute modulus values, but relative modulus changes have been shown. Other work in this laboratory has shown that this measurement is a true modulus measurement under certain circumstances, and this would be the substance for a later publication.

TABLE I

TRANSITIONS IN POLYCARBONATE AND POLYETHYLENE TEREPHTHALATE

Polyethylene Terephthalate

DTA	Tensile Modulus	Linear Expansion	Penetration
-130	-126	-125	-130
-62	---	---	---
-5	-7	---	---
---	33	---	---
77	76	86	83
---	155	130	---
207	---	200	---
225	242	252	243

Polycarbonate

DTA	Tensile Modulus	Linear Expansion	Penetration
---	-140	-120	-145
75	53	55	52
145	148	151	162
170	158	170	183

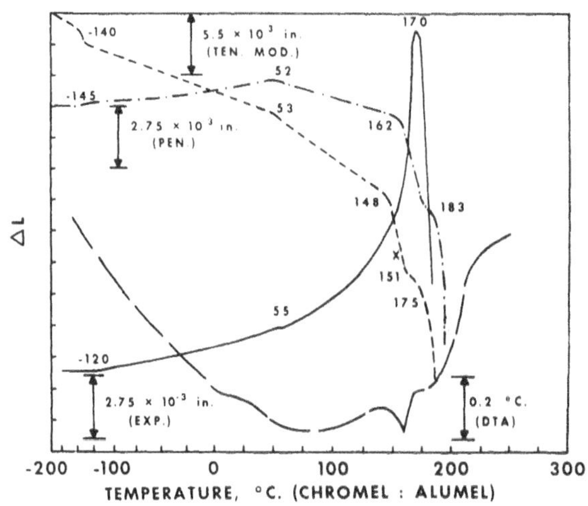

Fig. 5. Linear expansion (__), penetration (-.-), tensile modulus (---) and DTA (--) vs. temperature for polycarbonate.

2. Polyethylene Terephthalate

Though the work of Illers and Breuer (8) demonstrates well the dynamic behavior of PET, there has been no work published which shows the existing relationships between heat capacity, expansion, penetration and tensile strain. Figure 4 shows the relationship of the PET phase transitions observed by these four techniques, and these values are tabulated in Table I. There is a low temperature transition which exists at approximately -130°C.

Illers and Breuer (8) have reported maxima in loss modulus at -70, -105 and -165°C. by plotting the slope of the shear modulus vs. log frequency at 1 cycle/sec. as a function of temperature, but they only see a maximum in the loss modulus at -65°C. These are two large relaxations at -62 and -130°C. by tensile strain measurements, and only the latter transition has been detected by DTA, expansion and penetration. Some very small heat change is probably associated with all of the relaxations shown in Figure 4, but the detection of them by DTA could be limited by the lack of sensitivity of present DTA instruments. Nevertheless, the change in these mechanical modes of examination at -130°C. is detectable by DTA.

A number of small changes in measurement are listed in Table I, e.g., -5 (DTA) and -7°C. (strain), which are not indicated in Figure 4, for their magnitude of change is extremely low. All four techniques show a change in behavior around 70°C., and this phase transition has been termed the glass transition. A transition at 200°C. is confirmed by DTA and linear expansion, and the high temperature transition at 226 to 252°C. is that due to the polymer crystallites melting, usually identified as the polymer softening temperature.

The unassigned DTA endotherms as well as the change in expansion at 130°C. may be due to small areas of molecular order, not readily detectable by X-ray. Statton (15) has suggested that areas of molecular order of crystallinity be identified by their mode of measurement rather than expecting that all molecular order be confirmed by X-ray data.

3. Polycarbonate

Correlations by DTA, modulus, linear expansion and penetration for MERLON polycarbonate are shown in Figure 5 and tabulated in Table I. The marked changes in expansion and modulus at -120 to -140°C. correlate with the maximum in mechanical damping at -120°C. by Nielsen (13).

The change in behavior at 50°C. for all of the four techniques shown in Figure 5 confirms the existence of this previously unreported transition.

The relaxation occurring at 150°C. has been termed the Tg of polycarbonate (16). It is noteworthy that all four techniques also show a transition around 170°C. This former transition (150°C.) is very similar in expansion measurements to the melting of crystallites in crystallized PET in the region of 250°C. This expansion behavior is more typical of a first order transition, indicating that the 150°C. transition in PC may not be that of a Tg. The existence of the -130°C. transi-

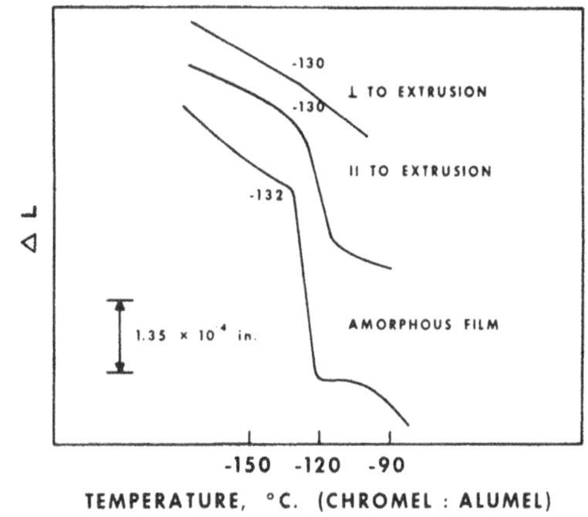

Fig. 6. The low temperature relaxation of polyethylene terephthalate for oriented and cast films.

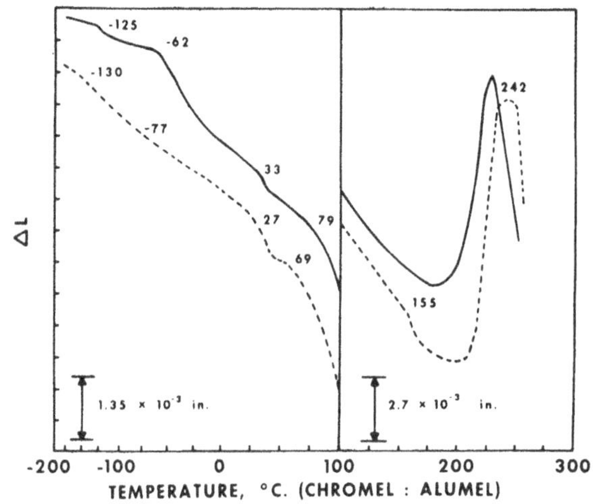

Fig. 7. The tensile modulus behavior of amorphous (__) and crystallized (---) polyethylene terephthalate.

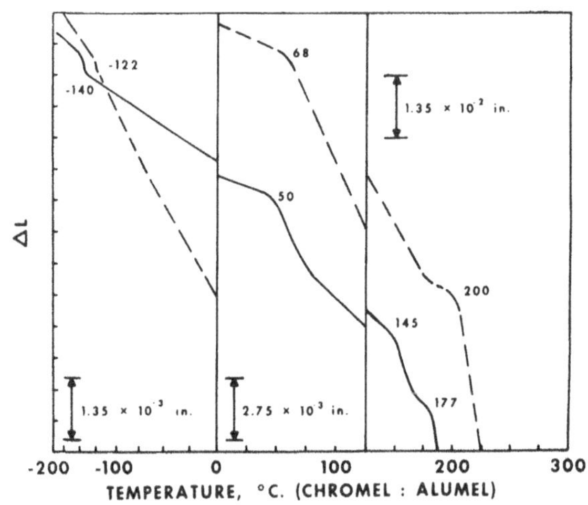

Fig. 8. The tensile modulus behavior of amorphous (__) and crystallized (---) polycarbonate.

tion in PET casts doubt on the assignment of the -140°C. transition in PC as due to rotation of the carbonate group (17).

4. Effect of Orientation on Low Temperature Transition in PET

The initial tensile strain data was gathered on extruded PET tape, and some amount of orientation in the film caused differences in the magnitude of modulus change for the -130°C. transition. Figure 6 shows the relative magnitude of the strain changes occurring for amorphous cast film, parallel to the direction of extruded film and perpendicular to the direction of extruded film. Under identical conditions, the largest change in strain occurred with the amorphous film, whereas the measurement perpendicular to the direction of extrusion showed the smallest change.

This difference in magnitude of strain change could be explained by the following considerations. There are many areas of small order within the extruded tape, and these act as tie points between the entangled chains. When the polymer is stretched in the direction of extrusion (orientation), the tie points hold the matrix together but allow the polymer chains to lengthen. When the polymer is strained perpendicular to extrusion, the oriented molecules must flex rather than stretch, and equivalent loading of the film in each of these two directions must cause a larger elongation in the direction of extrusion. The amorphous PET film has perhaps very few ordered areas or very small ordered areas which will not physically link the polymer chains, and on straining the bulk polymer, these polymer molecules are more readily elongated than the more crystalline extruded polymer. Hence, the amorphous polymer has the largest change in elongation at -130°C.

5. Effect of Crystallization on Modulus

a. Polyethylene Terephthalate

Crystallization by thermal or solvent techniques always induces some change in modulus behavior, depending upon the time, temperature or concentration of the desired technique. Figure 7 shows the small effect of thermal crystallization for PET heated 28 hours at 140°C. Cobbs and Burton (2) have shown that this condition causes a significant density increase in amorphous PET. These changes are measured on extruded tape in the direction of extrusion. One of the most notable changes is the decrease in intensity and shift of the -62°C. transition to -77°C. with small changes in the other detectable relaxations. The maximum at 225°C. has shifted to 242°C., indicating a further perfection of the molecular ordering of the polymer chains. The creation of the 155°C. transition on heating is not easily explained, but this transition could arise from some type of imperfect crystallization or ordering caused by the exothermic behavior at 140°C.

This behavior could well indicate the existence of small areas of order, even in "amorphous" polymer. The movement or creation of relaxations such as 155°C. (above exotherm

TABLE II

TEMPERATURES OF MODULUS CHANGES
DURING THE CRYSTALLIZATION OF MERLON POLYCARBONATE FILM IN CCl_4

Time, hrs.	First	Second	Third	Fourth
0	-140	53	148	158
1.0	-140	50	160	173
2.0	-140	67	155	182
8.0	-129	67	150	183
24.0	-122	68	---	200
72.0	-125	68	---	193

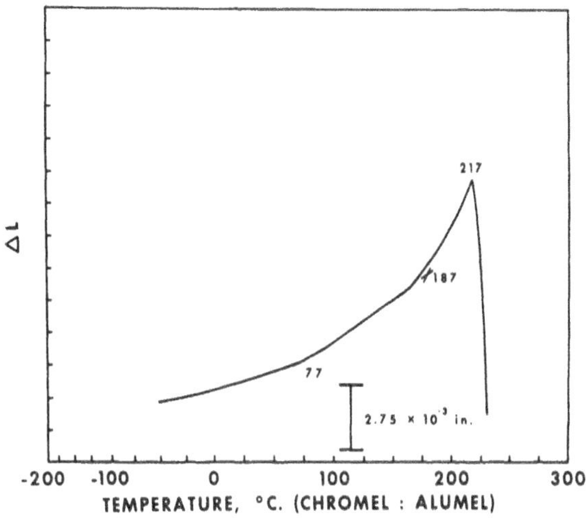

Fig. 9. Linear expansion as a function of temperature for crystalline polycarbonate.

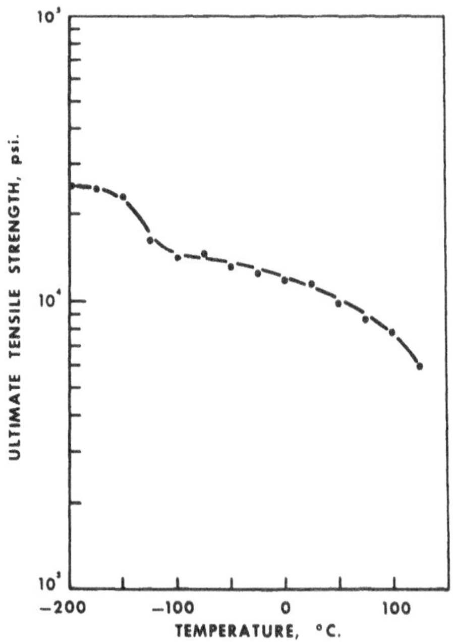

Fig. 10. Tensile strength as a function of temperature for linear amorphous polycarbonate.

at 140°C.) could illustrate that these small areas of molecular order become larger and temperature dependent according to their size as the polymer is heated at some T > Tg or treated with a "crystallizing solvent".

b. Polycarbonate

MERLON polycarbonate film, however, shows some very marked changes on crystallization either in stress cracking solvents or heating above 150°C. for a short period of time. Figure 8 shows the effect of crystallization in carbon tetrachloride (24 hours) on the apparent modulus behavior. These same effects have been produced by heating for 24 hrs. at 190°C. There is a significant shift in the low temperature transition from -145 to -122°C. as well as a shift of the 50°C. transition to 68°C. with a very significant increase in the magnitude of elongation. Crystallization causes the 145°C. transition to disappear, and it moves the 177°C. transition to 200°C. demonstrating the formation of more perfect areas of molecular order. This tendency to crystallization of PC by solvents, causing an increase in softening temperature and in tensile strength, suggests its application in fiber and film markets which are currently occupied by PET.

The changes in temperature of modulus decrease are given in Table II. The low temperature transition shifts from -145 to -129 after 8 hours with concurrent higher temperatures for the 50, 145 and 177°C. transitions. After 24 hours in CCl_4, there is no transition detectable at 150°C., but the other transitions have leveled out at -125, 68 and 200°C. On further treatment with CCl_4, no further changes are noted in temperature behavior of this thin film for several days of immersion.

However, the expansion characteristics of the 145°C. transition in Figure 5 for amorphous polycarbonate are moved to 187°C. as the polymer is crystallized with CCl_4 for several months (Figure 9). The polymer softening temperature is then raised from 170°C. to 217°C. with a shift in the 50°C. transition to 77°C. The similarity of behavior in expansion characteristics of the amorphous and crystalline polymer indicates similar melting processes occurring for each. It doesn't appear reasonable that one of these melting processes (amorphous) would be due to a Tg and the other that of a typical crystalline melting behavior. Furthermore, these expansion characteristics have been shown to be typical of first-order transitions (18).

While the 50°C. transition in the less ordered polycarbonate could be due to small areas of order (fringed micelle type), the low temperature transition at -145°C. corresponds with a significant change in tensile strength vs. temperature (Figure 10) (19).

ACKNOWLEDGMENT

The author wishes to thank the Instrument Products Division of the DuPont Co. for the quartz probe and quartz hook assemblies used in this work and Professor D. M. Cates of North Carolina State University for the amorphous sample of PET film. The permission for publication by the Mobay Chemical Co., the discussions with A. S. Morecroft, and the experimental help of O. L. Sheaffer and C. Crookshanks in obtaining these data are gratefully acknowledged.

REFERENCES

1. Kolb, H. J. and E. F. Izard, J. Appl. Phys., 20, 564 (1949).

2. Cobbs, W. H. and R. L. Burton, J. Polymer Sci., 10, 275 (1952).

3. Bateman, J., R. E. Richards, G. Farrow and I. M. Ward, Polymer, 1, 63 (1960).

4. Farrow, G., J. McIntosh and I. M. Ward, Makromol. Chem., 38, 147 (1960).

5. Gillham, J. K., and R. F. Schwenker, Jr., J. Polymer Sci., 6, 2, 59 (1966).

6. Cates, D. M. and S. T. Tarim, J. Polymer Sci., Appl. Polymer Sym., 1 (1966).

7. Schulken, R. M., Jr., R. E. Bay, Jr., and R. H. Cox, J. Polymer Sci., Appl. Polymer Sym., 6, 17 (1964).

8. Illers, K. H. and H. Breuer, J. Colloid Sci., 18, 1 (1963).

9. Mayhan, K. G., W. J. James and W. Bosch, J. Appl. Polymer Sci., 9, 3605 (1965).

10. Mitsuishi, Y. and M. Ikeda, J. Polymer Sci., A-2, 4, 283 (1966).

11. Ke, B., J. Appl. Polymer Sci., 6, 624 (1962).

12. Schwenker, R. F. and R. K. Zuccarello, J. Polymer Sci., C,6, 1 (1964).

13. Nielsen, L. E., "Mechanical Properties of Polymers", Reinhold Publishing Corp., New York, 1962, p. 187.

14. Schnell, H., "Chemistry and Physics of Polycarbonates", Interscience Publishers, New York, 1964, p. 148.

15. Statton, W. O., J. Polymer Sci., C, 18, 33 (1959).

16. Krum, F. and F. H. Mueller, Kolloid Z., 164, 81 (1959).

17. Reding, F. P., J. A. Faucher, and R. D. Whitman, J. Polymer Sci., 54, S-56 (1961).

18. Kauzmann, W., Chem. Rev. 43, 219 (1948).

19. Miller, G. W., Polymer Preprints, ACS Meeting, Chicago, 1967, p. 1072.

CALORIMETRIC STUDIES OF DRAWN POLYETHYLENE TEREPHTHALATE

by
J. L. Koenig and M. D. Mele
Division of Polymer Science
Case Western Reserve University
Cleveland, Ohio 44106

ABSTRACT

Dynamic differential calorimetry and density measurements have been made on drawn polyethylene terephthalate. The crystallinity and heats of fusion depend on the type of molecular rearrangements produced by the extent of draw and the temperature of draw. At low temperatures (near the glass transition) the molecules relieve strain by uncoiling with little sliding or alignment occurring. At temperatures above 100°C the molecules glide past one another to relieve strain with little uncoiling. Strain - induced crystals are formed in the samples drawn more than 160% at temperatures near the glass transition; at high strain, both chain alignment and uncoiling occur and produce crystallization at these temperatures. At higher temperatures and high strain, the uncoiling is too small to allow crystallization. Annealing a drawn sample under strain resulted in an increase in uncoiling which is proportional to the strain.

INTRODUCTION

A new approach to the study of drawn polymers is calorimetry(1). The proposed procedure for studying a polymer by calorimetry is to measure the heats of fusion and heats of crystallization of samples drawn different amounts and at different temperatures. The dependence of these heats upon draw, along with other data, should provide some insight into the changes that occur during drawing.

The purpose of this paper will be to determine the effects of draw ratio and draw temperature upon the heats of transition of PET. These values will be measured with the dynamic micro-calorimeter. Crystallinity will be evaluated through density measurements and orientation by x-ray diffraction patterns.

The results of these measurements will be used in an attempt to deduce the mechanism by which PET crystallizes and orients upon drawing. These results will also be correlated with the previous results obtained by infrared(2) and morphological studies(3).

Thermal Parameters for Undrawn PET

As a control for the drawn samples, a sample of the glassy PET was cut to a weight sufficient to give optimum results in the calorimeter. A second order transition, an exotherm and a endothermic peak appeared between the starting temperature (18°C) and final temperatures (280°C). The baseline was very smooth and had a consistent positive slope.

The second order transition occurred from 72°C to 80°C and was consistant with temperatures reported for the glass transition of amorphous PET[4]. The baselines before and after the transition were similar in slope.

The exotherm, identified as a cold crystallization peak, began at 108°C. The difference in temperature steadily increased and reached its maximum value at 129°C. The baseline after the peak was of slightly less slope than before, but remained positive. The best baseline appeared to be a straight line drawn between temperatures of 108°C and 147°C. The slope of this extrapolated baseline was the same as before the peak. The area of this peak was 1.27 ± 0.02 in^2. Considering the calibration constant of 70.1 ± 1.2 mcal/°C-min at 122°C, the heat of crystallization is found to be 1.05 ± 0.04 kcal/mole of repeat unit.

The endotherm began at 227°C and reached its minimum value at 259°C. There was no evidence of any shoulder on the peak. A baseline could be drawn between 227° and 275°C with its slope equal to the baseline before the peak. Because of the temperature range this peak may be assigned to the melting of crystalline regions of the sample[4]. The temperature of the minimum of the peak is 260°C ± 1.5°C. Since the weight is known and the area of the peak can be measured, a heat of fusion was calculated. With a calibration constant of 77.96 ± 1.33 mcal/°C-min at 232°C the heat of fusion is found to be 1.55 ± 0.04 kcal/mole of repeat unit. It should be noted that this value of the heat of fusion is for an annealed sample since the polymer crystallized (below 150°C) and from that temperature until the melting point at 227°C, it was annealing. This was indicated previously through the density measurements.

Thermal Parameters for Drawn PET

Strips of amorphous PET were clamped in the jaws of the Instron TT-C and drawn at 0.1 inches/minute. The temperatures of draw was adjusted to 60°C or 85°C. The polymer sample was stamped with parallel lines and after drawing the spacings were measured to give a per cent draw. The draw ranged from 75% to 300 ± 10%. The portion of the sample drawn 75% was in an unnecked but slightly drawn region. Portions of sample, drawn more than 75% were from the well formed necking region in samples drawn at both temperatures.

A feature of this thermogram of samples from >160°% at 60 and 85°C is the lack of a cold crystallization exotherm from ambient to 150°C. There is no trace of such a peak near 120°C where cold crystallization occurred in the undrawn sample. This was the case for all samples drawn over approximately 150%.

There is a definite endotherm similar to the endotherm in the undrawn sample. It begins at approximately 228°C and reaches a minimum value at 259°C. The peak returns to a constant baseline at 273°C. There is, however, a shoulder on the low temperature side oi the peak (257°C) that was not observed in the undrawn sample.

The thermogram of the samples drawn less than 160% at 60° and 85°C showed a cold crystallization exotherm at approximately 106°C and a maximum value at 129°C. The melting endotherm began at 226°C and had a peak at 260°C, similar to the undrawn sample. However, there is a slight trace of a shoulder at 259°C.

The heats of fusion of various samples of PET, as a function of the amount of draw, are fitted to a straight line with none of the points deviating more than 25%. Assuming a linear relationship, the data are expressed by following formula:

$$\Delta H_f(\text{kcal/mole}) = 1.53 + 1.10 \times 10^{-3} D \qquad (60^\circ\text{C and } 85^\circ\text{C})$$

where D is the per cent draw.

The temperature of the minimums on the melting endotherm were experimentally the same with a value of $259^\circ \pm 1.5^\circ$C.

The crystallinity of each sample was calculated from the density. The crystallinity does not increase above the value of an undrawn sample until the polymer has been drawn more than 160%. Thus strain-induced crystallization at 60°C and 85°C does not occur until the sample has been drawn more than 160% or until a neck has formed.

The samples were annealed in the calorimeter to 225°C at 10°C/min and the resulting crystallinities determined. Crystallinity increased upon annealing to 50%; crystalline content is a linearly increasing function of draw although the slope is very small $(1.55 \times 10^{-2} \frac{\%X}{\%D})$.

Some of the drawn samples were examined by x-ray diffraction to determine the crystallinity and the nature of the orientation. The samples drawn less than 160% were found to be unoriented and amorphous to x-ray diffraction. After annealing, these samples were crystalline but were still unoriented. However, samples drawn above 160% were semicrystalline and oriented. Upon annealing, the crystallinity increased and the orientation was perfected. These results agree with the relative crystallinities of the samples determined by density measurements.

To this point, samples have been studied that were drawn below or very near the glass transition temperature. The following set of data concern samples drawn well above the glass transition temperatures where chain mobility is greatly increased. The samples were drawn at 2 and 5 inches/minute with the sample temperature at 105°C and 114°C. Although the sample narrowed somewhat between the jaws of the Instron, there was no well defined necking region as in previous experiments. The per cent draw ranged from 78 to 360%. At a draw temperature of 105°C and 114°C, all samples produced similar thermograms. The thermogram is very similar to that for an undrawn polymer (and to samples drawn less than 160% at 60° and 85°C). The cold crystallization exotherm starts at 108°C, reaches a maximum at 127°C and returns to the baseline at 145°C. The baseline, after the peak, has a less positive slope than before the peak as in previous cases. The extrapolated baseline is drawn between 107° and 145°C. A heat of crystallization of 1.16 kcal/mole is found. The baseline, after the exotherm, is linear and with a slight positive slope until 224°C where melting begins. The minimum value is reached at 260°C. The baseline before and after the fusion endotherm are of identical slope and, therefore, it is easy to extrapolate a straight baseline across the peak. It is not necessary, in this case to use two fixed points in order to establish the desired baseline. The heats of fusion as a function of draw ratio at

105°C and 114°C were fitted with a smooth curved line instead of a straight line. It is, however, possible to approximate the curve by two straight lines. The formula for these lines are:

$$D<250\% \quad \Delta H_f(\text{kcal/mole}) = 1.55 + 2.00 \times 10^{-4} D \quad (105^\circ C \text{ and } 114^\circ C)$$

$$D>250\% \quad \Delta H_f(\text{kcal/mole}) = 1.22 + 1.53 \times 10^{-4} D \quad (105^\circ C \text{ and } 114^\circ C)$$

The slope for D greater than 250% is close to that of samples drawn at lower temperatures. The average peak temperature was 260° ± 1.5°C. The heat of crystallization as a function of the amount of draw are fitted by an assumed straight line with no point deviating more than 4%. The slope is positive by small $(2.8 \times 10^{-4} \frac{\text{kcal/mole}}{\%D})$.

Unannealed drawn samples did not increase in crystallinity as measured by density; in other words all samples had approximately 3% crystalline content. Upon annealing, all samples increased in crystallinity to the same level -- 40%.

Flat plate x-ray patterns confirmed the results of density measurements. All samples, no matter how much they were drawn at 105°C, appeared completely amorphous. Annealed samples were crystalline but there was no orientation (diffraction patterns appeared as rings and not spots).

The heats of crystallization for samples drawn at 114°C as a function of draw are unusual in that the slope of the line is negative. The points show greater scatter deviating as much as 5%.

Densities were measured and crystallinities calculated. For unannealed samples there is a slight, linear increase in crystallinity with draw in contrast to results at 105°C. Crystallinity of annealed samples also increased linearly with draw with no point deviating more than 2%.

Flat-plate patterns confirmed the crystallinity data. Unannealed, drawn samples showed a very slight amount of crystallinity and orientation upon annealing. Crystalline content increased greatly upon annealing but orientation was slight.

An important fact must be kept in mind when considering samples drawn at low temperatures (60° and 85°C). This fact is the low mobility for whole molecules and for aggregates of molecules such as the ball-like structures. When a sample is drawn at a low temperature, the strain may not be relieved by slipping of molecules past one another; the primary means to relieve the strain will be uncoiling of segments of molecules in the amorphous or tie-molecule regions of the sample. This rearrangement can only relieve the major portion of the strain. Previous results indicate this mechanism of elongation occurs up to approximately 160% draw. After this draw ratio has been reached, the ball-like areas must change in order to relieve strain. Therefore, above 160% draw, the strain will be "experienced" by the balls themselves; to relieve the strain, the molecules in these regions will uncoil and if in proper registry will crystallize.

No crystallization will occur below 160% draw because there is uncoiling and elongation in tie molecule regions. Above 160% draw, the chains in the ball-like regions uncoil to relieve the strain. The more strain imposed on the molecules the more uncoiling and subsequent crystallization[5]. The relationship between strain and crystallinity is nearly linear.

Since there is no measurable difference between strain induced crystals and heat induced crystals, all samples are nearly identical at the melting point whether they have been initially drawn enough to strain crystallize or not. The linear increase in the heat of fusion with draw is a result of greater trans content in the crystalline regions. As molecules were put under strain the gauche isomers are converted to trans isomer, and the more trans isomer the easier it is for the sample to crystallize.

The results relating to samples drawn at 105° and $114^{\circ}C$ can be interpreted in a similar manner. The most important difference is that the higher temperatures give greater mobility to the molecular chains. The strain imposed on the sample drawn at higher temperatures is relieved by rearrangement of the amorphous molecules. This is similar to drawing at lower temperatures except that the molecules in this case move more easily past one another and therefore may not uncoil as much. However, the more the sample is drawn the more tie molecules that will be drawn taut; and these tie molecules will transfer the strain to the ball-like regions in which they are incorporated.

In the ball-like regions the chains are also more mobile and, instead of relieving strain by conversion from gauche to trans isomers, the strain is partially relieved by movement of molecules past one another in the draw direction. The fact that less conversion to trans isomers is found in samples drawn at $105^{\circ}C$ than at $85^{\circ}C$ has been varified by infrared studies[2]. Thus there may be an increase in alignment of the chains upon drawing but, since there is little uncoiling (gauche isomers converting to trans amorphous isomers), there is little crystallization. Molecules that are incorporated in more than one ball-like structure may uncoil but there are not enough of these to produce a notable change in crystallinity. Thus annealed and unannealed densities do not depend on the amount of draw at higher temperatures.

CONCLUSIONS

The results that have been discussed are consistent with the following conclusions:

1. As a sample is drawn at lower temperatures (60° and $85^{\circ}C$) the strain imposed on it is relieved by uncoiling and some alignment of molecules; at temperatures over $100^{\circ}C$ strain is relieved by the molecules sliding past one another in the draw direction.

2. Strain induced crystallization will only occur in a PET sample when it is drawn more than 160% at temperatures near the glass transition.

3. Heat-induced crystals and strain-induced crystals have similar heats of fusion.

4. When a sample is drawn and annealed under strain, the molecules further uncoil by an amount proportional to the stress in order to relieve the strain.

REFERENCES

1. Peterlin, A., and G. Meinel, J. Applied Physics, 36, 3028 (1965).

2. Koenig, J. L., and S. W. Cornell, "Characterization of Uniaxially Drawn Polyethylene Terephthalate by Polarized Infrared Spectroscopy", Case Institute of Technology, Technical Report Number 42 (1967).

3. Yeh, G. S. Y., and P. H. Geil, "Crystallization of Polyethylene Terephthalate from the Glassy, 'Amorphous' State", Case Institute of Technology, Technical Report Number 44 (1967).

4. Brandrup, J., and E. H. Immergut (Ed.), Polymer Handbook, Interscience (1966).

5. Mandelkern, L., Crystallization of Polymers, McGraw-Hill, N.Y. (1964).

CRYSTALLIZATION OF POLY(ETHYLENE TEREPHTHALATE)

by

E. L. Lawton and D. M. Cates

Department of Textile Chemistry

North Carolina State University, Raleigh, North Carolina

INTRODUCTION

It has been reported that several endothermic melting transitions exist in differential thermal analysis thermograms of polyethylene.[1] Similar transitions are reported in differential scanning calorimetric thermograms of poly(ethylene terephthalate)[2] and bisphenol-A polycarbonate.[3] The appearance of separate melting endotherms in these polymers is evidence that almost completely discontinuous distributions of crystallite sizes or perfections can be induced in a polymer by appropriate thermal treatments.[4] At present, there is no evidence of the existence of polymorphic crystalline modifications in poly(ethylene terephthalate).

EXPERIMENTAL

The poly(ethylene terephthalate) polymer (PET) was a melt cast 0.8-mil film. The polymer had an intrinsic viscosity in tetrachloroethane-phenol at 30°C of 0.56, a diethylene glycol content of 1.44 mole percent, and a ratio of weight-average to number-average molecular weight of 1.6-1.7. Density, infrared, and x-ray diffraction measurements indicated that the film was amorphous and unoriented. Results reported in this study were reproduced using amorphous and unoriented PET film melt cast from polymer from a completely different commercial source (Goodyear PET chips).

Unrestrained samples were isothermally crystallized in a forced convection oven controlled to $\pm$ 1.0°C. Samples were liquid-crystallized by immersion in acetone at 25°C for 48 hours; this was followed by extraction of retained acetone in ethanol and drying under vacuum. The removal of imbibed acetone was carefully monitored by gravimetric and calorimetric techniques. Infrared spectra were determined on a Perkin-Elmer Spectrophotometer Model 337. Densities were measured by flotation in a carbon tetrachloride-heptane density gradient column at 25°C. Density measurements were also checked by the titration-flotation method of Moore and Sheldon.[5] X-ray measurements were made by the method of Dumbleton and Bowles.[6] Thermograms of samples were determined on a Perkin-Elmer Differential Scanning Calorimeter Model 1B using a heating rate of 40°C/min. Any temperature lags caused by the rapid programming rate were not corrected for. Quantitative fusion data were obtained in the following way. Tangents to the baseline were drawn on both sides of the transition. A straight line was used to connect the two sides at the points where the baseline departed from the tangent. The area enclosed by this straight line and the curve outlining the transition peak was measured with a planimeter. The calorimeter was calibrated with indium, tin, and lead standards.

Reported measurements are average values from three or more replicate samples. Mean deviations from average values did not exceed the statistical limits stated below. For samples isothermally crystallized at 150°C, 200°C, and 235°C, sufficient replicate determinations were made to

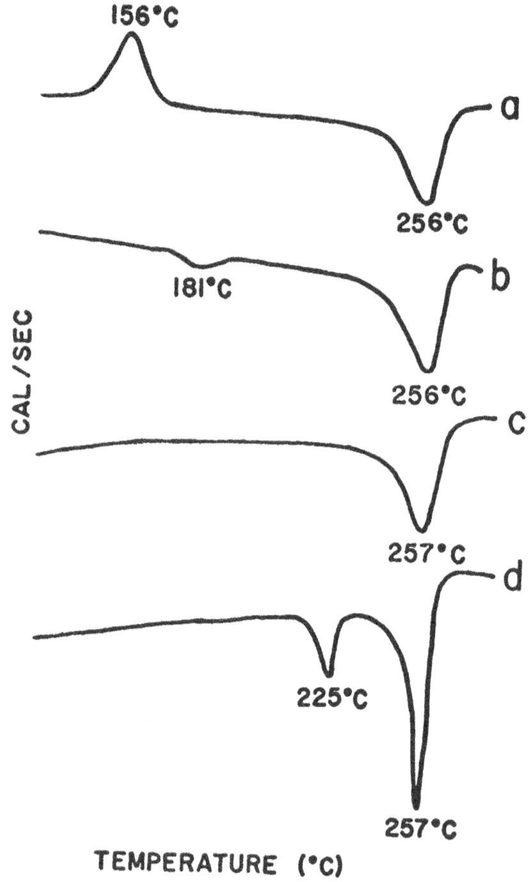

Fig. 1. Thermograms of PET crystallized by different treatments: a, amorphous sample, no treatment; b, one hr. at 150°C; c, in acetone; d, in acetone followed by 12 hr. at 195°C.

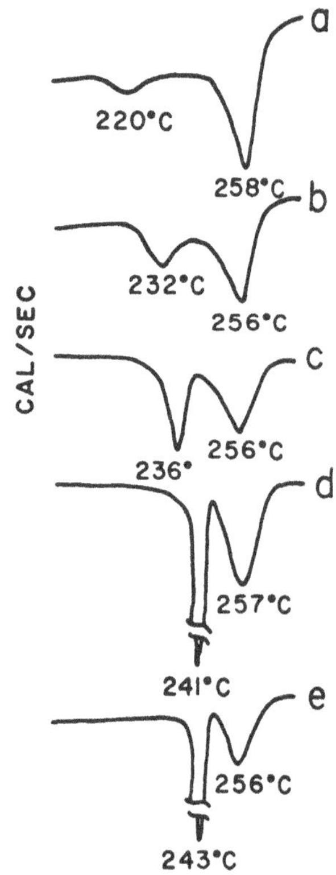

Fig. 2. Thermograms of PET previously crystallized at 200°C: a, one hr.; b, 12 hr.; c, 48 hr.; d, 96 hr.; e, 144 hr.

permit the data to be treated statistically. The statistical analysis indicated that the standard deviations of the data were as follows: infrared relative absorbances, $\pm$ 2 percent; densities, $\pm$0.1 percent; thermal transitions, $\pm$ 3 percent. The peak maxima of the thermal transitions were reproducible to $\pm$ 1°C.

RESULTS

Characteristic thermograms of PET crystallized in various ways are shown in Fig. 1. The thermogram of the amorphous sample, with a crystallization exotherm and a melting endotherm, shows that changes may occur in the sample while the temperature of the calorimeter is being raised. The extent of the changes that may take place in the calorimeter vary with the initial state of the sample and the rate of temperature rise. Thus an amorphous polymer undergoes more change during a run than a crystalline polymer. In this work, temperature was programmed to increase at the rapid rate of 40°C/min in order to minimize possible changes in the sample during scanning.

The major endotherm with a peak temperature near 257°C is obtained upon crystallization either by heat or by treatment with organic liquids (thermograms b and c in Fig. 1.). The lower endotherm, appearing in thermogram b with a peak temperature of 181°C, occurs only in those samples crystallized by heat. Thermogram d shows that PET which has previously been crystallized in acetone also exhibits the lower endotherm provided it is subsequently heat treated.

The enthalpy change and temperature at the maximum of the upper endotherm are designated respectively as ΔH^1_m and T^1_m. Corresponding values for the lower endotherm are designated as ΔH^2_m and T^2_m.

Both ΔH^2_m and T^2_m increase depending upon time and temperature of crystallization as shown by Table 1 and Fig. 2.

T^2_m clearly shifts to higher temperatures with increasing crystallization temperature, but ΔH^2_m exhibits only small variation over the temperature range 125 to 200°C (Table 1a). In this same range neither T^1_m nor ΔH^1_m appears to change.

When amorphous PET is crystallized at 150°C for varying times up to 48 hours, the enthalpies and peak temperatures of the two endotherms are found to be unchanged from their values after a crystallization time of one hour. At a crystallization temperature of 200°C, however, the intensity and peak temperature of the lower endotherm increase with time until the upper and lower endotherms begin to merge (Table 1b and Fig. 2.). The fusion data given in Table 1b for each individual peak was obtained by assuming each peak to be symmetrical. The thermograms suggest that upon prolonged crystallization crystalline regions melting near 257°C are transformed into crystalline regions melting near 243°C.

This conclusion is borne out by the thermograms of Figures 3 and 4. The material used for these thermograms was PET which had previously been crystallized in acetone and dried; the samples were then treated for the designated time and temperature in the calorimeter and quenched to room temperature. The thermograms were then run. (X-ray diffraction and infrared absorbance measurements showed that the acetone-crystallized sample was about 33 percent crystalline.)

Table 1. Heats of fusion of PET: (a) PET crystallized one hour at indicated temperatures; (b) PET crystallized at 200°C for indicated times.

Crystallization Temperature or Time	ΔH_m^2 mcal/mg	T_m^2 °C	ΔH_m^1 mcal/mg	T_m^1 °C
(a) 125°C	1.0	147	10.0	257
150°C	1.1	181	10.1	257
175°C	1.1	194	10.0	257
200°C	1.8	220	9.7	258
235°C	14.5	257	-	-
(b) 1 hr	1.8	220	9.7	258
12 hr	3.2	232	9.5	256
48 hr	5.4	236	8.1	256
96 hr	8.5	241	5.6	257

Thermogram d in Fig. 3 shows that after a few hours of treatment at 215°C only the lower endotherm can be resolved, its peak temperature being 249°C. The corresponding peak for treatment at 225°C occurs at 263°C (thermogram d in Fig. 4); after treatment for 9 hours at 235°C the peak occurs at 268°C (not shown).

When amorphous PET is crystallized in the calorimeter at 215°C or 225°C, one endotherm is observed at short times as a shoulder on the other endotherm. Crystallization times longer than a few hours, however, produce only a single peak – the lower endotherm (thermograms e in Figures 3 and 4). Crystallization of amorphous polymer at 235°C at times exceeding a few minutes produces only a single peak – the lower endotherm; but it is possible that the other endotherm can be detected at very short times.

A crystallinity index of PET, x, may be calculated from a specific property of the polymer using the relationship,

$$x = \frac{p - p_a}{p_c - p_a};$$

where p is a specific property, and subscripts "a" and "c" refer to the completely amorphous and crystalline polymer, respectively. The values below are those used in this paper.

	Density g/cm^3	Relative Infrared Absorbance ($973\ cm^{-1}/795\ cm^{-1}$)
Completely amorphous	1.335	0.56
Completely crystalline	1.455	2.92

The density of the completely crystalline polymer was taken as that of the crystalline unit cell.[6] The relative infrared absorbance of the completely crystalline polymer was extrapolated from the linear relationship that was found between absorbance of the 973 cm^{-1} band relative to the 795 cm^{-1} band

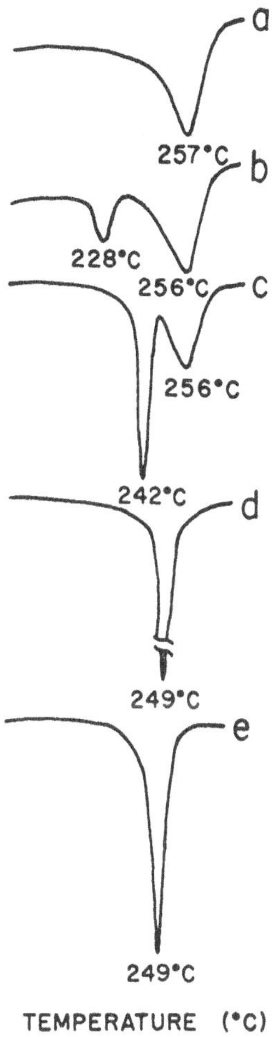

Fig. 3. Thermograms of PET crystallized first in acetone and subsequently treated in the calorimeter for varying periods at 215°C: a, no additional treatment; b, 10 min.; c, 2 hr.; d, 12 hr. Crystallization in the calorimeter of amorphous PET: e, 3 hr. at 215°C.

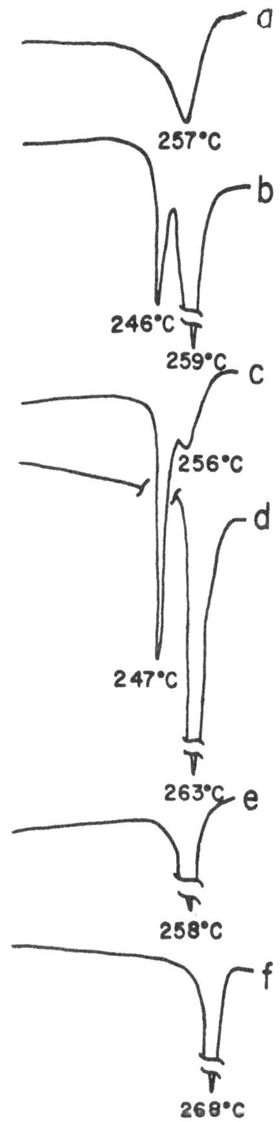

Fig. 4. Thermograms of PET crystallized first in acetone and subsequently treated in the calorimeter for varying periods at 225°C: a, no additional treatment; b, 30 min.; c, one hr.; d, 15 hr. Crystallization in the calorimeter of amorphous PET: e, 3 hr. at 225°C; f, 7 hr. at 235°C.

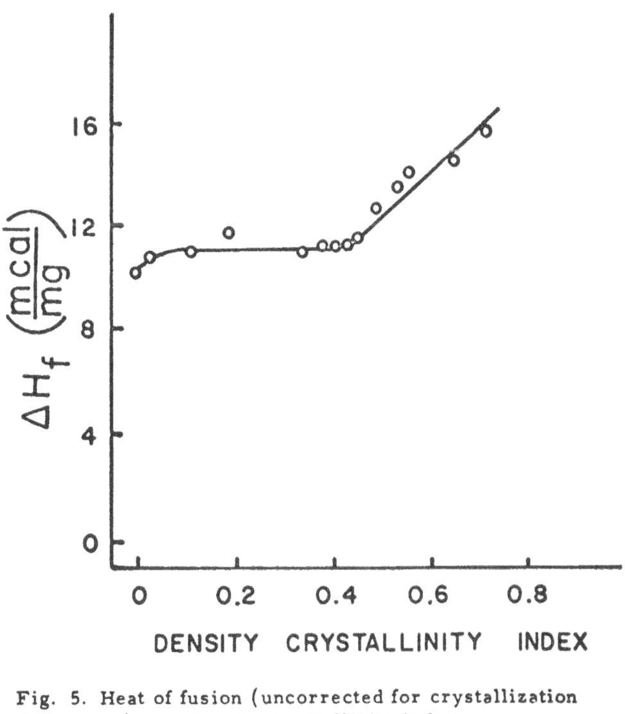

Fig. 5. Heat of fusion (uncorrected for crystallization during run) vs. density crystallinity index.

and density. The density and relative absorbance of the completely amorphous polymer were found experimentally.

Table 2 compares indexes of crystallinity which were calculated using the specific properties of density, relative infrared absorbance, and x-ray diffraction.

Heats of fusion are also given in Table 2 under the heading "Endotherm", and the values include ΔH_m^2 and ΔH_m^1 in the cases where both endotherms occur. The enthalpy values in the column labelled "Exotherm" represent "cold crystallization". The four samples which exhibited "cold crystallization" contained appreciable amorphous fractions prior to the runs in the calorimeter.

There are certain difficulties which tend to make the value for heat of fusion less accurate than might be desired. It has been suggested from DTA measurements[7] and adiabatic calorimetry[8] that, during differential heating of PET of low degrees of crystallinity, in addition to the "cold crystallization" exotherm there is continuous crystallization up to the melting temperature which is not readily detected in the thermogram baseline. A rapid rate of temperature rise was employed in order to minimize the additional crystallization taking place during a run. To the extent that such crystallization does take place, the enthalpy change (value under "Endotherm" in Table 2) that results from melting crystalline regions present in the polymer before differential heating was begun is overestimated.

The plot of the measured enthalpy change against density crystallinity index, given in Fig. 5, shows that the heat of fusion is greatly overestimated particularly in the least crystalline samples. This may be partly corrected for by considering the heat of cold crystallization. The heat of cold crystallization (value under "Exotherm" in Table 2), measured at the cold crystallization temperature, may properly be subtracted from the heat of fusion at the melt only after it has been corrected to the melting temperature by utilizing the temperature dependence of the difference in heat capacity between liquid and crystalline phases of PET. Such a correction might reduce the difference (1.9 mcal/mg) between the heat of cold crystallization and the "heat of fusion" of completely amorphous PET. It would be desirable also to correct for the crystallinity which develops in a specimen during a run without causing significant departure from the thermogram baseline. There does not seem to be any simple method by which this crystallinity can be taken into account. The heat of fusion values plotted in Fig. 5 have not been corrected in any way.

DISCUSSION

When molten PET is rapidly quenched, the polymer chains are frozen into metastable states (amorphous polymer) of higher free energy content than the equilibrium crystalline state at that temperature. Upon annealing amorphous PET at temperatures between the glass temperature (near 80°C) and the equilibrium melting point, the chain segments become sufficiently mobile to make irreversible transitions to more stable (normally semicrystalline) states.

The T_m^1 endotherm (peak at 257°C) observed in heat-crystallized PET is attributable to the fusion of crystalline regions. The T_m^2 endotherm, which appears when PET is crystallized above 125°C, might be attributed to

Table 2. Crystallinity indexes and enthalpy changes of heat-treated PET

Treatment	Crystallinity Index			ΔH ($\frac{mcal}{mg}$)		Optical Density to White Light
	IR	Density	X-ray	Exotherm	Endotherm	
amorphous	0.00	0.00	0.00	8.3(156°C) [a]	10.2	0.04
110°C, 20 min.	0.04	0.03		6.5(152°C) [a]	10.7	0.04
110°C, 60 min.	0.12	0.11		4.0(143°C) [a]	11.0	0.04
112°C, 25 min.	0.18	0.19		2.6(140°C) [a]	11.7	0.07
125°C, 60 min.	0.32	0.34			11.0	0.10
150°C, 60 min.	0.35	0.38	0.30 ± 0.03		11.2	0.14
150°C, 24 hr.	0.36	0.41			11.2	0.14
150°C, 48 hr.	0.36	0.43			11.2	0.14
175°C, 60 min.	0.36	0.41			11.1	0.16
200°C, 60 min.	0.46	0.45			11.5	0.16
200°C, 12 hr.	0.50	0.49			12.7	0.20
200°C, 48 hr.	0.54	0.53			13.5	0.20
200°C, 96 hr.	0.56	0.56			14.1	0.22
235°C, 60 min.	0.61	0.65	0.59 ± 0.03		14.5	0.24
150°C, 60 min. followed by 235°C, 120 min.	0.65	0.72			15.4	0.30

[a] Peak temperature

a solid to solid transition or to the fusion of a distribution of crystallites differing in certain ways from the T^1_m crystallites. The temperature at which the T^2_m transition occurs and the associated enthalpy change increase with temperature of crystallization. When the polymer is heat treated at a temperature of 200°C or greater, the thermograms show that the T^1_m endotherm disappears in favor of the T^2_m endotherm (Figures 2, 3, and 4). Moreover the transition temperature of the T^2_m endotherm can be made to move past that of the T^1_m endotherm, showing that the structure associated with the former endotherm is intrinsically more stable, although it is much slower to form at lower temperatures. The conclusion is thus reached that the two endotherms represent fusion of two crystalline distributions differing in size, perfection, or internal structure. Mitsuishi[9] has recently proposed that crystallites in PET exist both in a fringed micelle formation and a folded chain formation. Thus T^2_m could correspond to the melting of unfolded lamellae, and T^1_m to the melting of folded lamellae.

The T^2_m structure is very slow to form at low temperatures. When this structure melts below T^1_m during a run in the calorimeter, the melted chains along with any previously uncrystallized material are free, at the higher temperatures occurring during the run itself, to reform T^2_m crystals with a higher melting point. This cannot be detected in the thermograms produced under the conditions cited. However, if the sample is cycled under conditions allowing adequate time and temperature, i.e., the sample is heated to a temperature above T^2_m and below T^1_m, cooled, and then reheated, the T^2_m peak temperature moves upward and T^1_m remains constant.

A plot of heat of fusion against the density crystallinity index shows a distinct break near an index of 0.45 (Fig. 5). This index represents an upper limit which was exceeded in the present work only by using temperatures of crystallization equal to or greater than 200°C; this is also the temperature required to bring about significant changes in ΔH^2_m (Table 2). Mayhan et al.[10] report that the maximum density that can be obtained at different crystallization temperatures increases rather sharply above 200°C. Koenig and Hanson[11] report that an additional long period peak, detected by small-angle x-ray diffraction, appears at temperatures of 200°C and above. The structure responsible for the T^2_m endotherm does not develop at appreciable rates except at 200°C or higher; this structure may well account for these observations.

ACKNOWLEDGMENT

We wish to thank Dr. C. J. Heffelfinger, E. I. du Pont de Nemours and Company, Inc., Circleville, Ohio, who furnished the film, and Mr. B. B. Bowles, Chemstrand Research Center, Durham, North Carolina, who carried out the x-ray analysis.

REFERENCES

1. K. Hoashi and J. Mochyuki, Makromol. Chem., 100, 78 (1967).

2. N. J. Wakelyn and P. R. Young, J. Appl. Polymer Sci., 10, 1421 (1966).

3. R. P. Kambour, F. E. Karasz, and J. H. Doane, J. Polymer Sci., A, 4, 337 (1966).

4. A. P. Gray and K. Casey, J. Polymer Sci., B., 2, 381 (1964).

5. W. R. Moore and R. P. Sheldon, Polymer, 2, 315 (1961).

6. J. H. Dumbleton and B. B. Bowles, J. Polymer Sci., A-2, 4, 951 (1966).

7. M. A. Hughes and R. P. Sheldon, J. Appl. Polymer Sci., 8, 154 (1964).

8. C. W. Smith and M. Dole, J. Polymer Sci., 20, 37 (1956).

9. Y. Mitsuishi and M. Ikeda, Chem. High Polymers, 23, 310 (1966).

10. K. G. Mayhan, W. J. James, and W. Bosch, J. Appl. Polymer Sci., 9, 3605 (1965).

11. J. L. Koenig and M. J. Hannon, J. Macromol. Sci. (Phys.), B1(1), 119 (1967).

DIFFERENTIAL SCANNING CALORIMETRY OF POLYURETHANES

W. J. MacKnight, M. Yang and T. Kajiyama
Department of Chemistry and Polymer Science and Engineering Program
University of Massachusetts, Amherst, Massachusetts 01002

Introduction:

The urethane linkage results from the reaction of an isocyanate group and a hydroxyl group and has the structure $\left(\begin{array}{ccccc} & H & & O & \\ & | & & \| & \\ - & N & - & C & - O - \end{array} \right)$. Polymers containing this linkage are referred to as polyurethanes although the term has also been applied to polymers containing related linkages (such as the disubstituted urea linkage.) A large number of polyurethane polymers are known, ranging from thermoplastics to elastomers, and a few are commercially important. There are two principle synthetic routes to the polyurethanes.[1] In the prepolymer method, a low molecular weight polyester or polyether containing hydroxyl end groups is reacted with an excess of diisocyanate. The resulting isocyanate terminated species is called the prepolymer and is usually of sufficiently low molecular weight to be liquid at room temperature. The prepolymer is cured to an elastomeric product by reacting it with a diamine or a diol. Small amounts of triols and higher polyols are often used to introduce crosslinking.

The polyurethanes selected for the present study were synthesized by the method known as the "one shot" polymerization process. In this method, which is described in detail by Lyman[2] in a recent review, a monomeric diisocyanate is polymerized with a diol to give a high molecular weight product by step growth polymerization. Such polymers are often semicrystalline in nature with glass transitions above room temperature. In this work the glass transition temperatures (Tg's) and melting points (Tm's) of three series of polyurethanes were determined by Differential Scanning Calorimetry (DSC). The structures of the repeat units for each series are

(1) H series

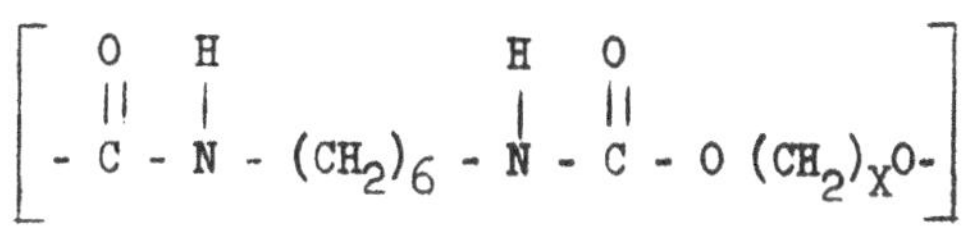

(2) D.P. series

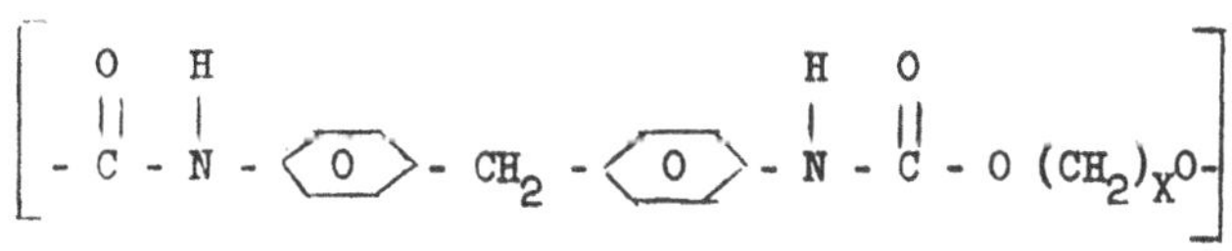

(3) 4 M series

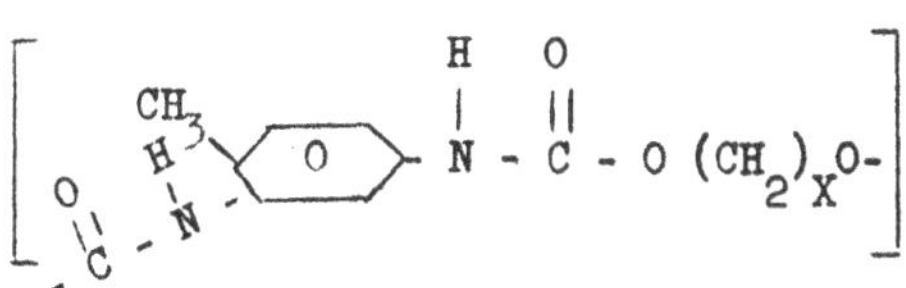

In all cases X = 2 - 10. The H series results from the polymerization of hexamethylene diisocyanate with various diols varying from ethylene glycol to 1 - 10 decanediol; the D.P. series results from the polymerization of 4 - 4' diphenylmethane diisocyanate and various diols; and the 4 M series results from the polymerization of 4-methyl-m phenylene diisocyanate and various diols. The H and D.P. series are semicrystalline while the 4 M series is entirely amorphous.

Experimental:

The monomers were obtained commercially and purified by the usual methods. Polymerization was carried out in dimethyl sulfoxide by the method of Lyman.[2] The polymers were isolated by precipitation into methanol and were solvent-cast into films. Drying of the films was accomplished by heating in a vacuum oven at 110° C for a period of one week. In the case of the D.P. and H series all the polymers were completely soluble in solvents such as dimethyl sulfoxide, dimethyl formamide, etc., showing that no side reactions leading to crosslinking took place during polymerization. In the 4 M series, however, it was found that the polymers, while undergoing considerable swelling in the above solvents, would not completely dissolve. On the other hand they could be compression molded into films at temperatures of 120° - 130° C. Thus it appears that the 4 M series is lightly crosslinked, probably by allophanate linkages. Allophanate formation is known to occur as a side reaction in urethane polymerizations and results from the reaction of a urethane group with an isocyanate group.[7]

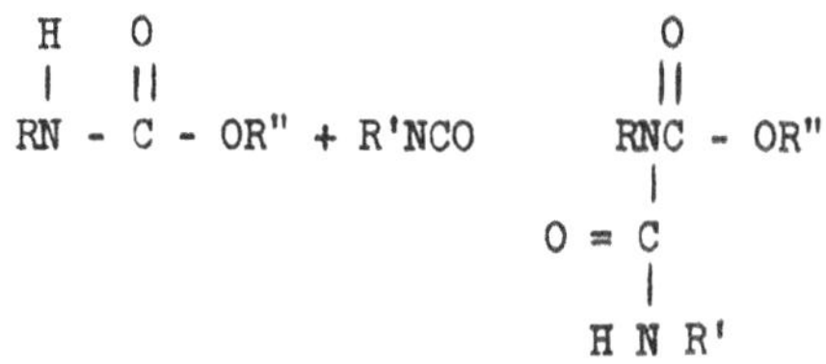

Why this should have occurred with this series and not with the other two is not entirely clear although it may have to do with the fact that the reactivity of an aryl isocyanate is decreased by the presence of an ortho or para methyl substituent.

The Differential Scanning Calorimeter used was the Perkin Elmer Model DSC 1-B. All scanning speeds were 10°C per minute unless otherwise noted and calibration of the calorimeter was accomplished using Induim, Tin and Lead. After preliminary runs to establish the approximate melting points, all samples were annealed in the calorimeter for 4 hours at 20° C below their melting points, slowly cooled to room temperature and then programmed up in temperature at 10° C per minute through the melting point. All experimental heats of fusion reported refer to samples with this thermal history. Tg's were taken as the midpoint of the observed transition while Tm's were taken as the point at which the melting endotherms returned to the base line.

Results and Discussion:

Glass Transitions

The Tg's for the three series are summarized in Table I. As expected, the D.P. series shows the highest Tg's, reflecting the influence of the "stiff" diphenylmethane group. In addition, the Tg's of the D.P. series decrease in regular fashion with increasing methylene sequence lengths (increasing X), in the repeat unit. This is due to the increased mobility introduced by the presence of these methylene sequences. It is possible to regard the phenomenon as arising from the fact that the polymer becomes more polymethylene-like with increasing methylene sequence length. The Tg's of many copolymers may be represented by the empirical equation

$$Tg = w_1 Tg_1 + w_2 Tg_2 \qquad (1)$$

Table I

Glass Transition Temperatures for Urethane Polymers (°C)

X	D.P.	H.	4 M
2	139	56	52
3	119	55	72
4	109	59	42
5	95	58	52
6	91	59	32
7	84	55	61
8	79	58	64
9	72	58	62
10	72	55	18

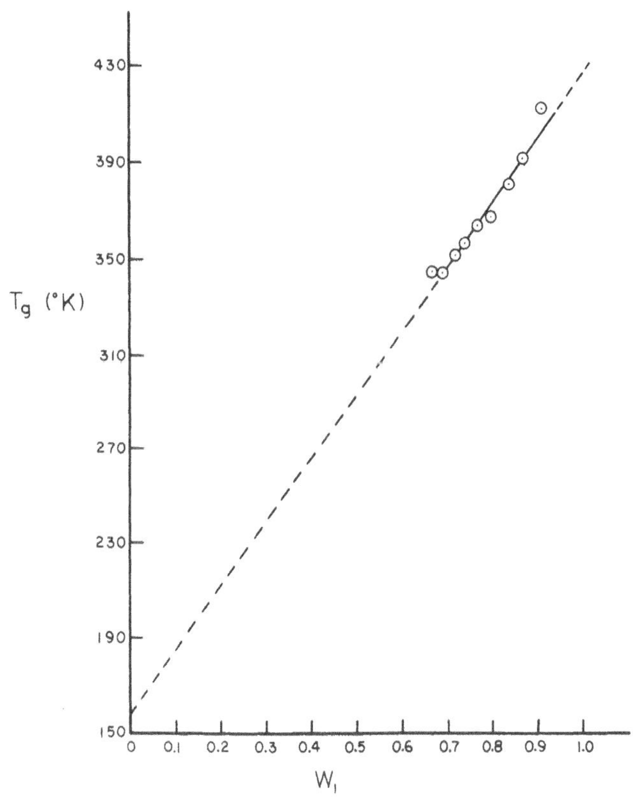

Fig. 1 T_g as a function of w_1 for the D.P. series

where Tg is the glass transition of the copolymer, Tg_1 is the glass transition of one homopolymer and Tg_2 is the glass transition of the other homopolymer. w_1 and w_2 are the weight fractions of homopolymers one and two in the copolymer. The D.P. series may be regarded somewhat arbitrarily as a copolymer of methylene and diphenylurethane. According to this view the repeat unit of homopolymer one is

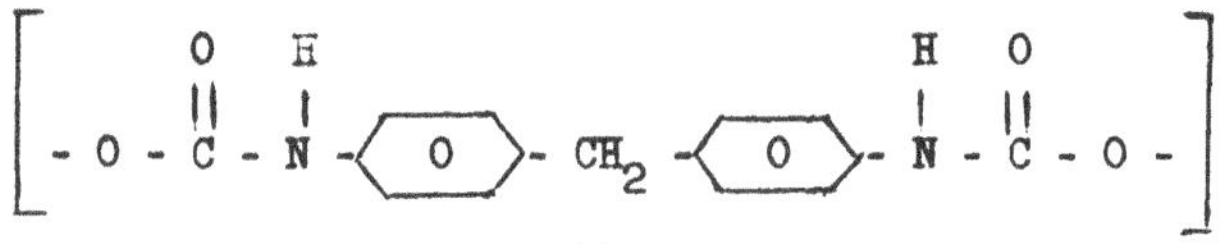

and that of homopolymer two is $\left[-CH_2-\right]$

Figure 1 is a plot of Tg versus w and it is seen that a fairly good straight line results with an extrapolated value of Tg_2 of -115° C. It has been established by a number of workers[3] that the Tg of polymethylene is approximately -120° C, in excellent agreement with the extrapolated value. It should be noted that the intermolecular hydrogen bonds which undoubtedly exist in these polymers seem to play a rather minor role in determining Tg, at least in the samples studied. This view is supported by the behavior of the H series in which the "stiff" diphenylmethane group is absent. The Tg's of this series remain essentially constant regardless of the number of methylene groups in the methylene sequence. In this case, the concentration of intermolecular hydrogen bonds decreases with increasing X, but the overall chain mobility does not change greatly since the diisocyanate unit already contains a methylene sequence consisting of six methylene units. The H series bears a very close structural relationship to the Nylons prepared by the condensation polymerization of hexamethylene diamine and various dicarboxylic acids. These Nylons have Tg's very close to those of the H series and their Tg's show relatively little variation with varying methylene sequence lengths in the repeat unit.[3]

The situation is quite different in the case of the 4 M series. Since the diisocyanate monomer is unsymmetrical, it can enter the chain in more than one configuration. This introduces sufficient disorder in the structure to make packing into a crystal lattice difficult and thus the 4 M series is amorphous, at least without prolonged annealing at elevated temperatures. It would appear that the same trends should be noted with the 4 M series as with the D.P. series since increasing the number of methylene units in the repeat unit should lead to increased chain backbone mobility in an analogous fashion to the effect in the D.P. series. Examination of Table I shows, however, that the Tg's for the 4 M series behave in an irregular manner. The reason for this is not entirely clear at this time. One possibility is the effect of water. All of the series are hydrophilic, but the effect is somewhat more pronounced in the 4 M series, probably because of the amorphous nature of this series. Caution was taken to exclude water but it is certainly possible that traces might remain. Water is an extremely effective plasticizer for Nylons so that the irregularities observed in the 4 M series may be due to a similar cause. A second possibility is that the various polymers in the 4 M series may have different concentrations of intermolecular hydrogen bonds. Changes in the level of hydrogen bonding are known to have significant effects on Tg's and a study has been made of this effect on a series of copolymers of various methacrylates containing hydroxyl groups.[4] It was found that Tg decreased markedly as the concentration of hydrogen bonding hydroxyl groups decreased. The situation in the 4 M series may be similar. An infrared study is presently in progress to attempt to obtain quantitative information about the hydrogen bonding in all the series investigated. If this explanation for the irregular Tg behavior is correct, it leads to the conclusion that intermolecular forces play a dominant role in determining Tg in the 4 M series while chain mobility is the important factor in both the D.P. and H series.

Table II

Melting Points (Tm) of Urethane Polymers (^{o}C)

X	D.P.	H
2	-*	166
3	241	163
4	248	182
5	192	157
6	200	171
7	198	151
8	201	162
9	194	154
10	194	161

*Data not available due to decomposition overlapping the melting point.

Table III

ΔHm's for Urethane Polymers (cal/g)

	D.P. Series	H Series		
X	(ΔHm)exp	(ΔHm)exp	(ΔHm)th	wt. % Crystallinity
2	-	9.23	43.5	21.2
3	11.2	9.72	43.1	22.6
4	13.2	16.8	45.2*	37.2
5	7.95	12.0	42.6	23.9
6	9.23	16.8	44.0	38.2
7	6.92	14.0	42.0	33.4
8	11.6	14.4	43.1	33.4
9	10.2	14.5	42.2	34.4
10	7.91	13.3	43.0	30.9

*This is the value for Nylon 6-6, assumed valid for H_4
ΔSm = 0.099 cal/g for H_4 and is assumed constant throughout the H series.

Melting Points and Heats of Fusion

The melting points of the D.P. and H series are collected in Table II. There are scattered references to the melting points of some of these polymers in the literature[5,8]. The melting point of H4 for example is quoted as 180° C[5] in excellent agreement with our value of 182° C. In both series it is seen that the Tm's decrease in a zig-zag fashion with increasing methylene sequence length in the repeat unit. The zig-zag phenomenon has been observed before, both in the case of Nylons containing even and odd numbers of methylene groups in the repeat unit and in the case of low molecular weight hydrocarbons of even and odd chain lengths. The effect is attributed to packing efficiency with the even numbered members packing more efficiently on the basis of geometrical considerations and thus melting higher than the odd numbered members.

The experimental enthalpies of fusion, (ΔH_m)exp., determined from the area under the melting endotherm, are presented in Table III. It would be possible to obtain weight per cent crystallinities for the various polymers from this data if the enthalpy of fusion of the 100% crystalline material were known. This data is not available at present, however. A rough estimate of the degree of crystallinity of the H series may be obtained by approximating the enthalpy of fusion of 100% crystalline H4 by the corresponding value for Nylon 6-6. This latter value is available[6] and an average of four determinations leads to 45.2 cal/g for the enthalpy of fusion of 100% crystalline Nylon 6-6. As already noted, H4 is structurally very similar to Nylon 6-6, so that the approximation is a reasonable one. The entropy of fusion (ΔS_m) for the 100% crystalline H4 may then be obtained from the familiar relationship

$$Tm = \frac{\Delta H_m}{\Delta S_m} \tag{2}$$

This is 0.099 cal g^{-1} deg^{-1}. If the additional assumption is introduced that ΔS_m is constant throughout the H series, the enthalpies of fusion for all the polymers in the H series may be obtained through eq. (2). The results are tabulated in Table III under the column headed $(\Delta H_m)_{th}$. Finally the weight per cent crystallinities are given as

$$\text{weight \% crystallinity} = \frac{(\Delta H_m)\text{exp}}{(\Delta H_m)_{th}} \times 100 \tag{3}$$

The constant ΔS_m approximation is, of course, equivalent to the statement that the melting point variations are entirely enthalpic in origin. This is almost certainly not strictly true but is probably satisfactory for a semi-quantitative approach to the problem.

Acknowledgments:

The authors are grateful to the National Science Foundation for partial support of this research and to Professor R. W. Lenz and F. E. Karasz for many helpful discussions.

References:

1. J. H. Saunders and K. C. Frisch, "Polyurethanes, Chemistry and Technology, Part I. Chemistry," Vol. XVI in the series "High Polymers," Wiley, New York 1962.
2. D. J. Lyman, Reviews in Macromolecular Chemistry 1, (1), 191 (1966).
3. L. E. Nielsen, Mechanical Properties of Polymers, Reinhold, New York, 1962, Chapter 2.

4. A. V. Tobolsky and M. C. Shen, J. Phys. Chem. 67, 1886 (1963).
5. R. Hill and E. E. Walker, J. Polymer Sci. 3, 609 (1948).
6. R. L. Miller in Polymer Handbook, J. Brandrup and E. H. Immergut, editors, Wiley, New York 1966, p. III - 17.
7. R. W. Lenz, Organic Chemistry of Synthetic High Polymers, Interscience, New York 1967, p. 187.
8. K. Saotome and H. Komoto, J. Polymer Sci. A1, (5) 119 (1967).

GLASS TRANSITION TEMPERATURE MEASUREMENT OF ELASTOMERS

John J. Maurer
Enjay Polymer Laboratories
Linden, New Jersey

Introduction

Analysis of the glass transition (Tg) process is a subject of wide and continuing interest both in the research and applied areas of polymer chemistry.(1-4) The advent of commercially available, high sensitivity DTA equipment has made it possible to easily undertake research in this area, and also to perform certain types of measurements which are not possible by other methods commonly used in Tg studies. This investigation was undertaken to examine critically the type of data about the Tg process which is obtained by DTA. Also of interest is the interpretation of certain characteristics of this process which appear in the DTA thermogram. This latter objective is attempted by comparison of the DTA data with those from linear expansion measurements. Throughout this work, attention was focused on polymers which, because of their elastomeric nature, have very low values of Tg.

Experimental

Differential Thermal Analysis

Reference: Glass beads; Sample Size: 0.03 gms; Thermocouples: Chromel/ Alumel; Atmosphere: Nitrogen; Heating Rate: 5 or 20°C/min; Cooling Rate: Varied; Temperature Range: Usually-100 to +200°C; Instrument: Dupont 900 DTA.

Linear Expansion

Measurements were performed using the apparatus previously described.(5) The samples varied in height, and were prepared by (a) molding 0.075" pads of gum rubber in a standard rubber press and plying discs of these pads to the required height; or (b) cutting gum samples to size and, where necessary, using silica powder to level the surface. This latter procedure has the advantage that it preserves the original thermal history of the sample, and minimizes flow problems.

Characteristics of the Tg Region of the Thermogram

In the first portion of this paper, the following points are considered: (1) the characteristic features of the Tg region of the thermogram, (2) how to measure the main Tg of elastomers from DTA thermograms, and (3) how these Tg values correspond to those obtained by other methods. This requires consideration of the influence of heating rate, since DTA is a dynamic measurement, and of heating vs cooling rates (thermal history) since relaxation phenomena are widely regarded as being involved in the Tg process. As shown in Figure 1, there are several easily recognizable regions of the Tg interval, which may vary depending on polymer type. Some details about each of these regions are presented below.

Region A: This region of the thermogram has been attributed to heat capacity differences between sample and reference, and used to calculate Cp for polymers by means of cup-type cells. It is clear from the present work with a thermocouple type DTA system that the size and general appearance of this region can vary extensively during replicate runs of the same polymer. This should be kept in mind when attempting qualitative or quantitative analyses of polymers via this feature of the thermogram.

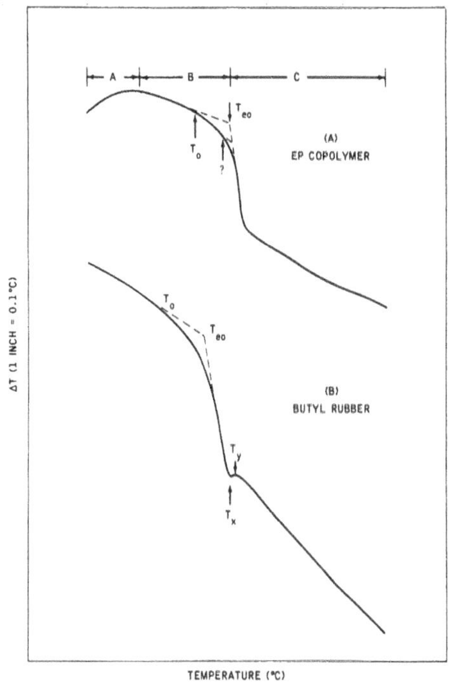

Fig. 1. Characteristics of DTA glass transition intervals.

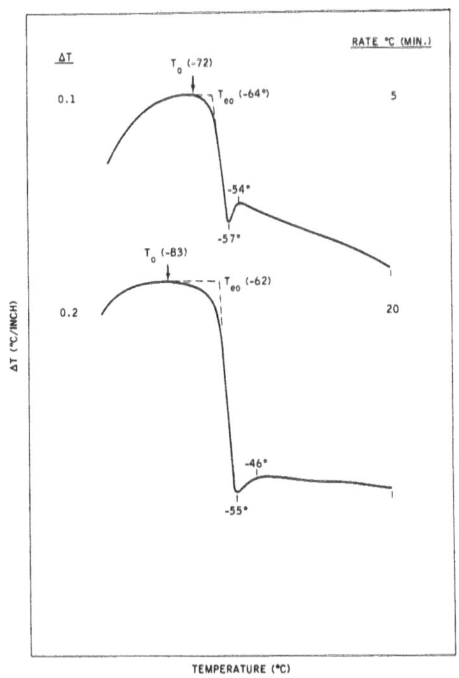

Fig. 2. Effect of heating rate on natural rubber T_g characteristics.

Region B: Tg values can be calculated from this region in two ways. The first is to select the Onset Temperature, To, as Tg. This is the temperature (generally beyond Region A) where the first endothermal departure from a baseline is noted. For some polymers To is difficult to estimate, because of its proximity to the termination of Region A. The second method is to locate the Extrapolated Onset Temperature, Teo, as shown in Figure 1(A) and use this as Tg. In using this method, one frequently finds that it is difficult to select a linear portion of the curve above Teo to use in the extrapolation method. Two such choices are shown in Figure 1(A); and several could be selected in cases such as Figure 1(B) due to the pronounced curvature of the thermogram in Region B. Experience has shown, however, that this problem causes only a few degrees variation in the estimation of Teo. An interesting point in connection with the selection of To is that there sometimes appears to be an easily recognizable inflection between To and Teo.

Region C: Significant variation in the appearance of this region of the thermogram can be observed as shown in Figure 1 (A and B). The minimum at Tx, which was observed in several elastomers, could be variously interpreted, based upon recent literature, as being (1) entirely relaxational in character,[3] (2) a consequence of "physical orientation" in the polymer above Tg,[6] or (3) as resulting from the inherent first order nature of glass transition behavior when treated as a freezing-in process.[7] Some comments about the relative applicability of these proposals to the present data will be considered subsequently.

Influence of Heating Rate

Figure 2 shows the influence of a four-fold variation in heating rate on the characteristics of the Tg interval for Natural Rubber (smoked sheet). The well-known observation that higher heating rates increase sensitivity but decrease resolution is shown here also. Thus, we note a much more intense transition at 20°C/min; whereas both Region B and the minimum following Teo broaden perceptibility. Two other features are also significant. First, Teo was increased by only 2°. Second, using the definition of To previously given we would estimate a To in the region of -83° with possibly another inflection near -68°. If -83° is To, then we have the case where an increase in heating rate leads to a decrease in transition temperature. We shall return to this point in the next section. For the moment, we may speculate that the increased sensitivity at the higher heating rate has detected a feature not observed in the 5°/min experiment.

Reproducibility of To and Teo For Elastomers

Results from triplicate determinations, at two heating rates, are shown in Table I which also allows comparison with literature Tg values. The following observations about the data seem significant:

(1) The To values obtained at 5°/min appear similar to literature Tg values from other techniques. Replicate DTA experiments of this type are therefore a potential means for obtaining Tg values for comparison with those from other methods.

As previously noted, considerable scatter can be encountered in To estimation. The accuracy of single determinations of To is thus highly questionable. There seems to be a tendency to estimate a lower value of To at the higher heating rate. This value is also lower than literature Tg values. If this is not an experimental artifact, the possibility of transitions or relaxations below Tg is suggested.[3]

TABLE I

Tg BEHAVIOR OF SELECTED ELASTOMERS

Polymer	Literature Tg (°C)	Heating Rate (°C/Min)	Temperatures As Defined in Figure 1			
			To	Teo	Tx	Ty
Butyl	-67 to -75	5	-73,-84, -78	-66,-66, -66	-57,-57, -57	-52,-54, -54
		20	-96,-83, -80	-65,-64, -64	-53,-52, -53	-47,-47, -48
EP Copolymer A	-55 to -60	5	-68,-73, -71	-59,-57, -56	Weak	Weak
		20	-88,-80, -98	-56,-56, -56	Weak	Weak
Neoprene W	-43 to -45	5	-47,-46, -45	-38,-38, -38	-32,-32, -32	-27,-28, -28
		20	-52,-52, -51	-38,-37, -37	None	None
Natural Rubber (Smoked Sheet)	-69 to -74	5	-69,-68, -72	-64,-64, -64	-57,-57 -57	-53,-54, -54
		20	-73,-83, -83	-61,-62, -64	-54,-55, -55	-47,-46, -50
SBR 1500	-59 to -64	5	-59,-56, -61	-52,-52, -54	-40,-44 -45	-38,-36, Weak
		20	-62,-65, -81	-50,-50, -51	-39(W), -39(W)	Very Weak

TABLE II

EFFECT OF POLYMER VARIATIONS ON Tg BEHAVIOR (5°C/MIN)

	IV	Wt. % C_2	To	Teo
EP Copolymer A (Gum)	3.4	47	-68,-73,-71	-59,-57,-56
EP Copolymer A (Gum Vulcanizate)	--	--	-80(?),-63(?) -83(?),-64(?)	-56 -56
EP Copolymer A (Gum + 10 phr Oil)	--	--	-61(?)	-56
EP Copolymer C	2.70	42	-57(?)	-54
EP Copolymer B	1.60	34	-59(?)	-50

(2) Teo can be precisely defined even at a high heating rate. It appears to be as acceptable as To for distinguishing between polymers of different type, and thus may be quite useful for identification or screening studies.

(3) Tx and Ty are as reproducible as Teo in several cases. They appear to be a function of polymer type, since for equal thermal history they do not appear in every case shown in Table I, and the influence of heating rate varies significantly with polymer type.

Effect of Polymer Variations on To and Teo

The influence of selected variations in sample characteristics on To and Teo is shown for EP Copolymer A in Table II. Although a 5° heating rate was used, To was difficult to estimate because of the nature of Region B. In the case of the vulcanizate, a possible weak inflection was noted near -80, in addition to a stronger one near -64°. The variation of Teo in EPR as a function of ethylene composition, oil content and crosslinking was small in accord with previous observations of this type.[4,8]

The Nature of Region C of the Thermogram

Limited experiments were performed to examine whether the cause of the minimum at Tx could be easily demonstrated in terms of the rate effects described in the literature.[2,9] The two experimental EP copolymers selected did not normally exhibit this minimum. (See Figure 3A and 4A) They were then subjected to low temperature annealing near Teo to determine whether this lead to a minimum as reported for PVC systems.[6] These experiments also involved controlled, slow cooling rates. As shown in Figure 3B and 4B, a sharp minimum appeared in one of the copolymers but not the other. The final part of this experiment consisted of annealing EP copolymer A near To rather than Teo. In this case, a minimum was induced as shown in Figure 4C. It seems clear from these experiments that polymers of the same basic type can vary significantly in the degree to which, under the same experimental conditions, this minimum can be induced. Consideration of these data and the relevant literature suggest two interpretations. One is that variations in polymer properties exist which cause very significant differences in the glass transition behavior of these polymers. In this event, a complete thermal history study would seem to be required in order to adequately describe the differences in relaxation behavior of different polymers. The other possibility is that, similar to the PVC data of McKinney and Foltz,[6] varying degrees of "physical orientation" are induced in these polymers during the annealing process.

Linear Expansion Measurements of Elastomers

The linear expansion behavior of the elastomers discussed in the DTA section of this paper was examined in an attempt to assist in analyzing the Tg region of the DTA thermogram. It was found that the expansion behavior was also quite complex as will now be shown. The specific objectives of this preliminary report of this work are: (1) to show the types of data which were generated by this technique, (2) to present some relationships between thermogram and expansion plot characteristics, and (3) to discuss several possibilities which could account for the observed complexity of the expansion data.

Influence of Thermal History on V-T Behavior

Variations in thermal treatments of polymers which could lead to anomalies in the Volume-Temperature curves subsequently obtained at rapid heating rates have been described by Marvin and McKinney as shown in Figure 5.[2] These will now be briefly considered in order to provide a

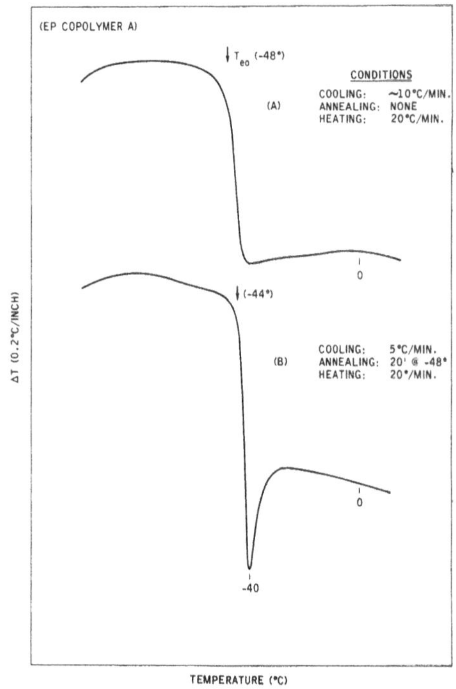

Fig. 3. Low temperature annealing can influence T_g region (EP copolymer).

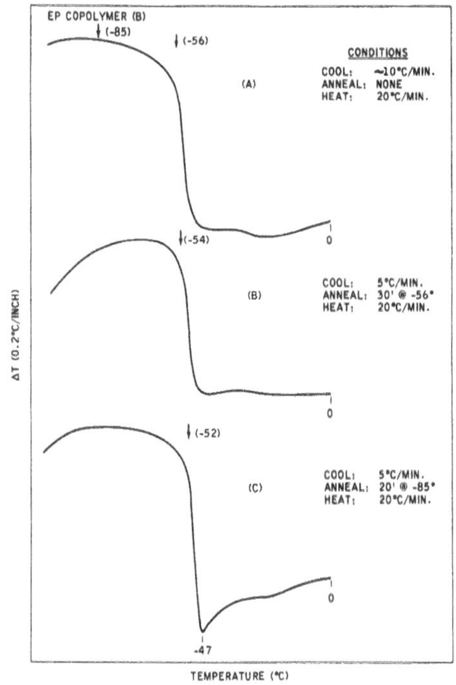

Fig. 4. Influence of low temperature annealing on EP copolymers is not uniform.

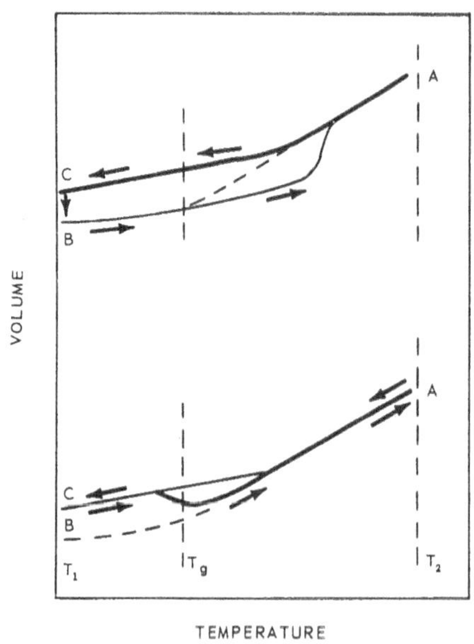

Fig. 5. Influence of thermal history on V-T curves* (*R.S. Marvin, J.E. McKinney in Physical Acoustics 2B, 1965).

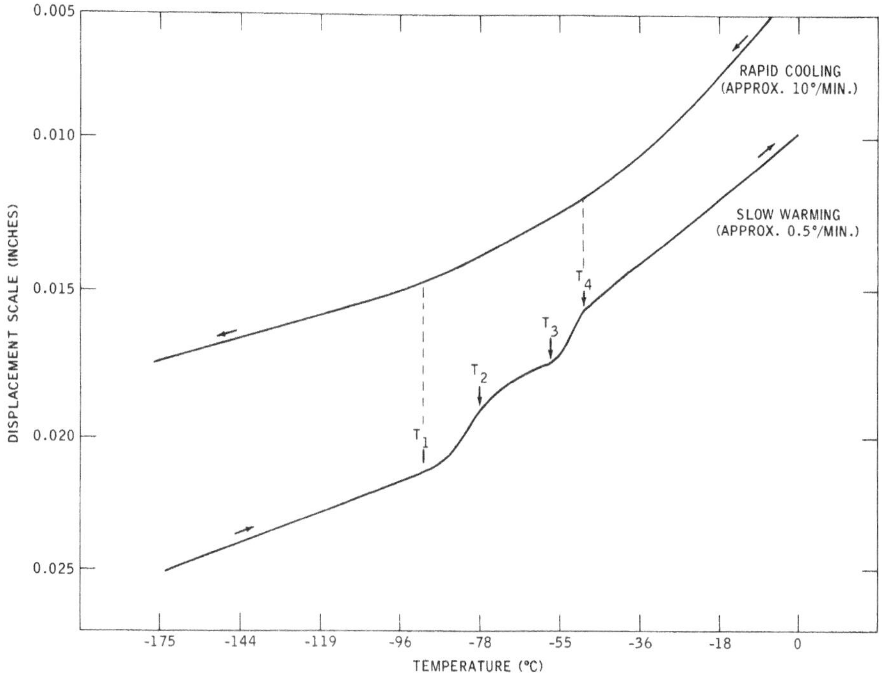

Fig. 6. Linear expansion behavior of EP copolymer A.

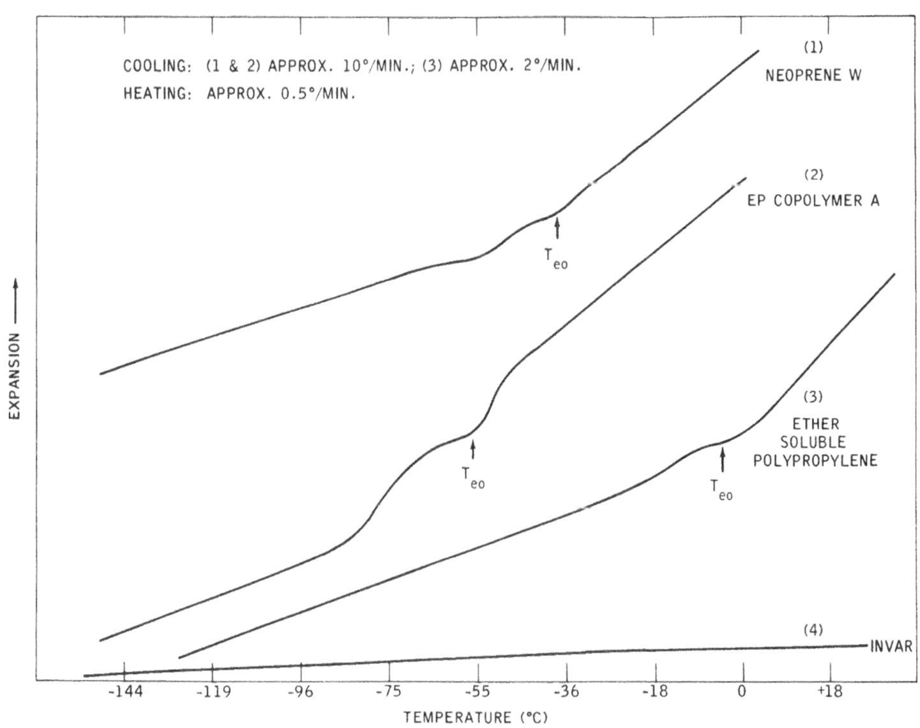

Fig. 7. Linear expansion behavior of different polymers.

reference point for discussion of the linear expansion data. The upper curves represent a system cooled rapidly from A-C, held for an extended time at T_1(C-B), and then rapidly heated. Note that the apparent Tg indicated by the cooling curve is higher than that of the equilibrium (dotted) curve, and the V-T curve (B-A) has an appearance similar to that observed in the melting region of polymers of low degree of crystallinity. The lower curves in Figure 5 represent a system cooled rapidly from A-C but not held at T_1 prior to heating. In this case rapid heating of the system leads to a minimum in the V-T curve commencing below Tg of the "equilibrium" curve.

Anomalous Behavior Encountered in Expansion Plots

The complex behavior encountered in the linear expansion study is illustrated in Figure 6 for EP Copolymer A. The thermal history utilized was fast cooling followed immediately by slow warming. These conditions are similar to the lower curves of Figure 5 except for the heating rate. Note that the cooling curve in Figure 6 exhibits a flat region in the T_1-T_4 temperature interval. Thus, the transition region is suggested as being broad. The warming curve could be interpreted in several different ways. First, one might consider that in the region T_1-T_2 expansion of a supercooled system occurs through T_2, followed by "collapse" of the system toward equilibrium dimensions between T_3 and T_4. A problem to be considered here is that the region T_3-T_4 resembles that presented in Figure 5 for a quite different thermal history. Other interpretations suggested by Figure 6 are (a) two transition regions of peculiar shape because of relaxation effects related to the thermal treatment employed, (b) two transition regions of a type anticipated by Ueberreiter for incompletely annealed samples.(7) This latter treatment considers the glass transition as a first order freezing-in process, and (c) two transition intervals, one being the main Tg region which is perturbed by relaxation effects and the other a lower temperature transition process such as those described by Boyer(3) and Wada.(10) The former author discusses transitions below Tg due to side group or subgroup motions in the polymer chains; the latter proposes that certain transitions below Tg represent the point where sufficient free volume change in the system has occurred to permit motion of chain ends.

Relationship of Thermogram and Expansion Plot Characteristics

Several types of elastomers were compared to determine whether the behavior shown in Figure 6 was specific to EP Copolymer A. As shown in Figure 7, the same general characteristics are observed, to different degrees, in Neoprene W, EP Copolymer A and an ether-soluble fraction of polypropylene. These curves, as well as that of the Invar system (Curve D), also indicate that the anomalous plots are not artifacts of the expansion unit itself. Comparison of several sets of corresponding DTA and expansion data reveal a rather striking similarity between characteristic temperatures from the two techniques as shown in Table III.

The temperatures compared in Table III are To, Teo and Tx for 5°/min. DTA runs (See Figure 1) and T_1, T_2, T_3, and T_4 for Expansion (See Figure 6). The agreement between the temperatures compared in the table is considered good considering the differences in experimental conditions. Some variations in the temperatures due to differences in thermal history, sample size, and thermocouple placement are to be expected. Further improvement in the comparison is expected to result when data for the SBR 1500 and polypropylene systems are also obtained using the rapid cool/slow heat treatment. In any event, the data shown in Table III suggest that the shape and characteristic temperatures of the DTA interval can apparently be directly related to dimensional changes in the sample via linear expansion measurements. This would be similar to the type of information obtained by Tool for inorganic glasses.(11)

TABLE III

COMPARISON OF EXPANSION AND DTA PROPERTIES OF ELASTOMERS

EP Copolymer A(1)		Natural Rubber(1)		Neoprene W(1)		SBR 1500(2)		Ether Soluble(2) Polypropylene	
Expansion	DTA(3)	Expansion	DTA	Expansion	DTA	Expansion	DTA	Expansion	DTA
T_1 (-89)	--	T_1(-90)	--	T_1(-53)	--	T_1(-78±1)	--	T_1(-34)	--
T_2 (-75)	To(-68 to -73)	T_2(-70)	To(-70±2)	T_2(45±1)	To(-46±1)	T_2(-64±2)	To(-59)	T_2(-15)	To(-11)
T_3 (-56±1)	Teo(-56 to -59)	T_3(-64)	Teo(-64)	T_3(-38±1)	Teo(-38)	T_3(-48±2)	Teo(-53)	T_3(-13)	Teo(-7)
T_4 (-50)	Tx(None)	T_4(-58)	Tx(-57)	T_4(-30±1)	Tx(-32)	T_4(-41)	Tx(-40)	T_4(None)	Tx (0)

(1) Cooled at 10°/min; warmed at 0.5°/min.

(2) Cooled at approximately 1°/min; warmec at 0.5°/min.

(3) All DTA runs at 5°/min heating rate.

The T_1 value and region of the expansion plot is quite interesting. As shown in Figures 6 and 7, this temperature clearly represents the beginning of a significant change in sample height and thus possibly of a transition region. A similar indication is not usually suggested by the DTA thermograms obtained at 5°C/minute. However, the To values detected in the 20°C/minute DTA runs may correspond to the "missing" T_1 values. Possibly the increased sensitivity of the higher heating rate experiment has enabled detection of the "T_1" region, but the accompanying reduced resolution does not generally permit detection of both "T_1" and "T_2" in the same DTA run. If this hypothesis is proven valid, then the use of a range of heating rates in DTA is suggested as a means for detecting transitions which occur below the main Tg.

Influence of Thermal History On Expansion Characteristics

The effect of variations in the thermal history on the anomalous expansion behavior is shown in Figures 8 and 9. In Figure 8, all three curves represent systems cooled at approximately 10°C/minute. Curve 1 is the previously shown EP Copolymer A heated at approximately 0.5°/minute. Curve 2 represents the behavior of the same polymer after "annealing" at -175°C for two hours prior to heating at 0.5°C/minute. The interesting feature of this system is the sharply increased intensity of the peak below the Tg region anticipated from DTA, and the weak T_3-T_4 region. In terms of relaxation effects, this system might be proposed to differ from Curve (1) because of the different degree of "equilibrium" achieved during the annealing process. However, it is interesting to note that based on relative peak intensities the annealed sample would seem to be further from an equilibrium condition than was Sample (1). This suggests an alternate interpretation, i.e., there are two transition regions whose relative intensities vary with thermal history. Curve (3) of Figure 8 shows the behavior of EP Copolymer B under identical conditions as for Copolymer A (Curve 2). Note that the expansion behavior differs significantly from Copolymer A, thus suggesting again that characteristics of the polymers themselves must also be considered in explaining and comparing the behavior observed during thermal analysis.

In the final set of experiments to be described, slow cooling was employed in an attempt to reduce possible relaxation effects in the expansion plots. Curve (3) of Figure 9 shows a sample cooled at 1.7°C/min and warmed at 1.5°C/min. Note that despite significant changes in thermal history, transition behavior is again suggested near -73 and -50°C, and the latter region still resembles melting behavior. A second step to achieve a system which was closer to the equilibrium state was to cool the sample very slowly (0.5°C/min) and subsequently heat it at the same rate. Here (Curve 2) we note only the transition near -50°C which again resembles melting behavior. The absence of the lower transition region may result from the fact that the sample was cooled only to -96°C. Alternate interpretations are that: (a) the conditions are unfavorable for detecting the lower transition, or (b) the "transition" is an artifact which disappears in this type of experiment.

The shape of the transition in Curve 2 is especially interesting because it resembles the type of behavior described by Ueberreiter for an incompletely annealed polymer.(7) An attempt was made to investigate this idea by slow cooling a sample and then holding it at -175°C for 2 hours. Subsequent slow warming of the sample led to Curve 1. This curve shows distinct regions near -80 and -50°C which could be interpreted as separate relaxation effects or as two transition regions which resemble first order behavior as detected by linear expansion.

In order to determine whether the shapes of the transition regions shown in Figure 9 are artifacts of the particular thermal histories employed, it would be desirable to obtain expansion data for samples cooled at extremely slow rates. This is not possible with the present unit. As an alternate

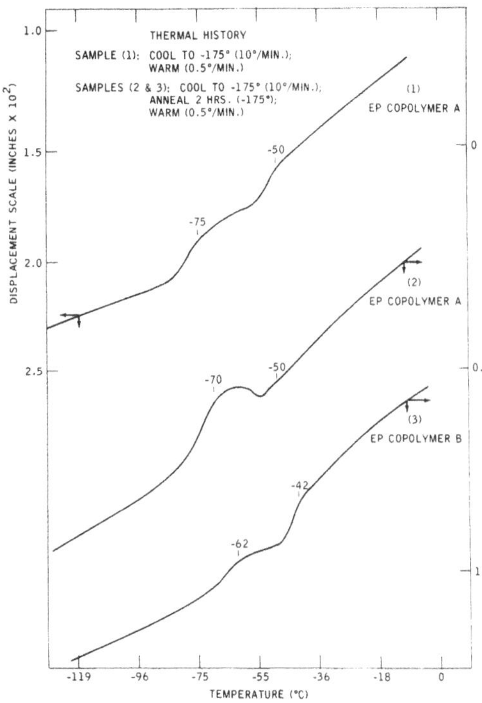

Fig. 8. Effect of fast cooling on EP copolymers.

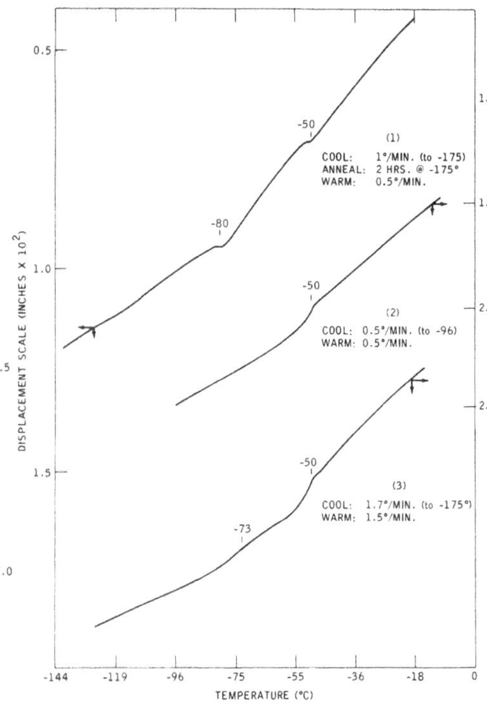

Fig. 9. Effect of slow cooling on EP copolymer A

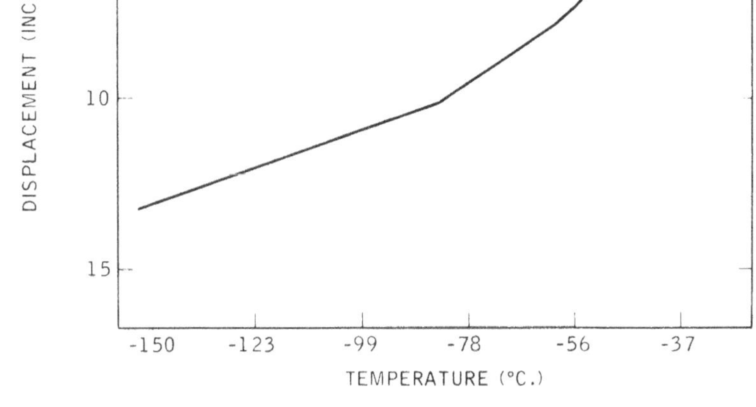

Fig. 10. Effect of low temperature annealing on EP copolymer A.

approach, the effect of prolonged low temperature annealing was evaluated. A sample of EP Copolymer A was stored at -70 to -78°C for 21 days, transferred rapidly to the expansion unit, cooled at 2°C/min to -180°C, and warmed at 0.5°C/min. The resulting expansion plot (Figure 10) indicates two clearly separated transition regions. The one at about -85°C has the appearance expected for a glass transition, whereas that at about -52°C has characteristics which suggest the melting of a polymer containing a low degree of crystallinity. The present data apparently do not permit an unequivocal determination of whether the -52° transition region could still be explained as a rate effect, or whether a degree of first-order-like characteristics must be admitted for this region--and possibly for the endothermal behavior observed at the termination of the DTA Tg interval as well.

The following general comments summarize the anomalous linear expansion data described above (a) the expansion plots illustrate complex behavior in the glassy region of the polymers which is, as anticipated, strongly influenced by thermal history; (b) further experiments, especially at much lower cooling rates, are needed to permit a complete analysis of the data; (c) for certain experimental conditions, there appears to be a good correlation between characteristic temperatures of DTA and expansion plots. Both methods suggest that a transition or relaxation region occurs below the main Tg. (d) One of the transitions observed in an EP Copolymer resembles first order behavior, but is difficult to distinguish from one type of anomalous glass transition interval due to rate effects.

Summary

An analysis of the glass transition (Tg) region of elastomers by DTA shows that a range of experimental conditions is needed to completely define the characteristics of a given sample. This is illustrated by the effect of heating rate and thermal history on (a) the values of Tg obtained from DTA by two methods, and (b) the development of an apparent endotherm at the end of the Tg interval. Linear expansion data on several of these polymers exhibits anomalous behavior which is a function of thermal history of the sample. Under certain conditions, there appears to be good correspondence between the DTA and expansion data for different types of elastomers. Both methods suggest that a transition or relaxation region occurs below the main Tg.

Acknowledgment

The author wishes to express his appreciation to Drs. N. Bekkedahl, J. E. McKinney and L. A. Wood of the National Bureau of Standards for helpful discussions of the linear expansion data.

References

(1) Rheologica Acta, 5, 1966, "Symposium on Glass-Rubber Transition".
(2) Marvin, R.S., and McKinney, J.E., "Volume Relaxations in Amorphous Polymers" in Physical Acoustics, II-B, Academic Press, New York, 1965.
(3) Boyer, R.F., Rubber Chem. and Tech., 36, 1303, (1963).
(4) Maurer, J.J., "Differential Thermal Analysis of Compounded and Vulcanized Elastomers", Division of Rubber Chemistry, Montreal, May 1967.
(5) Maurer, J.J. and Tsien, H.C., J. Applied Poly. Sci., 8, 1719 (1964).
(6) Foltz, C.R., and McKinney, P.V., J. Applied Poly. Sci., 11, 1189 (1967).
(7) Ueberreiter, K., Kolloid Zu Z Polymere, 216-217, 217 (1967).
(8) Maurer, J.J., Rubber Chem. and Tech., 38, 979 (1965).
(9) Wunderlich, B., Bodily, D.M., and Kaplan, M.H., J. Applied Phys., 35, 95 (1964).
(10) Wada, Y. et al, J. Phys. Soc. Japan, 14, 1064 (1959).
(11) Tool, A.Q., J. American Ceramic Soc., 29, 240 (1946).

THE TEMPERATURE DEPENDENCE OF THE GRÜNEISEN RATIO OF POLYMERIC MATERIALS DETERMINED BY THERMAL AND ULTRASONIC SOUND VELOCITY MEASUREMENTS

S. R. Urzendowski, A. H. Guenther and J. R. Asay
Air Force Weapons Laboratory
Kirtland Air Force Base
Albuquerque, New Mexico

The Grüneisen ratio, γ_G, often appears as a temperature independent factor in the thermal term of the Mie-Grüneisen equation of state for metals and refractory materials.[1,2,3,4]. This temperature independence is based upon the assumption made by Grüneisen[5,6] that γ_j defined as

$$\gamma_j = -\frac{d \ln \nu_j}{d \ln V} = -\frac{V}{\nu_j}\left(\frac{\partial \nu j}{\partial V}\right)_T \tag{1}$$

which leads to

$$\gamma_G = \frac{\beta \cdot B^T}{C_v \cdot \rho} = \frac{\beta \cdot B^S}{C_p \cdot \rho} \tag{2}$$

where each frequency, ν_j, is assumed to be the same for all frequencies and is strictly a volume dependent function expressed by β, the volume coefficient of thermal expansion, ρ, the density, C_v and C_p, the heat capacities at constant volume and pressure, respectively, and B^T and B^s, the isothermal and adiabatic bulk moduli.

Since γ_j values are known to vary in a solid, equation (2) may be related to an Einstein varying frequency, ν_i, so that γ_j becomes γ_i

$$\gamma_i = -\frac{d \ln \nu_i}{d \ln V} = \frac{V}{\nu_i}\left(\frac{\partial \nu i}{\partial V}\right)_T \tag{3}$$

and is now known as the thermal formulation, γ_T, of the Grüneisen law. Under these conditions, γ, as defined by equation (2) is regarded as a temperature dependent integrated value which often departs from constancy as a result of the temperature dependence of the corresponding frequencies.

The temperature regions where the Grüneisen approximation is reduced to the existence of a purely volume-dependent characteristic temperature for the vibrational or thermal free energy of the solid has been the topic of much discussion. Rice et al[7] have derived P-V isotherms at 0° K for a number of metals from shock compression data utilizing a constant Grüneisen ratio. Experimental investigation of metallic solids by Grüneisen,[5,6] Born[4], and Davies[8], have verified the validity of the parameter over specified temperature ranges. Recently Bijl and Pullan[9] explained the deviations in γ for copper, aluminum and iron in terms of a constant Poison's ratio, σ, and its effect on logitudinal and transverse velocity. The temperature dependence of the Grüneisen ratio for alkali halides was first noted by Born[4] wherein low temperature, γ_o values were approximately 30% smaller than the value at high temperatures. Barron[1] discovered a similar variation with temperature for argon-type face-centered latices which he defined in terms of the actual lattice spectrum and the corresponding γ_j. Blackman[10] explained deviations from Debye's specific heat law by analyzing the frequency spectrum according to Born - von Karman[4] lattice dynamics and explained variations in the Debye characteristic temperature, θ.

In this study, ultrasonic and thermal measurements were used to calculate the Grüneisen ratio for several polymeric materials according to equation (2). Since the Grüneisen law does not hold through transitional regions, variations in γ were assumed to result from the transition and/or relaxation phenomena observed by the thermal expansion, heat capacity and sound velocity measurements. For the nontransitional regions, the temperature dependence of the Grüneisen ratio was related to average frequency values, $\left(\frac{\nu}{\nu_0}\right)$, calculated from the thermal equation.

THEORY

The thermal expansion of a solid is directly related to the motion of the atoms, hence, β may be used to derive information on the frequency spectrum of lattice vibrations.

The quantum expression for the energy of an oscillator defined by the coefficient of thermal expansion, β, is fundamentally related to the compressibility, X, or bulk modulus, B, the Grüneisen ratio, γ_j, and E_j, the thermal energy of the jth oscillator by

$$\beta = \frac{X}{V} \sum_{j=1}^{N} \gamma_j \frac{d\,E_j}{d\,T} \tag{4}$$

The parameter, E_j, is related to the heat capacity C_v, by

$$C_v = \sum_{j=1}^{N} \frac{d\,E_j}{d\,T}$$

and the Grüneisen ratio γ_j is related to the frequency of the jth mode of vibration by equation (1).

Grüneisen assumed that all γ_j's are equal and independent of temperature hence, equation (2) was assumed to be independent of temperature.

Barron[1] introduced a temperature dependent γ_T, (Equation 3) which redefines the Grüneisen ratio of equation (2) such that $\gamma_G = \gamma_T$. For Barron's derivation, the equation of state defined as

$$P = -\frac{d\,U}{d\,V} + \frac{1}{V} \sum \gamma_i \frac{h\nu_i}{(e^{h\nu_i/kT} - 1)} \tag{5}$$

was differentiated as a function of temperature giving

$$\left(\frac{\partial P}{\partial T}\right)_V = \frac{1}{V} \sum \gamma_i \frac{k(h\nu_i/kT)^2 e^{(h\nu_i/kT)}}{(e^{h\nu_i/kT} - 1)^2} \tag{6}$$

where U is the internal energy, and h and k refer to Planck and Boltzman constants, respectively.

Use of the identity, $(\partial P/\partial T) = \beta B$ and

$$C_i = \frac{k(h\nu_i/kT)^2 e^{(h\nu_i/kT)}}{(e^{h\nu_i/kT} - 1)^2} \tag{7}$$

relates the coefficient of thermal expansion and the bulk modulus, B, to the heat capacity, C_i of a harmonic oscillator of Einstein frequency, ν.

Equation (5) may now be rearranged to give equation (2) with γ_T based on Einstein frequencies.

Shauer[11] has demonstrated that the introduction of dT into equation (3) yields

$$\frac{1}{\nu_i}\frac{d\nu_i}{dT} = -\gamma_i \frac{1}{V}\frac{dV}{dT} = -\gamma\beta \tag{8}$$

which gives a direct relationship between the lattice vibrational frequencies, the Grüenisen ratio and the thermal expansion.

Since γ_i is unmeasureable, equation (9) was converted by C_i to

$$\sum C_i \frac{1}{\nu_i}\frac{d}{dT} \Big/ \sum C_i = -\gamma\beta \tag{9}$$

which defines the mean temperature coefficient of the Einstein lattice vibrational frequencies for non-transitional regions in terms of thermodynamic properties, and the observed sections of the $\gamma(T)$ curve may be explained on the basis of Debye frequencies as proposed by Barron.[1]

For the temperature range $T \ll 0.2\theta$, where $kT \ll h\nu_i$, only long waves are excited so that the solid behaves like a continuum. For such a solid the vibrational properties are described fully by θ, the characteristic Debye temperature and $\gamma_i = \gamma_o$ so that

$$\gamma_o = -\frac{d\log\theta}{d\log V} = -\frac{V}{\theta}\frac{d\theta}{dV} \tag{10}$$

and

$$\frac{1}{\theta}\frac{d\theta}{dT} = -\gamma_o\beta$$

Integration of (10) yields

$$\theta = \theta_o \exp\left(-\gamma_o \int_{T_o}^{T} \beta\, dT\right)$$

or, as the exponent is small,

$$\frac{\theta}{\theta_o} = 1 - \gamma_o \int_{To}^{T} \beta \, dT \tag{11}$$

When $T \stackrel{\sim}{\sim} C.2\theta$, the integration is not meaningful, so the average temperature coefficient of the frequencies may be obtained from equation (10).

In the temperature range $T \geqslant \theta$ the solid behaves classically for C_i=k and the high temperature limit, γ_∞ is defined by

$$\frac{1}{3N} \sum_i \frac{1}{\nu_i} \frac{d\nu_i}{d\,T} = - \gamma_\infty \beta$$

which upon proper integration and substitution yields (12)

$$\left(\prod_i \frac{\nu_i}{\nu_{i,o}} \right)^{\frac{1}{3N}} = 1 - \gamma_\infty \int_{To}^{T} \beta \, dT$$

where ν_i is the frequency of the ith mode at temperature T, and $\nu_{i,o}$ the frequency at a certain reference temperature T_o.

EXPERIMENTAL TECHNIQUES

PROCEDURE - Linear thermal expansions between -100° and 200°C were made on machined samples (0.25 in. x 0.25 in.) by means of a du Pont 940 amp Thermo mechanical analyzer[12]. Details of the experimental procedure were previously described.[13,14,15] After proper application of chromel-alumel thermocouple corrections, individual temperature determinations agreed to within 0.2°C and to within 2.0 to 4.0°C of values reported in the literature. The total probable error for the expansion measurements calculated as the square root of the summation of all errors associated with each component of the instrument was ± 2.5%[8,9].

Heat capacity data and latent heat of phase transitions were obtained from sample plugs weighing approximately 30 to 40 mg, by means of a du Pont 900 Differentia Thermal Analyzer with calorimetric plug-in attachment. Base-line corrections were applied for the empty calorimeter as described in references 13,15. The present limits of accuracy in the determination of ΔH are within 3-5%. The heat capacity values are accurate to approximately ± 2.0% and define all transitions present in the polymer. Negative heat capacity values listed in Tables IV-VI are the result of extrapolating about 0° C and do not imply negative heat capacities.

Ultrasonic sound velocities were measured by the phase comparison technique as originated by McSkimin[16]. Measurements were made at this laboratory by Asay, et al.[13,17,18] and were used to calculate Poisson's ratio and the adiabatic bulk modulus necessary for the solution of the Grüneisen ratio by equation (2). Details of the experimental techniques are given in the above references.

TABLE I Grüneisen Ratio, γ ,

For Nylon 6, Teflon and Plexiglas

T, °C	Nylon 6	Plexiglas	Teflon
0	0.94	1.13	
10	0.84	0.88	
20	0.91	0.80	
30	1.05	0.79	0.39
40	1.82	0.76	0.44
50	3.30	0.74	0.44
60	2.43	0.70	0.47
70	1.62	0.70	0.46
80	1.23	0.66	0.58
90	0.97	0.65	0.71
100	0.90	0.65	0.93
110	0.82	0.64	1.03
120	0.75	0.64	0.92

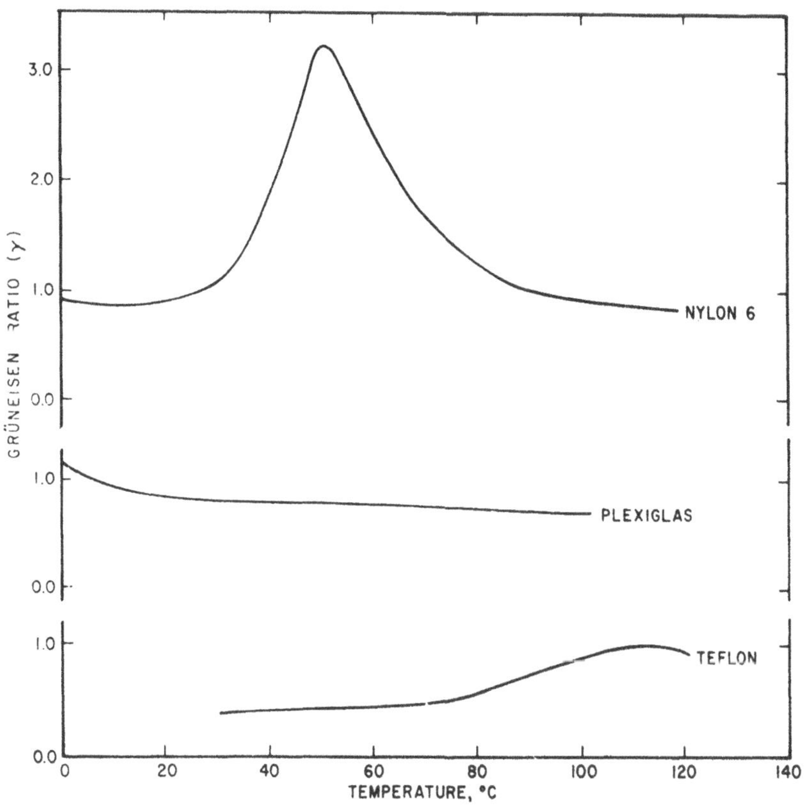

Figure 1.

MATERIALS STUDIES - Solid samples of Nylon 6, Teflon and Plexiglas were used for the experimental studies.

The Nylon 6 sample (du Pont de Nemours) had a calculated density of 1.14 g/cm^3 at 20°C. The polymer is a linear condensation product of caprolactam, [$\left[NH_2 - (CH_2)_5 - CO \right]$].

Polytetrafluorethylene, Teflon, typical of products currently made by du Pont de Nemours Co., has a density of 2.186 gcm^3 at 20°C. Elemental analysis of the sample as presented by specification data, indicated 76.0% fluorine and 24.0% carbon.

Polymethylmethacrylate, Plexiglas, density 1.19 gcm^3 at 20°C was a product of Rohm & Haas Co.

EXPERIMENTAL RESULTS

The thermal Grüneisen ratios, γ_G, for Nylon 6, Teflon and Plexiglas, determined according to equation (2) under constant pressure conditions, are listed in Table I and Figure 1. The temperature derivatives, $(\partial\gamma/\partial T)_p$, were obtained by differentiating each term of equation (2) with temperature so that

$$\left(\frac{\partial\gamma}{\partial T}\right)_p = \gamma\left[\frac{1}{\beta}\left\{\left(\frac{\partial\beta}{\partial T}\right) + \beta^2\right\} + \frac{1}{B^s}\left(\frac{\partial B^s}{\partial T}\right) - \frac{1}{C_p}\left(\frac{\partial C_p}{\partial T}\right)\right] \tag{13}$$

where, β, B^s, and C_p refer to the volume coefficient of expansion, the adiabatic bulk modulus and the heat capacity, respectively.

The experimentally determined linear coefficient of expansion, (α), data (Table II - Figure 2) were used to calculate the volume expansion as defined for isotropic materials in which $\beta = 3\alpha$. The temperature derivative, $(\partial\beta/\partial T)$, was obtained from a linear least squares fit of the original expansion data.

Table III lists the adiabatic bulk moduli, and Poisson's ratios which were obtained from ultrasonic sound velocity measurements by

$$B^s = [\rho\,(u_\ell^2 - 4/3 u_t^2)] = -\frac{1}{3}\,(C_{11} + 2C_{12}) \tag{14}$$

and

$$\sigma = \frac{u_\ell^2 - 2u_t^2}{2\,(u_\ell^2 - u_t^2)} = C_{12}/(C_{11} + C_{12}) \tag{15}$$

where u_ℓ and u_t define the longitudinal and transverse velocities and the c's refer to the elastic coefficients of the cubic system.

Density, ρ, values were obtained from a MacLaurin expansion of the value obtained at room temperature.

TABLE II Linear Coefficient of Expansion, (α), For Nylon 6, Teflon and Plexiglas ($\alpha \times 10^5$,in/in°C)

T,°C	Nylon 6	Teflon	Plexiglas
-100	5.02	7.80	2.40
- 90	5.30	8.66	2.56
- 80	6.24	9.42	2.81
- 70	6.25	8.96	2.87
- 60	5.93	9.12	3.24
- 50	6.60	9.36	3.39
- 40	6.25	9.72	3.41
- 30	5.90	9.79	3.21
- 20	5.97	10.05	4.42
- 10	7.08	10.24	4.69
0	8.01	10.60	5.84
10	7.82	13.09	5.31
20	9.06	21.62	5.53
30	11.01	9.55	6.23
40	23.98	9.16	6.70
50	40.12	8.99	7.16
60	31.90	10.06	7.63
70	23.08	10.13	8.09
80	18.54	13.05	8.56
90	15.36	16.48	9.02
100	15.12	22.89	9.49
110	14.89	26.48	13.62
120	14.64	25.02	26.90
130	14.41	26.99	18.55
140	14.15	24.14	17.88
150	13.96	22.39	17.17
160	13.70	20.31	17.10
170	16.08	20.26	17.17
180	25.58	19.07	17.24
190	35.08	18.97	17.24
200	34.59	18.97	17.24

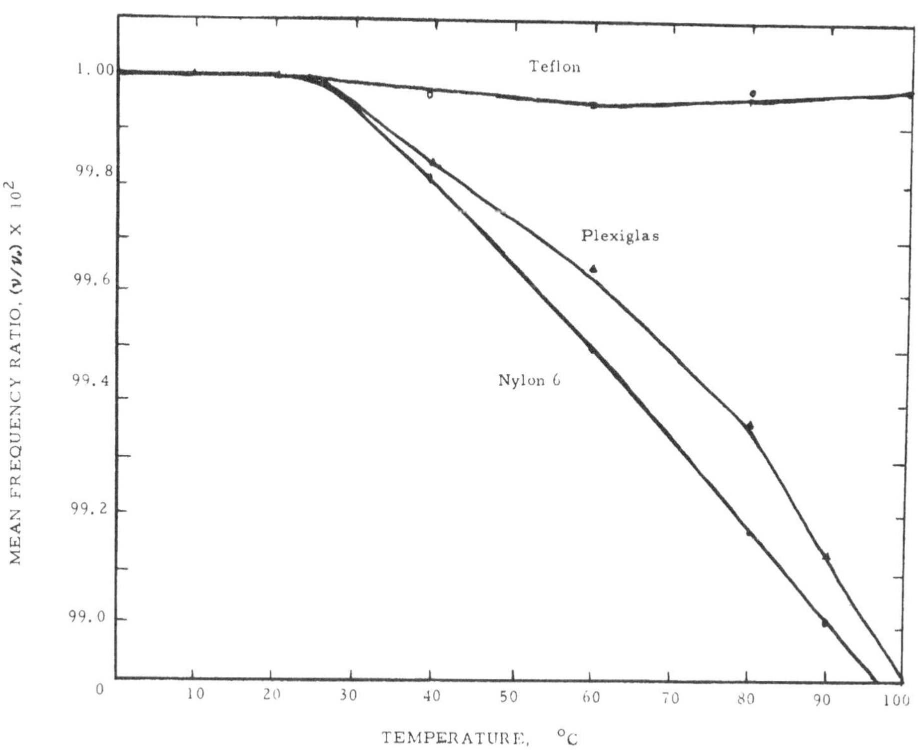

Figure 2.

Linear least square polynomial expressions for the heat capacity, are given in Tables IV, V, and VI. The temperature variation of the heat capacity, $(\partial C_p/\partial T)$, was obtained directly from the polynomial equations.

A close interrelationship is observed between the temperature dependence of the elastic moduli and the thermal data as evidenced from the similar major and minor transitional regions defined and verified by the heat capacity, thermal expansion, bulk modulus, Poisson's ratio, and DTA studies[13] of the polymers. (See Tables I-VI and Figures 1-3). Nylon 6 - A pronounced glass transitional region was evident at Tg = 50°C followed by a fusion peak temperature at T_f = 225°C. Nielsen and Coffman, et al,[20] report a similar glass transitional temperature for the polyamide determined from dynamic mechanical measurements. They attribute the transition to motion of large chain segments in the amorphous region. The high temperature thermal studies showed decomposition of the polymer at 345°C with a double decomposition peak occuring at 385°C.

The linear heat capacity equations (Table IV) may be used to define various phase changes occurring within the polymer in terms of the enthalpy, ΔH. For the solid glassy phase the enthalpy becomes

$$(H_{0^\circ C} - H_{-100^\circ C}) = 0.3716\ T + 5.0 \times 10^{-4} T^2$$

The rubbery state of the polymer, where $Tg < T < T_f$ may be expressed

as

$$(H_{200^\circ C} - H_{50^\circ C}) = 0.4039\ T + 7.60 \times 10^{-4}\ T^2$$

whereas the liquid region is defined by

$$(H_{350^\circ C} - H_{270^\circ C}) = 0.5318\ T - 0.85 \times 10^{-4}\ T^2$$

The heat of fusion, ΔH_f, as calculated from the planimeter measurement of the fusion peak area was ΔH_f = 15.2 cal/g. The value is in good agreement with fusion data presented by Dole and Wunderlich.[21]

Velocity determinations[13,17,18] on the nylon polymer were made at frequencies of 1.5, 4.33, 3.0 and 10.0 mc./sec. The frequency-dependent deviation appears to be no greater than the deviation due to experimental error, indicating that the material is not dispersive in this frequency range. Teflon - The transition observed on the DTA trace[13] at 23°C was verified by the expansion, heat capacity and ultrasonic data. The DTA scan showed an additional break at 30°C which was previously confirmed by Quinn, et al[22] and from heat capacity measurements reported by Furukawa, et al[23]. X-ray studies by Rigby and Bunn[24] and Bunn and Howells[25] indicate no new well-defined crystal structure above 20°C. The assumed change in lattice dimensions resulting from the thermal stress, changes the equilibrium order and leads to anelastic effects. The original velocity data [17,18] verify the 20° - 30°C transition by the sharp inflection point observed on the velocity curve. Determinations made at frequencies of 1.5, 3.0 and 4.33 Mc indicate the velocities measured in Teflon appears to be more frequency-sensitive than those observed in Nylon 6 or Plexiglas.

TABLE III Ultrasonic Data for the Polymers

B^s - Adiabatic Bulk Modulus, Dyne/cm σ - Poisson's Ratio

	Nylon 6		Teflon		Plexiglas	
T,°C	$B \times 10^{10}$	$\sigma \times 10^1$	$B \times 10^{10}$	$\sigma \times 10^1$	$B \times 10^{10}$	$\sigma \times 10^1$
0	6.99	3.99			5.86	3.27
10	6.80	3.98			5.75	3.23
20	6.61	3.98			5.63	3.21
30	6.43	4.00	3.32	1.17	5.52	3.19
40	6.24	4.03	3.15	4.31	5.41	3.19
50	6.05	4.08	3.04	4.36	5.30	3.20
60	5.86	4.14	2.92	4.41	5.18	3.21
70	5.67	4.22	2.84	4.45	5.07	3.24
80	5.49	4.37	2.78	4.49	4.96	3.28
90	5.30	4.41	2.71	4.32	4.84	3.33
100	5.11	4.52	2.57	4.55	4.75	3.39

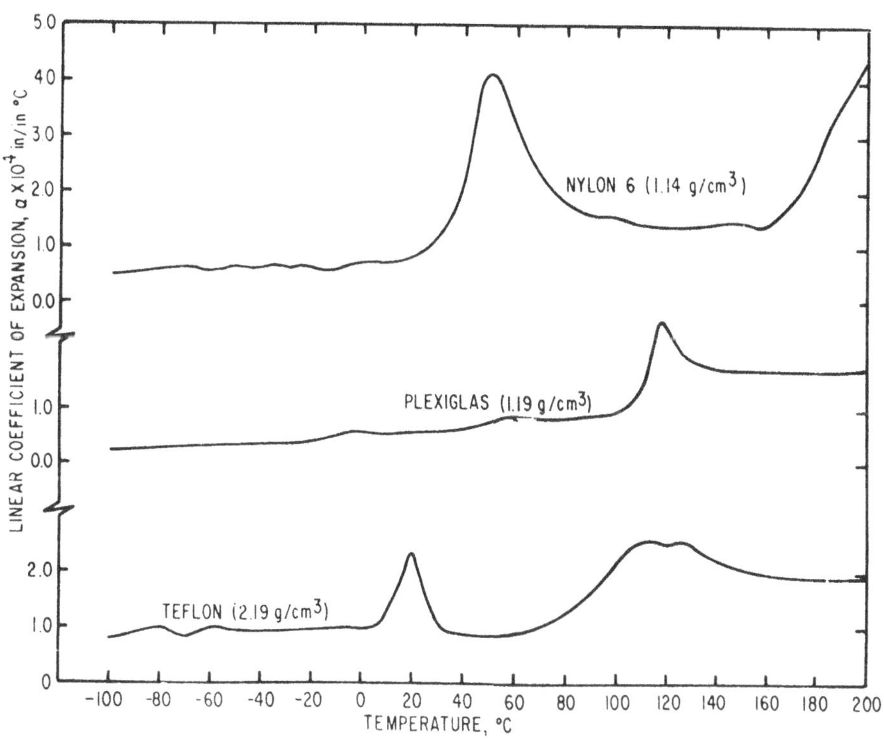

Figure 3.

The heat of transition, ΔH_{Tr}, calculated from the planimeter measurement of the peak area was 1.80 cal/g. The difficulty in determining the boundaries of the 30° transitional region made it impossible to calculate the corresponding enthalpy. Marx and Dole[26] ascribe a value of 1.96 cal/g to that at the 30°C temperature range.

Thermal expansion measurements (Table II - Figure 2) verified the 20°C transition but the region at 30°C was indistinct. It was assumed that the heat generated during the machining of the sample was instrumental in distorting the 30°C transitional region.

The secondary transition at 120°C was attributed to a crankshaft type of motion about the chain axis[27].

The melting transition was clearly defined by the sharp endothermal peak observed at 329°C (Table V). This peak was followed by an endothermal process with a major endothermal decomposition beginning at 530°C.

The calculated enthalpy of fusion, ΔH_f=10.1 cal/g when added to the ΔH_{Tr} at 20°C becomes 12.0 cal/g and is in good agreement with published values obtained from calorimetric studies.[28,29]

Plexiglas - Thermal studies showed a broad, irregular transition occurring from 0° to 60°C whereas the glass temperature was well-defined at 115°C. The transitions were more clearly defined by the expansion curve (Figure 2). The absence of a fusion peak on the DTA thermogram[14] verified the non-crystalline character of the sample.

Nielsen[30] reported a secondary transition at 50°C from dynamic mechanical measurements. A similar transition was reported by Becker[31] and other investigations[32,33,34,35] who attribute the transition to the motion of the ester group (-C = O - OCH_3) which becomes mobile at a lower temperature than the main chain. The longitudinal and transverse sound velocity waves showed a change in slope at this temperature interval which indicates a change in molecular forces.

Decomposition of the polymer was gradual with the DTA baseline deviation beginning at 150°C and terminating in an endothermal peak at 375°C. Fox, et al.,[36] and Cowley and Melville[37] verified the extensive depolymerization beginning at 150°C.

DISCUSSION

Examination of the graph of the Grüneisen ratio versus temperature (Figure 1) illustrates maximum deviations occurring at transitional regions which suggests relaxation and/or translational phenomena. Since γ in this region is no longer a vibrational-volume dependent function as defined by equations (1) and (2), the Grüneisen law does not apply and maximum deviations with temperature are expected. Exclusive of the previously postulated transitional regions, the Grüneisen ratios (Table I - Figure 1) are defined by equations (2) and (3) and slight variations of γ with temperature may be explained by a detailed analysis of the frequency spectrum of the cubic lattices.

The close relation between the lattice vibrational frequencies and the Grüneisen ratio as defined by equations (1) and (3) suggests similar interrelationships between the thermal expansion, heat capacity and compressibility, as defined by equation (2). Since deviations in β are greatly affected by anharmonic effects associated with lattice vibrations, the many theoretical considerations extending from recent publications, to the classic papers of Mie,[3,4] Nernst,[3,4] Lindeman,[4] and Grüneisen[5,6] support the idea that the volume dependence of the moments

Table IV Heat Capacity Date for Nylon 6 (C_p, cal/g°C)*

Temp. Range, °C	a	b x 10^3	Av. Error of C_p x 10^2
-100 to 0	0.3716	1.00	0.20
0 to 50	0.3640	2.74	1.78
50 to 200	0.4039	1.52	0.90
200 to 224	-4.9492	0.28	10.01
224 to 265	5.5194	-0.19	5.22
265 to 350	0.5318	-0.17	0.62

*C_p = a + bT

TABLE V Heat Capacity Data for Teflon (C_p. cal/g°C)*

Temp. Range, °C	a	b x 10^3	Av. Error of C_p x 10^2
-100 to 0	0.1915	33.72	0.13
0 to 20	0.1809	1.31	0.52
20 to 30	0.0904	5.81	0.38
30 to 45	0.3679	-3.56	0.61
50 to 290	0.2003	17.00	0.30
300 to 329	-0.1538	1.35	0.27
345 to 355	0.9874	-1.49	0.81
355 to 425	0.5390	-26.98	0.33

*C_p = a + bT

TABLE VI Heat Capacity Data for Plexiglas (C_p, cal/g°C)*

Temp. Range, °C	a	b x 10^3	Av. Error of C_p x 10^2
-100 to 45	0.1974	1.76	1.17
45 to 78	0.1578	3.05	0.56
78 to 112	0.2878	1.42	0.54
112 to 130	-0.0328	4.27	1.20
130 to 150	0.6376	-0.60	0.21
150 to 166	0.4264	0.82	0.38
166 to 200	0.4249	0.92	0.59

*C_p = a + bT

of the frequency spectrum can be related to the thermal expansion.

With the assumption that $T \gtrsim \theta$, equation (12) was used to estimate the frequency ratios which are plotted in Figure 3 as a function of temperature. Values used for the volume coefficients of expansion through major transitional regions were obtained from low temperature linear extrapolations. Similar extrapolations were used to obtain the high temperature limits of the Gruneisen ratio which were 0.90, 0.55 and 0.75 for Nylon 6, Teflon and Plexiglas, respectively.

Small overall variations observed for the frequency ratios (Figure 3) suggest minor deviations of the frequencies between 0° and 100°C temperature range; consequently, no complicated reorganization of structure is expected outside of the major transitional regions described above.

The gradual deviations in the ratio for Nylon 6 and Plexiglas above $\sim 30^{o}$C are associated with the onset of the liquid phase wherein the frequencies are expected to decrease. The decrease in the bulk moduli and Poisson's ratios (Table III) verifies onset of liquifaction.

The overall constancy observed for Teflon (Figure 3) suggests that $\nu = \nu_0$ and the strength of the polymer is not greatly affected by temperature within this range. These results agree with the experimental results obtained for the Grüneisen ratio wherein minor deviations in γ were observed for non-transitional regions.

CONCLUSION

This work has shown that anomalies in polymeric materials are interrelated. From a phenomenological point of view, the temperature dependence of the elastic moduli, the heat capacities, the thermal expansions and the ratios of the mean vibrational frequencies are all equivalent and may be related to the Grüneisen ratio. Further detailed structural relationships require experimental velocity and thermal measurements at low temperatures as well as theoretical calculations of the interchain contribution to the heat capacity at constant volume.

ACKNOWLEDGMENTS

The authors wish to express their appreciation to Leonie Boemer and Mark Middleton for their assistance in obtaining the data reported here.

References

(1) T. H. K. Barron, Phil. Mag., 46,720 (1955); Ann. Phys. 1, 77 (1957).

(2) B. Dayal, Proc. Indian Acad. Sci., A20, 138 (1944).

(3) J. C. Slater, Introduction to Chemical Physics, New York, N. Y., 1939.

(4) M. Born, Phys. Z., 13, 297 (1912); M. Born and T. Von Karman, 13, 297.

(5) E. Grüneisen, Handbuch der Physik, Vol. 10, p22, Springer, Berlin (1926)

(6) E. Grüneisen, Ann. Physik, 39, 258 (1912).

(7) T. Rice, R. McQueen and J. M. Walsh, in "Solid State Physics," F. Seitz and D. Turnbull, Ed., Academic Press, Inc., New York, N.Y., 1958, Vol. 6.

(8) R. O. Davies, Phil. Mag., 45, 290 (1954).

(9) D. Bijl and H. Pullan, Phil. Mag., 45, 290 (1954); Physica, 21, 285 (1955).

(10) M. Blackman, Proc. Roy. Soc., A148, 365 (1934); Phil. Mag., 3, 831 (1958); Proc. Phys. Soc., 74, 17 (1959).

(11) A. Schauer, Can. J. Phys., 42, 1957 (1964).

(12) E. I. duPont deNemours, Differential Thermal Analysis Instrument Manual, Wilmington, Del., 1965.

(13) J. R. Asay, S. R. Urzendowski, and A. H. Guenther, AFWL-TR-67-91, Air Force Weapons Laboratory, Kirtland Air Force Base, New Mexico, 1967.

(14) S. R. Urzendowski, J. R. Asay, and A. H. Guenther, submitted for publication in J. Appl. Phys.

(15) S. R. Urzendowski, J. R. Asay, and A. H. Guenther, submitted for publication in J. Appl. Phys.

(16) H. J. McSkimin, J. Acoust. Soc. Am., 22, 413 (1950).

(17) J. R. Asay, A. J. Dorr, N. D. Arnold and A. H. Guenther, AFWL-TR-65-188, Air Force Weapons Laboratory, Kirtland Air Force Base, New Mexico, 1966.

(18) N. D. Arnold, J. Appl. Poly. Sci., 10, 731 (1966).

(19) L. E. Nielsen, Mechanical Properties of Polymers, Reinhold Pub. Co., New York, N. Y., 1962, p. 33.

(20) D. D. Coffman, G. J. Berchet, W. R. Petersen and P. W. Spanagel, J. Poly. Sci., 2, 139 (1957).

(21) M. Dole and E. Wunderlich, J. Poly. Sci., 24, 139 (1957).

(22) R. A. Quinn, Jr., D. E. Roberts and R. N. Work, J. Appl. Phys., 22, 1085 (1951).

(23) G. T. Furukawa, R. E. McCoskey and G. J. King, J. Res. Nat. Bur. Stand., 49, 273 (1952).

(24) H. A. Rigby and C. W. Bunn, Nature, 164, 583 (1949).

(25) C. W. Bunn and E. R. Howells, Nature, 174, 549 (1954).

(26) P. Marx and M. Dole, J. Am. Chem. Soc., 77, 4771 (1955).

(27) W. J. Roff, Fibers, Plastics and Rubbers, Butterworth, Inc., London, England, 1956.

(28) H. W. Starkweather and R. H. Boyd, J. Phys. Chem., 64, 410 (1960).

(29) J. C. Siegel, L. T. Muus, T. P. Lin, and H. A. Larsen, J. Poly. Sci., 2A, 391 (1964).

(30) L. E. Nielsen, Soc. Plastics Eng. J., 16, 525 (1960).

(31) G. W. Becker, Kolloid Z., 140, 1 (1955).

(32) K. Schmieder and K. Wolf, Kolloid Z., 134, 149 (1953).

(33) P. Heydemann and H. D. Grucking, Kolloid Zeit., 1, 16 (1963).

(34) K. Deutsch, E. A. W. Hoff and W. Reddish, J. Poly. Sci., 13, 565 (1954).

(35) J. Koppelmann, Kolloid Z., 164, 31 (1959).

(36) R. B. Fox, L. G. Isaacs and S. Stokes, J. Poly. Sci. Part A, 1, 1079 (1963).

(37) P. R. E. Cowley and H. W. Melville, Proc. Roy. Soc., London, A 210, 153 (1955).

THERMAL PROPERTIES OF INORGANIC POLYMER SYSTEMS

by

Mark B. Myers

Research Laboratories
Xerox Corporation, Rochester, New York

The liquid structures of S and Se have been proposed (Ref. 1,2,3) to be equilibrium mixtures of linear polymer molecules and eight-membered monomer rings. The experimental phenomena observed in S which stimulated Gee's (1) proposal and its formulation by Tobolsky and Eisenberg (Ref. 2,3) were the temperature dependence of the viscosity and heat capacity in the region of 159 °C and the solubility or lack of solubility of various forms of S in solvents such as CS_2. Further, the structures of the highly molecular crystalline phases of S and Se have strongly influenced thought on the possible molecular forms in the liquid.

This model which has worked so successfully with S has been applied to liquid Se by Eisenberg and Tobolsky (Ref. 3,4). In the Se case, the experimental evidence supporting a polymer-monomer structure are the high liquid viscosities, the crystallization behavior, differential solubilities and infrared spectroscopy (5). Additional experimental evidence supporting a polymeric view of the structures of S and Se are the elastic moduli-temperature curves (6) and the glass transition phenomenon. It should be noted that the primary stimulus for Gee's proposal; i.e. the thermally induced polymerization transformation of the S monomer rings, has not been experimentally observed for the proposed Se monomer rings.

This paper reports on studies of the relationships between thermal properties and the possible liquid and corresponding glass structures in the S-Se, As-S, and As-Se binary systems. The thermal properties studied include the glass transition temperature, Tg, as a function of composition and its relationship to T_f the liquidus temperature; the variation of T_λ, the polymerization temperature, with composition; and a comparison of the ΔH_f and ΔS_f of the crystalline phases. Investigations of the systematic variation of such thermal properties can both further define the structural concepts of the end members S and Se, and can indicate possible changes in equilibria and structure in the copolymer systems S-Se, As-S, and As-Se.

EXPERIMENTAL

The transition temperatures were observed by the change of C_p as indicated by differential thermal analysis using a DuPont Model 900 differential thermal analyzer and a Perkin-Elmer DSC-1B scanning calorimeter. Most of the C_p and H_f measurements were made employing the DSC-1B. With heating rates of 10°/min., the transition temperature was interpreted as the break in the baseline of the thermogram trace. For practical reasons, this was normally identified by extrapolating straight lines above and below the departure from baseline.

If the sample to be studied was a glass, the powdered material was placed in the sample configuration and heated above the liquidus temperature, then rapidly cooled to room temperature with the subsequent measurements being made at defined times after cooling to insure similar thermal histories.

The samples were prepared by weighing 99.9999 percent pure elemental As, S, and Se obtained from Gallard Schlesinger and vacuum sealing them in silica glass ampuls. The constituents were reacted at 600 °C for 24 hours and air-cooled to room temperature.

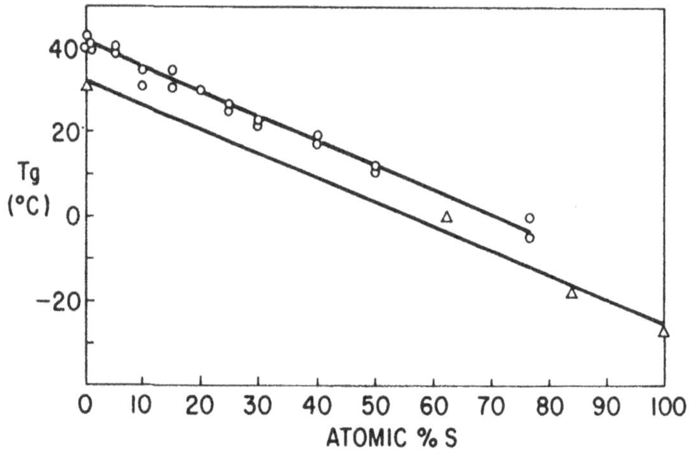

FIGURE 1. Variation of the Tg's with S concentration with S-Se system. Circles denote the DTA data of this investigation, and triangles denote elastic moduli-temperature data of Tobolsky et al.

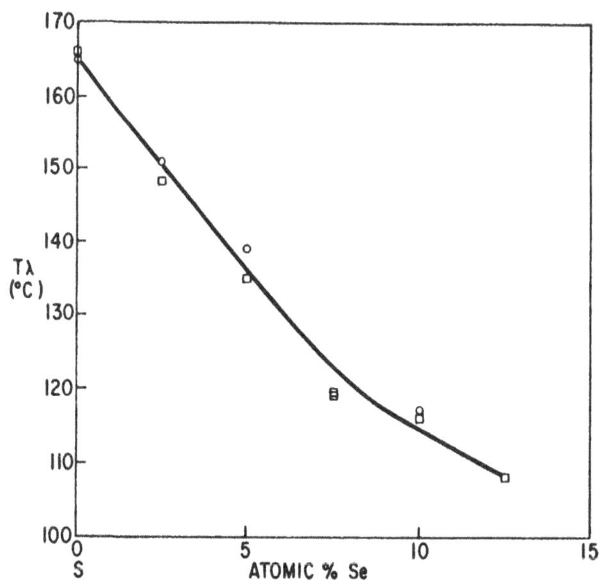

FIGURE 2. The variation of the polymerization transformation, T_λ, in the S-Se system. Circles denote DTA data and squares denote DSC data. Heating rates were 10°/min.

THE S-Se COPOLYMER SYSTEM

The comelting of S and Se results in a copolymer system in which copolymer molecules are considered to be dynamic equilibria with monomer rings. In such a system the relative monomer-polymer concentrations and chain lengths are not only dependent on temperature but also on the relative S and Se concentrations. From copolymerization considerations (Ref. 7), the addition of Se to S should decrease the monomer concentration, decrease the polymer chain length, and lower the onset of the polymerization transition.

The thermal properties of the S-Se mixtures are considered to be reasonably consistent with this interpretation. The variation of Tg with composition is shown in Fig. 1, and it is seen that Tg decreases linearly with S content. The increased monomer concentration associated with the S content is considered to have a plasticizing influence which produces a less rigid structure with a corresponding decrease in Tg. The data in Fig. 1 denoted by circles is from the investigation by DTA. This data is consistent with the earlier Tg's of Tobolsky et al (Ref. 6) determined from elastic moduli-temperature curves.

The variation of T_λ with composition is shown in Fig. 2, and it is seen that there is a rather sharp but uniform decrease with increasing Se content. A simple treatment with a van't Hoff relation employing the enthalpy for polymerization (8) for S suggests that at temperatures below T_λ the Se not only replaces a S monomer concentration equal to the amount of Se added, but it also has a polymerizing effect on the S in the solution. This is supported by the diminution and eventual disappearance of an observable dynamic polymerization phenomenon with increasing Se concentration to 15 atomic percent.

THE As-S AND As-Se COPOLYMER SYSTEMS

1. Molecular Configurations

In order to ascertain the possible relationships existing between the molecular structure of the copolymer systems and the thermal properties, evidence for the limiting molecular configurations in the systems will be considered. The most important of these are the structures of the highly molecular crystalline phases and their melting behavior. The six compositions in which unique crystal structures form in the composition ranges studied are listed in Table 1 and indicated in Figs. 3 and 4.

The structures of S and Se are well known and will not be further discussed except they are considered to fuse to liquids composed of rings and chains as previously discussed. As_2S_3 and As_2Se_3 crystallize in the orpiment structure. This structure consists of parallel sheets or layers of $(AsS_{3/2})_n$ molecules with van der Waal bonding between. The layers consist of arsenic trisulfide linked puckered 12-membered rings. The melting of these crystals may be envisioned as the disordering or floating apart of the layers, then the break up of the layers into large aggregations of linked chains or rings.

The crystal structure of α As_4S_4 is the realgar structure. It is composed of 8 membered puckered molecules held together with van der Waal forces. A fact seldom recognized in current work on this material is that As_4S_4 undergoes a polymorphic transformation at 273 °C to a high temperature β form. The structure of As_4Se_4 is less certain but it is considered to be isostructural with As_4S_4 or to have a very similar structure. An important difference between As_4S_4 and As_4Se_4 relating to the structure of the liquid is their melting behavior as indicated in Figs. 3 and 4. As_4S_4 melts congruently to a liquid of the same composition, whereas under equilibrium conditions As_4Se_4 decomposes at 264 °C to a liquid and crystalline As_2Se_3.

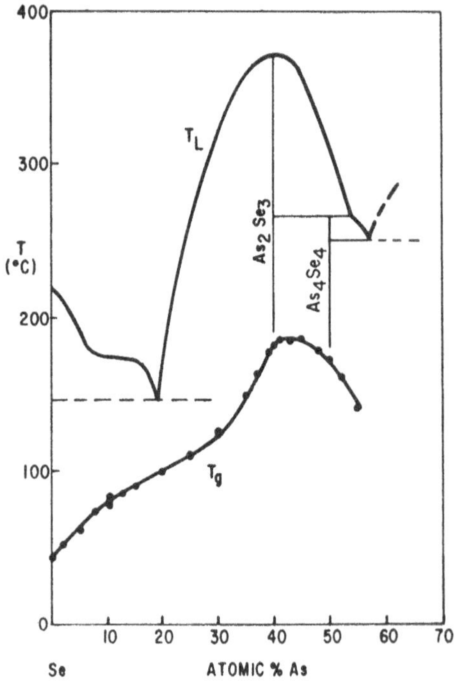

FIGURE 3. The glass transition temperatures and phase diagram (Ref. 10) of the system Se-As.

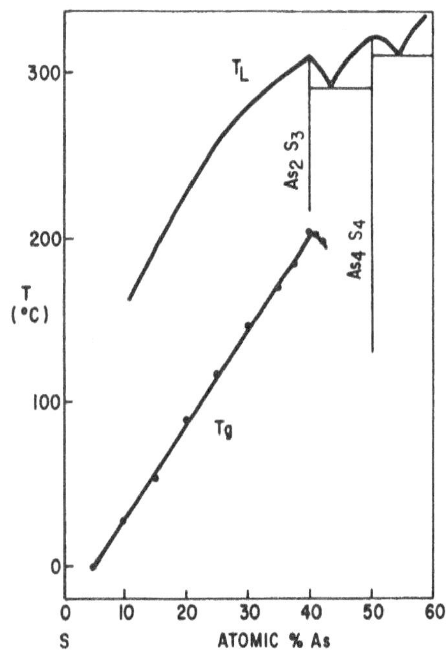

FIGURE 4. The glass transition temperatures and phase diagram (Ref. 9) of the system S-As.

The inability to quench glasses with more than 43% As in the As-S system and the extension of glass formation to 55% As in the As-Se system is most certainly related to this difference in melting behavior.

In Table 1, the similarity in the thermal properties of S and As_4S_4 is noted; both have polymorphic transformations and melt with relatively low enthalpies and entropies of fusion. By comparison, the crystalline phases Se, As_2S_3 and As_2Se_3 undergo greater rearrangement when melting to associated liquids and have correspondingly higher enthalpies and entropies of fusion.

2. Thermal Properties of the Copolymers

The addition of As to either Se or S liquids is considered (5) to produce a copolymer with the trivalent As forming branch points. This branching produces a more rigid network structure with a corresponding increase in Tg. If this branching is a continuous linear process, it is expected that Tg will increase monotonically with As content.

The glass transition temperatures and liquidus temperatures for the As-Se system are shown in Fig. 3. Tg does not increase monotonically with As content; in fact, a considerable amount of structure in the curve is observed. A broad maximum peaking at 43% As exists in the vicinity of compound formation, and there are two inflection points in the curve between Se and As_2Se_3 at 10% and 30% As. The correspondence between the broad maximum and the liquidus curve is quite apparent. It also appears that the inflection in Tg at low As content might be associated with the tendency of the liquid to dissociate as indicated by the shape of the liquidus.

Before attempting to relate the above observations to the variations in glass structure in the As-Se system, it is important to consider the As-S system. The glass transition temperature, as indicated in Fig. 4, increases monotonically with As content to 40% As and goes through a sharp maximum then decreases with higher As content. This difference in the behavior of Tg in the As-S and As-Se system in the region of compound formation is consistent with melting behavior and glass formation. The linear increase in Tg to 40% As content was previously observed by Tobolsky and co-workers (5); however, the composition region between the two compounds was not investigated.

The structural interpretation given (Ref. 10) to explain the variation of Tg with composition in the As-Se system is one consisting of modifications of the limiting polymer structures of Se and As_2Se_3 where there is a region of As branched Se chains, mixtures of branched chains and modified $AsSe_{3/2}$ networks, and aggregrates of modified $AsSe_{3/2}$ networks. Specifically considering Fig. 3, the relationship of Tg to these regions is the following. The addition of excess Se or As to the As_2Se_3 composition will break down the continuity of the $AsSe_{3/2}$ linked networks into smaller aggregrates with the excess Se or As residing at bridge or terminal points. This will cause a decrease in Tg. Moreover, this decrease should be greater for an equal amount of the divalent Se than an equal amount or trivalent As. As indicated in Fig. 3, the maximum in Tg at 43% instead of 40% As would suggest that the excess As initially produces further network formation before breaking down the continuity of the structure. The addition of sufficient Se to As_2Se_3 will eventually reach a composition where considerable Se-Se bonding is produced and the glass will have a mixed $AsSe_{3/2}$ linkage and As branched Se chain behavior. It is proposed that this corresponds to the break in the Tg curve near 30% As. Tg varys less with composition in between 10 and 30% As because two polymer structures are being modified. Finally, at lower concentrations of As between 0 and 10%, the glass consists essentially of As branched Se chains.

TABLE I. SUMMARY OF THERMODYNAMIC TRANSITIONS FOR CRYSTALLINE PHASES

Phase	Str. Unit Stable below T_f	Transformation	T $(^{\circ}C)^{-1}$	ΔH $(kcal/gm\ at)^{-1}$	ΔS $(cal/gm\text{-}at^{\circ})^{-1}$
S	S_8 rings	$\alpha - \beta$*	96	0.090	0.25
		f*	119	0.293	0.75
As_2S_3	$AsS_{3/2}$ linked sheets	f	322	1.60	2.69
As_4S_4	As_4S_4 molec.	$\alpha - \beta$	273	0.123	.23
		f	310	0.991	1.67
Se	Se chains	f*	217	1.25	2.56
		f	217	1.48	3.02
As_2Se_3	$AsSe_{3/2}$ linked sheets	f	365	2.01	3.14
As_4Se_4	As_4Se_4 molec.	d	264-280		

*NBS cir 500
Other values from present study.

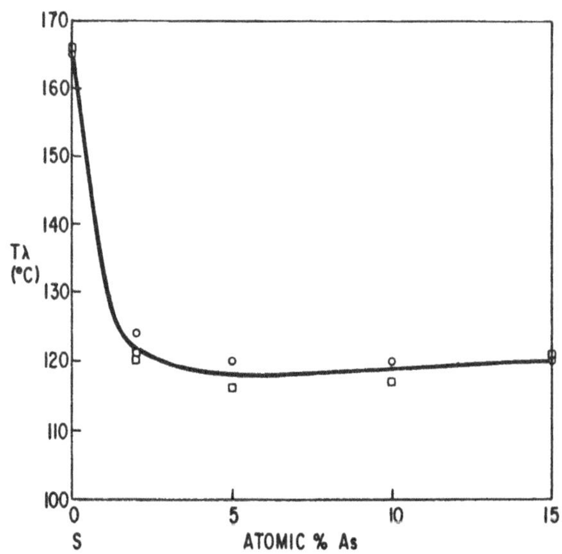

FIGURE 5. The variation of the polymerization transformation, T_{λ}, in the S-As system. Circles denote DTA data and squares denote DSC data. Heating rates were 10°/min.

The sharp maximum and linear behavior of Tg in the As-S system lead to a different structural interpretation. The glass structure is considered to mixtures of modified $AsS_{3/2}$ linked networks and S monomer rings or As_4S_4 molecules, depending on composition. Specifically considering Fig. 4, the relationship of Tg to these regions is the following. Adding excess As to the As_2S_3 composition leads to the formation of a small molecular species such as As_4S_4 and the break down of the continuity of the $AsS_{3/2}$ linked networks. This is indicated by the decrease of Tg and the limited glass formation. Initially additions of S to the As_2S_3 composition have the same effect that Se has on As_2Se_3; however, when a sufficient amount of S has been added to As_2S_3 to introduce a large amount of S-S linkages the Tg will be below the polymerization temperature of S. The excess S not included in S modified $AsS_{3/2}$ linked networks will be in the form of monomer rings not polymer chains. These monomer rings are considered to have a plasticizing influence on the polymer with a corresponding decrease in Tg. However, it must be remarked that it may be misleading to attribute the decrease in Tg completely to the plasticizing influence of S monomer rings. The characteristics and continuity of the $AsS_{3/2}$ networks are continuously modified by some excess S which interacts with the polymer rather than forming monomer.

The variation of T_λ with the addition of As is shown in Fig. 5. If the concept of the As-S glass structure being a mixture of aggregrates of $AsS_{3/2}$ linkages and S monomer rings is consistent with observation, it is necessary that T_λ lie above Tg for glasses with high S content. This is observed as indicated by comparison of Figs. 4 and 5. In contrast with the uniform variation of T_λ found in the S-Se copolymers, it is seen in Fig. 5 that there is a large decrease in T_λ with small additions of As and near invariance with further addition. An explanation for this, subject to verification by spectroscopic techniques, is that small additions of As decrease the initial monomer concentration significantly forming $AsS_{n/2}$ ($n \gg 3$) aggregrates. Further additions have less effect on the monomer concentration and interact primarily with these aggregrates shifting their composition toward $AsS_{3/2}$.

Finally, the structural interpretations based on the variation of Tg in the As-Se and As-S systems are essentially the same as those proposed by Nemilov and Petroskii (11) to explain the variation of liquid viscosity in the As-Se system. More recently similar interpretations have been proposed by Carrel and Wilder (Ref. 12) in studies of sintering phenomena in partially vitreous As-S-Se systems.

CONCLUSION

The study of the relationships between the thermal properties of inorganic polymers is found to be generally consistent with previous structural concepts and useful for providing insights where prior concepts do not exist. The linear variation of Tg in the S-Se is associated with variations of concentrations of linear polymer chains and monomer rings, and in the As-S it is associated with modification of $AsS_{3/2}$ networks and S ring concentrations. The nonlinear variation of Tg in the As-Se system is explained by variations of the limiting structures of As branched chains and $AsSe_{3/2}$ networks.

REFERENCES

1. G. Gee, Trans. Faraday Soc. 48, 515 (1952).
2. A. Tobolsky and A. Eisenberg, J. Am. Chem. Soc. 81, 780 (1959).
3. A. Eisenberg and A. Tobolsky, J. Polymer Sci. 46, 19 (1960).
4. R. Keezer and M. Bailey, Mat. Res. Bull. 2, 185 (1967).
5. G. Lucovsky and R. Keezer, Bull. Am. Phys. Soc. 11, 812 (1966).
6. A. Tobolsky, G. Owen and A. Eisenberg, J. Colloid Sci. 17, 717 (1962).
7. A. Tobolsky and G. Owen, J. Polymer Sci. 59, 329 (1962).
8. G. Gee, "Inorganic Polymers" Chem. Soc. Spec. Publ. No. 15, 67 (1961).
9. M. Hansen, "Constitution of Binary Alloys" McGraw Hill, New York (1958).
10. M. Myers and E. Felty, Mat. Res. Bull. 2, 535 (1967).
11. S. Nemitov and G. Petrovskii, Izv. An. SSSR Su. 28, 1283 (1964).
12. M. Carrell and D. Wilder, J. Amer. Ceram. Soc. 50, 604 (1967).

THERMAL EFFECTS IN ION-EXCHANGE REACTIONS: HEATS OF ION EXCHANGE

by

G. E. Boyd
Oak Ridge National Laboratory, Oak Ridge, Tennessee

Ion-exchange reactions between aqueous electrolyte solutions and organic ion exchangers are temperature dependent, and therefore they must be accompanied by the evolution or the absorption of heat. However, the magnitudes of these thermal effects generally are small so that they are measured best with a calorimeter rather than estimated by application of the van't Hoff equation to the variation of the thermodynamic equilibrium constant with temperature. Calorimetrically determined ion-exchange reaction enthalpy changes are useful further in that they may be combined with free energies computed from equilibrium measurements to obtain the change in entropy. The sign and magnitude of the enthalpy and entropy changes afford an insight into the factors which govern the selective uptake of one ion in aqueous solution over another. For example, the entropy change frequently suggests that ion-solvent interactions are of central importance, while the enthalpy change may indicate that "site binding" or ion-pair formation by the ion exchanger governs the observed selective ion absorption.

The thermal effects accompanying ion-exchange reactions are of importance to analytical and preparative chemistry. The presence of an ion-exchange reaction in a system may be detected by the heat evolved or absorbed and this fact has been used as the basis of a very general method for monitoring ion-exchange column chromatographic separations. In one practical embodiment of this concept, liquid from an ion-exchange column is led through a small, thermally insulated cell containing two thermistors embedded in a thin bed of the ion exchanger (Fig. 1). The electrical resistance of these detectors responds rapidly to the minute temperature changes generated in the cell as the composition of the exchanger adjacent to the thermistor varies when a chromatographic band moves off the column and through the cell.

The temperature dependence of the partition of ions between a solution and an ion exchanger suggests that heat instead of chemicals might be applied in ion-exchange column separations. This possibility appears to have been realized recently in a novel method for separating mixtures wherein a temperature gradient is applied to a cylindrical bed of ion exchanger and a solution of the mixture is pumped alternately back and forth through the bed (Fig. 2). Separation takes place at the expense of thermal instead of chemical energy as in the conventional operation of ion-exchange columns.

Chromatographic efficiency in ion-exchange column separations is enhanced by operation at temperatures well above room temperature. This improvement results from the diffusional nature of the rate process governing the attainment of equilibrium. As the temperature is increased, the chromatographic bands, or elution peaks, become sharper and the overlap between adjacent bands is reduced. However, the separation also depends on the differences between the equilibrium distribution coefficients for the solutes being fractionated: if these differences increase as the temperature increases, the separation attainable will be improved markedly, as with the rare earths (Fig. 3). If, however, the difference becomes smaller with increasing temperature, the separation may become poorer, as with the alkali metals, in spite of the sharpening of the bands (Fig. 4). These variations may be predicted from a knowledge of the enthalpies of the ion-exchange reactions occurring in the column. The measurement of these enthalpies will now be considered:

The Measurement of Thermal Effects in Ion Exchange Reactions: Estimates derived from the temperature coefficients of equilibrium constants reveal that the enthalpy changes in most ion-exchange reactions are usually of the order of 1-3 kcal/equiv so that sensitive calorimetric methods are required for their measurement. A solution-type calorimeter (Fig. 5) has been employed in the work performed at ORNL. The silvered Dewar vessel contained

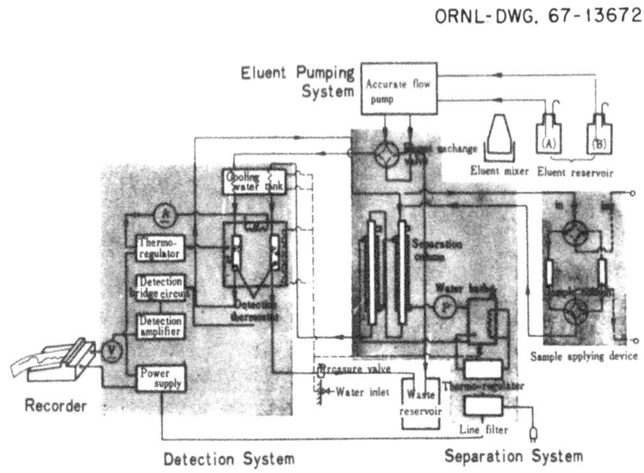

THERMAL REACTION ENERGY DETECTOR FOR LIQUID CHROMOTOGRAPHY

Fig. 1

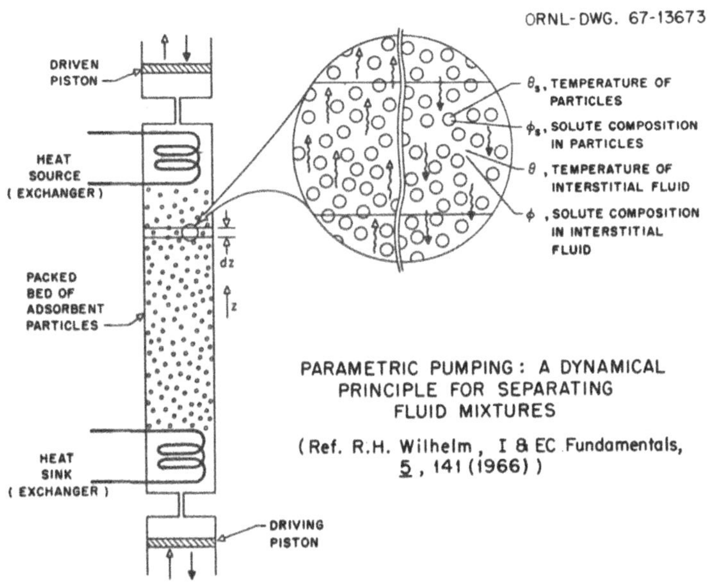

PARAMETRIC PUMPING: A DYNAMICAL PRINCIPLE FOR SEPARATING FLUID MIXTURES

(Ref. R.H. Wilhelm, I & EC Fundamentals, 5, 141 (1966))

Fig. 2

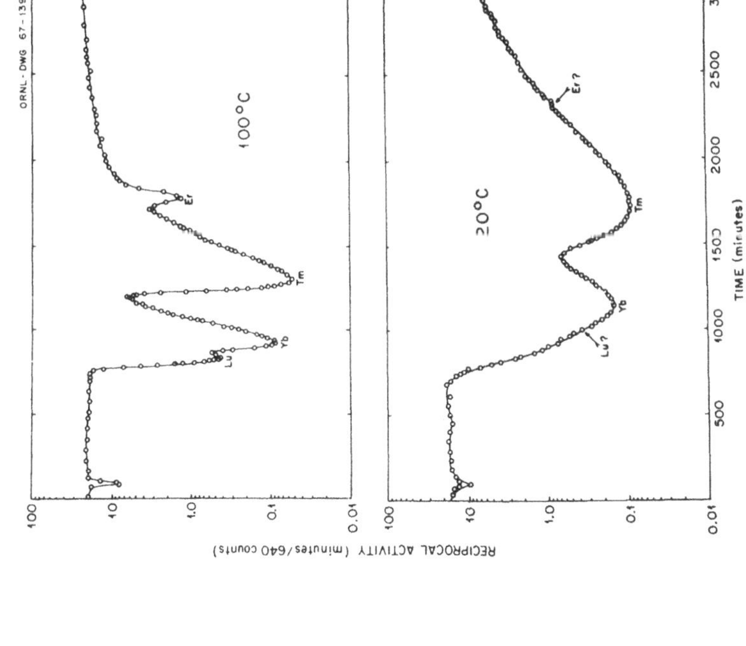

EFFECT OF TEMPERATURE ON THE SEPARATION OF THE YTTRIUM GROUP RARE EARTHS
(Ref. J. Am. Chem. Soc., 69, 2800 (1947))

Fig. 3

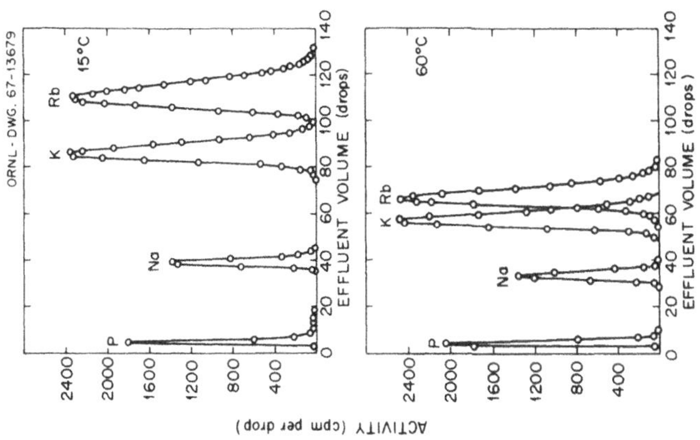

DEPENDENCE OF THE CHROMATOGRAPHIC SEPARATION OF THE ALKALI METALS ON TEMPERATURE
(R. Dybczynski, Institute of Nuclear Research, IBJ 809/VIII/C, 1967)

Fig. 4

approximately 500 ml of dilute electrolyte of either 0.01 or 0.1 N concentration (*i.e.*, 5.0 or 50.0 meq of exchangeable ion). The sealed pipet was filled with 1.0-10 meq of ion exchanger of predetermined initial ionic composition. When a steady-state temperature drift was reached inside the stirred solution in the Dewar, the ion-exchange reaction was initiated by opening the pipet and allowing the exchanger to mix with the external solution. Almost always the thermal effect (*i.e.*, heat absorbed or evolved) was complete within 5-6 minutes. The energy equivalent of the calorimeter was determined by heater ratings in which an accurately determined current was passed for a known time through the heater resistance. Temperature changes were sensed with a 10 K thermistor which was connected electrically in the arm of a Wheatstone bridge. The unbalanced current from the bridge was amplified and recorded as a function of time with a 100 mv L&N recorder. A temperature sensitivity of approximately 10 microdegrees, corresponding to 0.005 cal, was realized. The heat of opening of the pipet was 10 ± 5 millicalories exothermic.

On completion of the calorimetric measurements, the ion exchanger was recovered quantitatively and its ionic composition was determined by analysis. The total meq of exchanger employed was determined by titration of the acid form. The meq of exchange reaction was computed from the initial and final compositions of the exchanger and total meq involved. The uncertainty in the amount of reaction was ± 0.02 meq. The composition of the equilibrium mixed electrolyte solution in the calorimeter also was determined by analysis so that the purity of the calorimetric reaction could be established by material balances. Heats of ion exchange are expressed as defined calories per *equivalent* (1 cal = 4.1840 absolute joules).

Several criteria were applied to establish that the calorimeter performed reliably in a given measurement: (a) the precision of triplicate energy equivalent determinations did not exceed ± 0.2%; (b) the heater resistance as calculated from the measured current and voltage drop remained constant; (c) the heat leakage constant of the calorimeter was small and constant. The over-all performance of the calorimeter system was checked periodically by measurements of the heat of solution of KCl(c) in water and of tris(hydroxymethyl)aminomethane (THAM) in 0.100 N HCl.

Isothermal heat of ion-exchange measurements, Q, on the potassium-sodium ion reaction in cross-linked polystyrenesulfonic acid type (PSSA) exchangers and on the bromide-fluoride ion reaction in cross-linked strong-base anion exchangers are shown in Figs. 6 and 7. The observed heats are plotted as chords against the equivalent fraction of the preferred ion taken up by the exchanger. The curve defining the differential heat of exchange, $\Delta\bar{H} = (dQ/dx)$, is drawn through the mid-points so that the areas subtended above and below the chord are equal. It may be observed that the larger the cross-linking, the more $\Delta\bar{H}$ varies with the composition. Heat is not necessarily evolved on the absorption of a preferred ion, as may be seen for the Li^+/Cs^+ ion-exchange reaction with cross-linked polymethacrylic acid (PMA) exchanger (Fig. 8).

The integral or total heat of exchange, ΔH, for the conversion of the exchanger from one ionic form to another is given by the area under the differential heat, composition curve

$$\Delta H = \int_0^1 \Delta\bar{H}\, dx_{M_2}$$

where x_{M_2} is the equivalent fraction of the preferred ion on the exchanger.

The enthalpy change, $\Delta H°$, for the (hypothetical) standard-state cation-exchange reaction

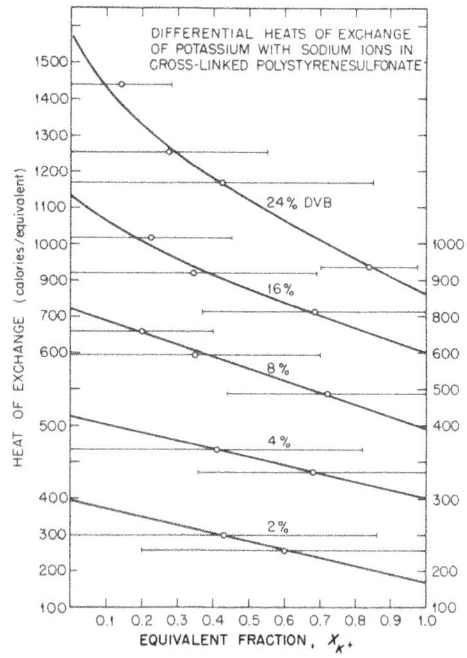

Fig. 6

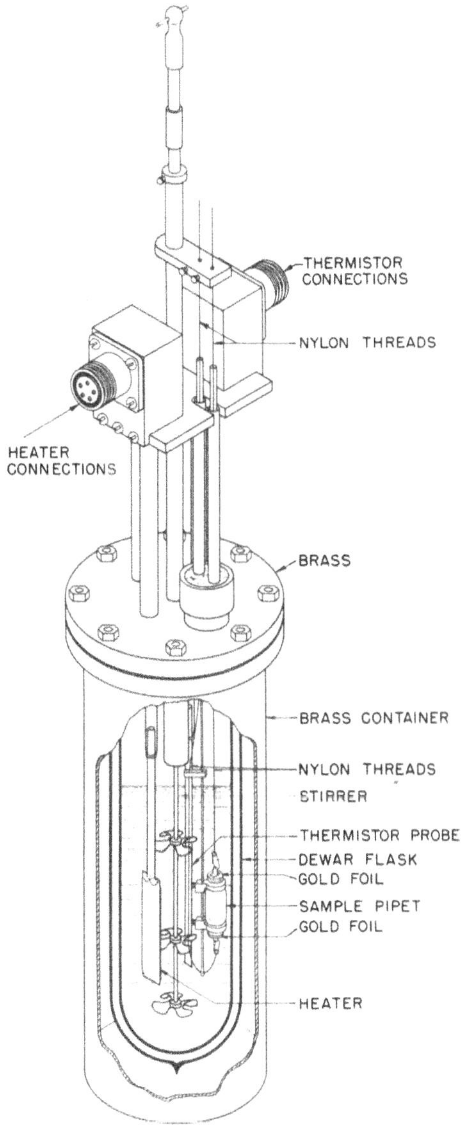

Fig. 5

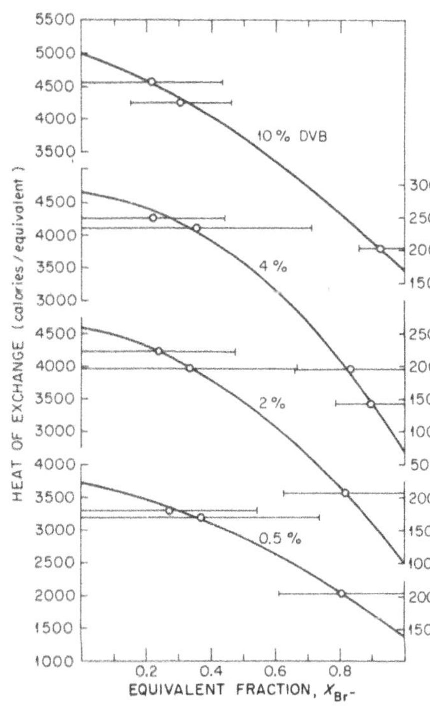

DIFFERENTIAL HEATS OF EXCHANGE OF BROMIDE WITH FLUORIDE ION IN STRONG-BASE ANION EXCHANGERS DERIVED FROM CROSS-LINKED POLYSTYRENE

Fig. 7

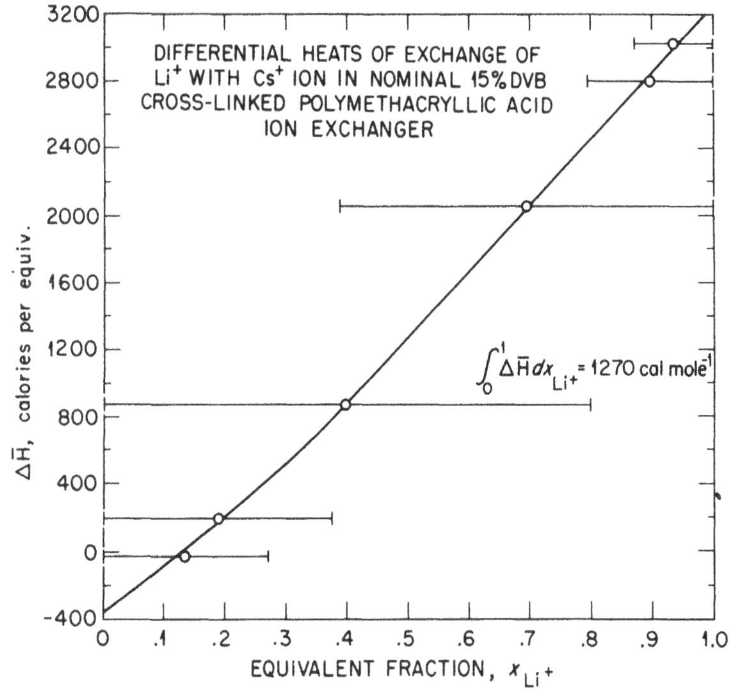

Fig. 8

Table I. Standard Free Energies, Enthalpies, and Entropies at 298.2°K for the Exchange of the Alkali Metal Cations with Sodium Ion in Cross-linked Polystyrenesulfonates.

($\Delta G°$ and $\Delta H°$ in kcal equiv^{-1}; $\Delta S°$ in cal equiv^{-1} deg^{-1})

Nominal Cross-linking % DVB	Na^+/H^+			Na^+/Li^+			K^+/Na^+			Cs^+/Na^+		
	$-\Delta G°$	$-\Delta H°$	$-\Delta S°$	$-\Delta G°$	$-\Delta H°$	$-\Delta S°$	$-\Delta G°$	$-\Delta H°$	$-\Delta S°$	$-\Delta G°$	$-\Delta H°$	$-\Delta S°$
0.5	0.00_5	0.26	0.9	0.06	0.33	0.9	0.12	0.23	0.3	0.23	0.36	0.4
2	0.05	0.46	1.4	0.07	0.65	1.9	0.17	0.29	0.4	0.25	0.56	1.0
4	0.14	0.86	2.4	0.23	1.08	2.8	0.20	0.41	0.7	0.29	0.62	1.1
8	0.23	1.18	3.0	0.36	1.46	3.6	0.26	0.55	1.0	0.33	0.77	1.5
16	0.28	1.25	3.3	0.46	1.70	4.0	0.36	0.88	1.8	0.37	1.19	2.7
24	0.48	1.40	3.1	0.54	1.83	4.2	0.41	1.17	2.6	0.45	1.48	3.5

Ref.: J. Phys. Chem., *68*, 590 (1964).

Table II. Standard Free Energies, Enthalpies, and Entropies at 298.2°K for the Exchange of Silver with Sodium Ion in Cross-linked Polystyrenesulfonates.

Nominal Cross-linking % DVB	$-\Delta G°$ kcal equiv^{-1}	$\Delta H°$ kcal equiv^{-1}	$\Delta S°$ cal equiv^{-1} deg^{-1}
1	0.30	0.00	1.0
4	0.56	0.04	2.0
8	0.83	0.11	3.1
16	1.23	-0.12	3.7
24	1.47	-0.34	3.8

Ref.: J. Phys. Chem., *70*, 2295 (1966).

$$M_1R(\text{a}=1, \text{ equil with } 0.1\ N\ M_1\text{Cl}) + M_2\text{Cl}(\text{aq, a}=1) \rightleftharpoons M_2R(\text{a}=1, \text{ equil with } 0.1\ N\ M_2\text{Cl}) + M_1\text{Cl}(\text{aq, a}=1)$$

between cations M_1 and M_2 is given by

$$\Delta H^\circ = \Delta H + \Delta\phi_L$$

where $\Delta\phi_L$ is the relative apparent molal heat content difference (*i.e.*, heat of dilution difference), $\phi_L(M_2\text{Cl}) - \phi_L(M_1\text{Cl})$, at 0.01 N (usually negligible) or at 0.1 N. An analogous treatment applies to anion-exchange reactions. The computation of ΔH° is valuable as standard enthalpies of ion exchange for many additional exchange reactions may then be derived by application of Hess' Law.

The standard free energy change for the standard-state reaction is given by

$$\Delta G^\circ = -2.3RT \int_0^1 \log D_\circ \, dx_{M_2}$$

where $D_\circ$ is the experimentally determined mass law concentration product ratio corrected for activity coefficients in the equilibrium mixed electrolyte solution.

Experimental Results and Their Qualitative Features: Standard free energies, enthalpies, and entropies of exchange of the alkali metal cations with sodium ion in cross-linked polystyrenesulfonate are summarized in Table I. The decrease in the free energy in all of these reactions was a consequence of the *enthalpy decrease* which was of sufficient magnitude to counterbalance the negative entropy changes which occurred. The enthalpy and entropy changes were the largest when the exchanging ions differed most in their hydration. It also may be noted that ΔH° and ΔS° increased with the cross-linking of the ion exchanger. This change was relatively more rapid when the cross-linking was light.

The exchange reaction of Ag^+ ion differed significantly from the behavior shown by other singly charged cations against sodium ion. The selective absorption of silver was accompanied by an entropy increase (Table II) for all cross-linkages, while the enthalpy change was small and was either positive or negative. The strongly positive ΔS° of the reaction was largely responsible for the highly selective uptake of Ag^+ ion, although with the most heavily cross-linked preparations the lowering of the enthalpy and the increase in the entropy both contributed to the lowering of the free energy.

The larger *n*-alkyl quaternary ammonium ions likewise contrast with the alkali metal ions in their exchange reactions with Na^+ ion (Table III). Measurements could be made only with the most lightly cross-linked PSSA exchanger because of the inability of these ions to enter the three-dimensional copolymer network. The large positive ΔS° in the absorption of tetra-*n*-butylammonium ion is noteworthy. The increase in selective uptake of the quaternary ammonium ions with size appeared to be an "entropy effect" as the exchange reaction became increasingly *endothermic*.

The effect of the cross-linking of the ion exchanger on the thermodynamic quantities in the tetramethylammonium-sodium ion-exchange reaction is shown in Table IV. The quaternary ammonium ion was preferentially taken up by the two most lightly cross-linked exchangers because with these the enthalpy lowering was sufficient to more than compensate for the entropy decrease when Na^+ ion was displaced from the exchanger. Increasing the cross-linking additionally, however, ultimately caused the reaction to be endothermic; the free energy then increased and sodium became the preferred ion. The decrease and reversal in sign of ΔH° may be explained in terms of the work to expand the molecular network of the exchanger when the large tetramethylammonium ion enters it.

Table III. Standard Free Energies, Enthalpies, and Entropies at 298.2°K for the Exchange of the Alkali Metal and Tetra-*n*-alkylammonium Ions with Sodium Ion on Nominal 0.5% DVB Cross-linked Polystyrenesulfonates.

Cation	$-\Delta G°$ kcal equiv^{-1}	$\Delta H°$ kcal equiv^{-1}	$\Delta S°$ cal equiv^{-1} deg^{-1}
H^+	-0.00_5	0.26	0.9
Li^+	-0.06	0.33	0.9
K^+	0.12	-0.23	-0.3
Cs^+	0.23	-0.36	-0.4
Me_4N^+	0.28	-0.56	-1.0
Et_4N^+	0.33	-0.50	-0.2
Pr_4N^+	0.37	0.56	3.1
Bu_4N^+	0.46	2.21	9.0

Ref.: J. Am. Chem. Soc., *89*, 6038 (1967).

Table IV. Thermodynamic Quantities at 298.2°K for the Exchange of Tetramethylammonium with Sodium Ion in Cross-linked Polystyrenesulfonate Exchangers.

Nominal Cross-linking % DVB	$-\Delta G°$ kcal equiv^{-1}	$\Delta H°$ kcal equiv^{-1}	$\Delta S°$ cal equiv^{-1} deg^{-1}
0.5	0.28	-0.56	-1.0
2.0	0.09_5	-0.36	-0.9
4.0	-0.06	-0.19	-0.9
8.0	-0.27	+0.04	-0.8

Rev.: J. Am. Chem. Soc.,*89*, 6038 (1967).

Table V. Standard Free Energies, Enthalpies, and Entropies at 298.2°K for the Exchange of Zinc with Sodium Ion in Cross-linked Polystyrenesulfonates.

Nominal Cross-linking % DVB	$-\Delta G°$ kcal equiv^{-1}	$\Delta H°$ kcal equiv^{-1}	$\Delta S°$ cal equiv^{-1} deg^{-1}
2	0.34	0.79	3.8
4	0.41	0.97	4.6
8	0.41	1.43	6.2
16	0.37	1.86	7.4
24	0.23	2.26	8.3

Ref.: J. Phys. Chem., *71*, 2214 (1967).

The role of cation charge in determining the changes in thermodynamic properties accompanying an ion-exchange reaction is elucidated by the data for the zinc-sodium reaction (Table V). Heat was absorbed during the exchange of Zn^{++} ion which was preferentially taken up because of the increase in entropy in the reaction. The balance between the increase in $\Delta H°$ and in $\Delta S°$ was such, however, that a maximum occurred in the zinc ion selectivity with cross-linking. Thus, the thermodynamic mass law constant increased until a nominal 8% DVB cross-linking was reached and then decreased as the cross-linking became heavier. The behavior shown in the zinc-sodium ion exchange is typical of many reactions between multivalent and univalent cations in that heat is absorbed and the entropy increases on the preferential absorption of the more highly charged ion.

The nature of the ion-exchange group attached to the polymeric network strongly influences the selectivity sequence among the alkali metal cations. An illustration of this is afforded by the cesium-lithium ion-exchange reaction (Table VI). The entropy change for the absorption of lithium ion is positive and of almost the same magnitude for all the cation exchangers. The reaction is endothermic in all exchangers, but with the polystyrenesulfonate, $\Delta H°$ is sufficiently large to cause the free energy to increase so that cesium becomes the preferred ion. With the polymethacrylic and the methylenephosphonic acid exchangers the entropy increase is more than sufficient to counter-balance the enthalpy increase, and the free energy decreases for the uptake of lithium ion.

The exchange reactions of the singly charged halide anions in cross-linked strong-base anion exchangers are analogous in many respects to the reactions of the alkali metal cations with one another in that an enthalpy decrease and an entropy increase accompanies the absorption of the preferred anion (Table VII). When the hydrations of the respective anions differed, the magnitudes of $\Delta H°$ and $\Delta S°$ were relatively large. However, in contrast to the ion-exchange reactions of the alkali metal cations with one another, as the cross-linking of the exchanger increased the standard entropy became less negative. The relatively large magnitude of $\Delta H°$ and $\Delta S°$ for the very lightly cross-linked exchanger is noteworthy also, and here there is a dramatic contrast with the exchanges of the alkali metal cations with one another in nominal 0.5% DVB PSSA exchanger. The strongly differing degrees of hydration of the ion-exchange groups in these two types of exchangers may be a factor.

Many reactions in aqueous solution occur with a change in heat capacity, and ion-exchange reactions are no exception. Heats of ion exchange are usually temperature dependent and, further, the $\Delta C_p°$ of exchange itself may vary with temperature. Measurements of the heat of exchange at various temperatures permit estimates of $\Delta C_p°$ which should be much more accurate than values derived from measurements of the temperature dependence of equilibrium constants. An example is afforded by heat of exchange measurements on the tetrabutylammonium-sodium ion reaction (Fig. 9). A relatively large, negative $\Delta C_p°$ was found which appeared to be nearly temperature independent over the range investigated. In the light of these measurements, in general, the heat of ion exchange must be expressed by the equation

$$\Delta H_T° = \Delta H_{298}° + AT + 1/2\ BT^2 + \ldots$$

Discussion: An immediate use for measurements of standard enthalpies of ion exchange is in the computation of the temperature variation of ion-exchange equilibria with the van't Hoff equation

$$d\ \ell n\ K_a/dT = \Delta H°/RT^2$$

With this relation it is possible to compute K_a at temperatures other than that at which it was measured assuming $\Delta H°$ may be taken as approximately constant. The magnitudes of the calorimetrically measured $\Delta H°$ are such that for

Table VI. Role of the Ionogenic Group in Determining the Thermodynamic Property Changes in the Reaction:

$$CsR + Li^+ \rightleftharpoons LiR + Cs^+$$

Type of Exchanger	Nominal Cross-linking % DVB	$-\Delta G^o$ kcal equiv^{-1}	ΔH^o kcal equiv^{-1}	ΔS^o cal equiv^{-1} deg^{-1}
RSO_3^-	16	-0.83	2.89	6.7
$RCOO^-$	15	0.62	1.22	6.2
$RPO_3^=$	5.5	0.72	1.20	6.4

Table VII. Standard Free Energies, Enthalpies, and Entropies at 298.2°K for the Exchange of the Halide Ions in Cross-linked Strong-base Anion Exchangers.

($\Delta G°$ and $\Delta H°$ in kcal equiv^{-1}; $\Delta S°$ in cal equiv^{-1} deg^{-1})

Nominal Cross-linking % DVB	Br^-/F^-			Br^-/Cl^-			I^-/Br^-		
	$-\Delta G°$	$-\Delta H°$	$-\Delta S°$	$-\Delta G°$	$-\Delta H°$	$-\Delta S°$	$-\Delta G°$	$-\Delta H°$	$-\Delta S°$
0.5	1.15	2.79	5.5	0.33	1.18	2.9	0.79	2.25	4.9
2	1.27	3.26	6.7				0.84	2.04	4.0
4	1.56	3.30	5.8	0.57	1.33	2.5	0.88	2.02	3.8
10	2.23	3.58	4.5	0.70	1.36	2.1	1.02	1.70	2.3
24				0.81	1.31	1.7		1.65	

Ref.: J. Phys. Chem., *70*, 2507 (1966).

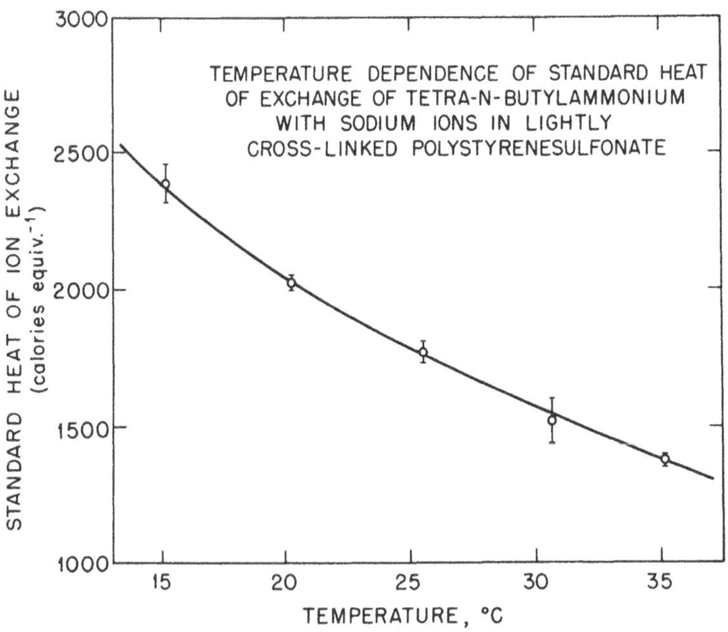

Fig. 9

accurate equilibrium constant measurements the temperature should be controlled to ± 1° or better. When the free energy decrease for the exchange reaction is determined by the enthalpy change, as is true for the alkali metal cations and for the halide ions, the larger the selectivity coefficient, the greater its temperature coefficient. Because of the dependence of ΔH° on temperature the magnitude of the enthalpy change may vanish and change sign, thereby signaling a reversal in the preferred uptake of one of the exchanging ions. These considerations show that the lyotropic series for selective ion uptake depends on temperature and that there is nothing unique about the ionic sequence at room temperature.

An argument similar to the foregoing leads to the conclusion that the usual dependence of selectivity coefficients on the composition of the equilibrium ion exchanger will become smaller as the temperature is increased and ultimately the system may become "ideal."

In the exchange reactions between unequally charged ions, the entropy increase almost always causes the selective uptake of the more highly charged species. The lowering of the enthalpy, however, appears to be cause for the preferred absorption of the alkali metal and the halide ions in their reactions, suggesting that ion binding with the exchanger may dominate occasionally over ion-solvent interactions.

Selected Bibliography -- Organic Ion Exchangers:

G. E. Boyd, F. Vaslow, and S. Lindenbaum, "Heats of Exchange of the Alkali Metal Cations in Cross-linked Polystyrenesulfonates," J. Phys. Chem., *68*, 590 (1964).

S. Lindenbaum and G. E. Boyd, "Thermodynamic Quantities in the Exchange of Li^+ with Cs^+ Ion on Cross-linked Polymethacrylate Ion Exchangers," ibid., *69*, 2374 (1965).

F. Vaslow and G. E. Boyd, "Thermodynamic Properties in the Exchange of Ag^+ with Na^+ Ions in Cross-linked Polystyrenesulfonate Cation Exchangers," ibid., *70*, 2295 (1966).

F. Vaslow and G. E. Boyd, "Heats of Exchange of Halide Ions in Cross-linked Strong-base Anion Exchangers," ibid., *70*, 2507 (1966).

K. E. Becker, S. Lindenbaum, and G. E. Boyd, "Thermodynamic Quantities in the Exchange of Li^+ with Cs^+ Ion on Cross-linked Phosphonic Acid Cation Exchangers," ibid., *70*, 3834 (1966).

G. E. Boyd, J. W. Chase, and F. Vaslow, "Structural Effects on the Enthalpy and Entropy of Dilution of Aqueous Solutions of Quaternary Ammonium Halides at 25°. Comparisons with Strong-base Anion Exchangers," ibid., *71*, 573 (1967).

G. E. Boyd and A. Schwarz, "Heat of Exchange of I^- with Cl^- Ions on a Unifunctional Weak-base Type Anion Exchanger at 25°," ibid., *71*, 1355 (1967).

G. E. Boyd, F. Vaslow, and S. Lindenbaum," Thermodynamic Quantities in the Exchange of Zn^{++} with Na^+ Ions in Cross-linked Polystyrenesulfonate Cation Exchangers at 25°," ibid., *71*, 2214 (1967).

G. E. Boyd, F. Vaslow, A. Schwarz, and J. W. Chase, "Thermodynamic Properties at 25° of Aqueous Solutions of *p*-Ethylbenzenesulfonic Acid and Its Alkali Metal Salts. Comparisons with Cross-linked Polystyrenesulfonate Cation Exchangers," ibid., *71*, 3879 (1967).

G. E. Boyd and Q. V. Larson, "The Binding of Quaternary Ammonium Ions by Polystyrenesulfonic Acid Type Cation Exchangers," J. Am. Chem. Soc., *89*, 6038 (1967).

O. D. Bonner and J. R. Overton, "The Effect of Temperature on Ion Exchange Equilibria. IV. The Comparison of Enthalpy Changes Calculated from Equilibrium Measurements and Calorimetrically Measured Values," J. Phys. Chem., *65*, 1599 (1961).

E. H. Cruickshank and P. Meares, "Thermodynamics of Cation Exchange. I. Determination of the Heats and Free Energies of Exchange by Resins," Trans. Faraday Soc., *53*, 1289 (1957).

E. H. Cruickshank and P. Meares, "II. Comparison Between Resins and Concentrated Chloride Solutions, ibid., *53*, 1299 (1957).

E. H. Cruickshank and P. Meares, "III. Thermodynamic Properties of Resins Containing a Mixture of Cations," ibid., *54*, 174 (1958).

D. S. Flett and P. Meares, "IV. Uni-, bi-, and tervalent Ions on Dowex-50," ibid., *62*, 1469 (1966).

D. S. Flett and P. Meares, "V. The Excess Enthalpies of Some Mixed Resinates of Dowex-50," J. Phys. Chem., *70*, 1841 (1966).

W. R. Heumann and D. Patterson, "Heats of Potassium-Lithium Ion Exchange on Sulfonated Polystyrene Resins," Canadian J. Chem., *44*, 2139 (1966).

J. S. Redhina and J. A. Kitchener, "Thermodynamics of Ion Exchange Processes. System H, Na and Ag with Polystyrene Sulfonate Resins," Trans. Faraday Soc., *59*, 515 (1963).

USE OF MICROCALORIMETRY FOR DETERMINATION OF ORDER IN POLYMERS

Samuel H. Maron and Frank Filisko
Division of Polymer Science
Case Western Reserve University
University Circle
Cleveland, Ohio

INTRODUCTION

Knowledge of the degree of order in polymers is of great importance in understanding the morphology and the physical properties of such materials. Such order has been determined heretofore by x-rays, density, specific heat, and infrared measurements, but such methods are either inapplicable or inaccurate at low degrees of order. However, Maron and Daniels (1) have shown recently that both high and low degrees of order in polymers can be determined to a high degree of accuracy from a combination of heat of solution and heat of dilution measurements. For application of this method the instrument most suitable is the Calvet microcalorimeter, particularly when modified in the manner to be described. The purpose of this talk is to describe this calorimeter and the modifications made in it, and to show how the data obtained with its aid can be used to ascertain polymer order.

THEORY OF NEW METHOD

According to the Maron theory (2), when w_2 grams of amorphous polymer are dissolved in w_1 grams of solvent, then the heat of mixing per gram of polymer is given by

$$\Delta H_m = -\frac{RT^2\lambda\; w_1 v_2}{M_1\; w_2} \tag{1}$$

where R is the gas constant, T the absolute temperature, v_2 the volume fraction of polymer in the solution formed, M_1 the molecular weight of the solvent, and λ is the temperature coefficient of the interaction parameter μ at constant v_2; i.e.,

$$\lambda = \left(\frac{\partial\mu}{\partial T}\right)_{v_2,\underline{P}} \tag{2}$$

However, if crystallinity of any type of order above that corresponding to the amorphous polymer is present in the sample, then the observed heat of mixing per gram of polymer, $\Delta H'_m$, will be greater than that given by Eq. (1). The difference between the two will be the heat of fusion of the polymer per gram, ΔH_f. Consequently we obtain

$$\begin{aligned}\Delta H'_m &= \Delta H_f + \Delta H_m \\ &= \Delta H_f - \frac{RT^2\lambda\; w_1 v_2}{M_1\; w_2}\end{aligned} \tag{3}$$

ΔH_f is calculable from Eq. (3) once λ is known. The value of λ for a given polymer-solvent system at any temperature and concentration v_2 can be obtained from osmotic pressure data or from heats of dilution. We shall consider here only the latter. According to the Maron theory (2) the heat of dilution per gram of polymer, ΔH_d, when we go from an initial volume fraction v_2 to a lower volume fraction v_2' is given by

$$-\frac{\Delta H_d \rho_2 V_1^\circ}{RT^2} = (\lambda' - \lambda) + \lambda v_2 - \lambda' v_2' \qquad (4)$$

where ρ_2 is the density of the polymer, V_1° the molar volume of the solvent, and λ' the value of λ at v_2'. Now experience shows that λ is either a constant or, at worst, a linear function of v_2 at any given temperature. We may write in general, therefore, that

$$\lambda = a + b\ v_2 \qquad (5)$$

where, for a given system and temperature, a and b are constants. Substitution of Eq. (5) into Eq. (4) leads then to

$$X = -\frac{\Delta H_d \rho_2 V_1^\circ}{RT^2\ (v_2 - v_2')} = (a-b) + b\ (v_2 + v_2') \qquad (6)$$

Consequently, a plot of X vs. $(v_2 + v_2')$ should be linear, and from the straight line both constants can be found. Once these constants are known λ follows from Eq. (5), and so we get

$$\Delta H_f = \Delta H'_m + \frac{RT^2 \lambda\ w_1\ v_2}{M_1\ w_2} \qquad (7)$$

$$= \Delta H'_m + \frac{RT^2 w_1 v_2}{M_1\ w_2}\ (a + b\ v_2) \qquad (8)$$

When λ is independent of v_2, as is frequently the case, then b = 0, and the equations simplify to

$$X = -\frac{\Delta H_d \rho_2 V_1^\circ}{RT^2\ (v_2 - v_2')} = a \qquad (9)$$

a constant, and

$$\Delta H_f = H'_m + \frac{RT^2 a w_1 v_2}{M_1 w_2} \qquad (10)$$

It should be observed that in Eqs. (6) and (8), the working equations for application of the method, the molecular weight of the polymer does not appear, and only the density of the polymer is needed.

PRELIMINARY TEST OF NEW METHOD

Limited published data on heats of dilution and solution conform the validity of Eqs. (6) and (8), and thereby also the theory on which they are based. Heat of dilution data, plotted in the manner called for by Eq. (6), yield lines which are straight within the accuracy of the data. Again, combination of heat of solution data with λ values obtained by osmometry yield, in line with Eq. (8), heats of fusion which are independent of weight of sample taken, v_2 of solution, or solvent used, as they should be. This may be seen from Tables 1 and 2, where there are given heats of fusion per gram calculated from the heat of solution data obtained with the same sample of polyisobutylene dissolved in benzene and chlorobenzene (3,4). The benzene solutions yield a value of $\Delta H_f = 2.44 \pm 2.1\%$ cal/gm, while the chlorobenzene solutions give $\Delta H_f = 2.55 \pm 2.3\%$. If we use the value of ΔH_f given in the literature (5) for the calculated heat of fusion of perfectly crystalline polyisobutylene, namely, 2,870 cal per monomer unit, then the percentages of crystallinity observed in the sample studied are 4.76% from the benzene solution data and 4.99% from the chlorobenzene solution data. The difference between the two results is, therefore, 0.23% at a crystallinity level of ca. 5%. These results are considerably better than any which can be obtained by use of presently available methods.

Table 1
Heats of Fusion for Polyisobutylene at 25°C Calculated from Heats of Solution in Benzene
(Sample PIJ-4; M_2 = 63,000)

Final v_2	Wt. of Polymer Dissolved (g.)	ΔH_f (cal./g.)
0.00714	0.0460	2.44
0.0139	0.0735	2.50
0.0169	0.0822	2.48
0.0191	0.0868	2.49
0.0230	0.0807	2.40
0.0249	0.0775	2.39
0.0279	0.0886	2.53
0.0286	0.0947	2.38
0.0347	0.1260	2.37
0.0367	0.1434	2.35
0.0377	0.0903	2.48
0.0593	0.2847	2.41
	Average	2.44 ± 2.1%

Table 2
Heats of Fusion for Polyisobutylene at 25°C Calculated from Heats of Solution in Chlorobenzene
(Sample PIJ-4; M_2 = 63,000)

Final v_2	Wt. of Polymer Dissolved (g.)	ΔH_f (cal./g.)
0.00559	0.0338	2.42
0.00914	0.0504	2.50
0.0108	0.0594	2.53
0.0121	0.0662	2.59
0.0151	0.0835	2.48
0.0161	0.0892	2.59
0.0178	0.1375	2.58
0.0218	0.1225	2.51
0.0226	0.1227	2.51
0.0278	0.1554	2.54
0.0286	0.1605	2.60
0.0379	0.2156	2.72
	Average	2.55 ± 2.3%

The above results show also that it should be possible to use this method with accuracy at levels of order much lower than 5%. Since the error is only small at the 5% crystallinity level, then it should be readily possible to go down to 1% order or less with considerable certainty.

THE MODIFIED CALVET MICROCALORIMETER

To obtain the kind of accuracy indicated above requires a calorimeter of very high precision and certain specific characteristics. An instrument which meets admirably all requirements is the Calvet Microcalorimeter (6), whose operation is practically isothermal. It can measure total heats evolved or absorbed over long periods of time with an accuracy of 0.001 calorie or better. The remainder of the paper will be devoted to a description of the calorimeter, the modifications made in it and its operation, also some preliminary results obtained with this instrument will be presented.

REFERENCES

1. C. A. Daniels, M.S. Thesis, Case Western Reserve University, 1967.
2. S. H. Maron, J. Polymer Sci., 38, 329-342 (1959).
3. C. Watters, H. Daoust, and M. Rinfret, Can. J. Chem., 38, 1087 (1960).
4. M. A. Kabayama and H. Daoust, J. Phys. Chem., 62, 1127 (1958).
5. J. Brandrup and E. H. Immergut, "Polymer Handbook," Interscience Publishers (1966).
6. E. Calvet and H. Prat, Recent Progress in Microcalorimetry, Pergamon Press, Oxford, England (1963).

DIFFERENTIAL THERMAL ANALYSIS AND CALORIMETRY OF WAXES

Robert G. Craig, John M. Powers, and Floyd A. Peyton
School of Dentistry, University of Michigan
Ann Arbor, Michigan 48104

Thermal analysis of waxes used in dentistry has been conducted using simple techniques such as differential plots from cooling curves[1,2] and comparison of the temperature of a wax and a silicone oil on cooling.[3] More recently the authors[4] reported a differential thermal analysis (DTA) study of dental waxes and their components. Infrared absorption studies of the phase transitions in waxes has been published by Martin, Johnson, Cannon, and O'Neal[5] and by Ludwig[6], and the latter showed the orthorhombic-hexagonal and hexagonal-liquid transitions of paraffin wax could be determined from the infrared spectra and an analysis of paraffin and microcrystalline waxes in mixtures could be determined by this method. Lange and Jochinke[7], Lorant[8], Kawasaki, Komizu, and Uchida[9], and Currell and Robinson[10] reported differential thermal analysis of a variety of commercial waxes and esters. Lange and Jochinke[7] concluded the height and position of the peaks were influenced by many factors and thus quantitative analysis of waxes by DTA in most instances was not possible. The later study by Currell and Robinson[10] showed microcrystalline and polyethylene waxes were characterized by an endotherm at 475 to 480° C and the area of this peak could be used to estimate the amount of these waxes in mixtures; the melting curves were not unequivocal but for a few exceptions the paraffin waxes had at least two endotherms and the microcrystalline and polyethylene waxes only one. Kawasaki, et al.[9] studied commercial paraffin waxes and the DTA curves had one premelting and one melting endotherm; both of these peaks were broad and the premelting peak was at a lower temperature than the comparable pure n-paraffin. They also reported heats of premelting of 14 to 22 cal gm^{-1} and heats of fusion of 36 to 44 cal gm^{-1} for the commercial paraffin waxes. Barmby, Bostwick and Huston[11] used x-ray diffraction, DTA, and dilatometric techniques to study the transitions of paraffin wax. They found waxes melting below about 60° C had two solid-solid transitions, those melting from 60 to 75° C had one such transition and those melting above 75° C exhibited no solid-solid transition. In the low melting waxes their data indicated the lower transition point was a first-order change involving a change in the a and b dimensions of the orthorhombic lattice while the higher transitions was a second-order change involving the transformation from an orthorhombic to a hexagonal lattice structure. The heat of transition for one of the fractions which melted between the C_{23} and C_{24} melting points, was reported as 17.4 cal gm^{-1}. Mazee[12] reported heats of transition for $C_{23}H_{48}$ and $C_{24}H_{50}$ waxes as 16.6 and 21.0 cal gm^{-1}.

Numerous papers have been written on the molecular structure and crystal habit of waxes. Hatt and Lamberton[13] discussed the structure of paraffin and ester waxes and proposed the solidification of hydrocarbons involved first the formation of a vertical rotating form (hexagonal) followed by a change to an opage vertical form (orthorhombic) and then to an opague tilted form (mono-or triclinic). The heats of transition were approximately 50% of the total heat of fusion. Mixtures of hydrocarbons still had these changes but they were surpressed. Edwards[14] reported the crystal structure of purest hydrocarbons. Normal hydrocarbons C_{18} and higher crystallized in a close-packed hexagonal lattice with molecules rotating on their long axes which were perpendicular to the basal plane. He reported hydrocarbons above $C_{40}H_{82}$ when pure, showed no transition point and crystallized directly from the melt in the low temperature orthorhombic lattice. Odd numbered hydrocarbons have an orthorhombic low temperature form with the long axis perpendicular to the basal plane. The lower even numbered paraffins, including $C_{24}H_{50}$, on cooling formed a triclinic lattice, while the higher even numbered hydrocarbons, $C_{28}H_{58}$ and higher, transformed on cooling from hexagonal to a low temperature form with the long axis even farther from the perpendicular and were monoclinic. Small amounts of impurities

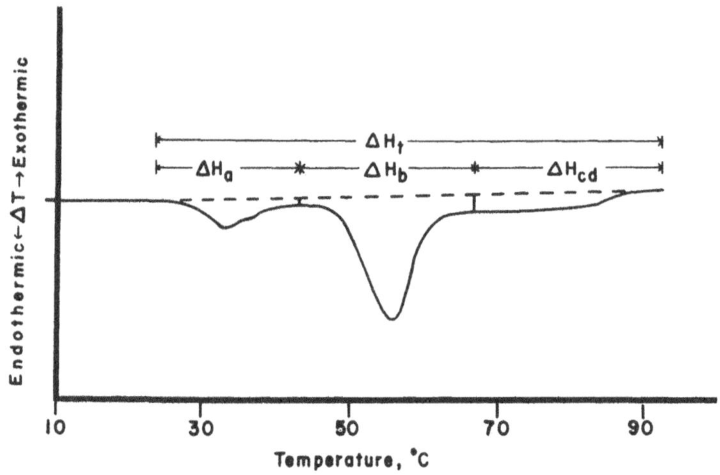

Figure 1 - Area determination of a calorimetric curve for a binary mixture of paraffin and carnauba wax.

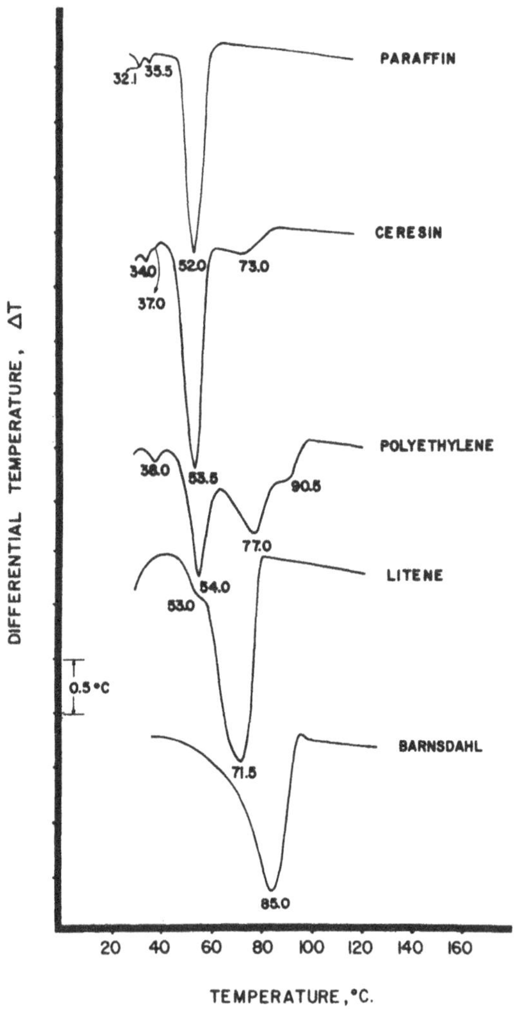

Figure 2 - Thermograms of some hydrocarbon waxes in nitrogen.

(other members of the series) can cause an inclined lattice to become orthorhombic or can depress the transition point or increase the range when the hexagonal lattice is stable. This effect appears to be related to the ability of the hexagonal and orthorhombic lattices to accommodate the formation of solid solutions. Broadhurst[15] has collected, summarized, and interpreted information published on the solid phase behavior of n-paraffins. Those from C_{22} to C_{30} had heats of transition that were 50% or more than the heats of fusion. The hexagonal phase had a high degree of rotational freedom around the chain axis; the orthorhombic transition took place easily from the hexagonal when chain rotation ceased as a result of cooling. It was proposed the shift from the triclinic to monoclinic above C_{26} was a result of end group packing. Recently, Kawasaki and Uchida[16] reported x-ray diffraction studies of the premelting of commercial paraffin waxes.

It was the purpose of this paper to summarize the results of differential thermal analysis and calorimetric analysis of waxes and mixtures of waxes. The transition temperatures and heats of transition are reported and the interactions between waxes in mixtures are evaluated.

Experimental

The waxes were used as purchased from the supplier* and were blended as indicated for the mixtures. Differential thermal analysis was conducted from 20 to 200° C using a du Pont 900 DTA. The analyses were carried out in an atmosphere of nitrogen at a heating rate of 20° C/min. Thermograms of paraffin wax also were obtained at heating rates of 10 and 5° C/min., in order to provide better resolution of the solid-solid transitions. Calorimetric analysis was evaluated from 0 to 100° C in a nitrogen atmosphere using a du Pont 950 Calorimetric Cell at a heating rate of 5° C/min. The cell was calibrated using pure metals having known heats of fusion over the temperature range of interest. The areas above the calorimetric curves used in the calculation of the heats of transition were determined as shown in Figure 1. The determination of the areas for the individual components of binary mixtures of waxes was the most difficult as a result of incomplete resolution of the various peaks. For example, in Figure 1 the determination of ΔH_T presents no difficulty, but the accurate determination of ΔH_a, ΔH_b, and ΔH_{cd}, representing the solid-solid transitions of paraffin, the melting transition of paraffin; and the melting transitions of carnauba wax, is difficult. The closest return to the baseline appears to be the best choice realizing, that the melting peak of paraffin contributes to the area left and right of the vertical lines and the solid-solid transition of paraffin and the melting transition of carnauba contributes about an equivalent amount to the area of the melting transition of paraffin.

In all instances, the values reported are means of five separate replications. Statistical analysis of the data was accomplished using analysis of variance[17] (ANOVA) and where significant differences were observed the means were ranked using Duncan's Multiple Range Test.[18] Regression analysis and correlation[19] was used to evaluate the interaction of the heats of transition on composition and waxes.

Results

DTA curves for five hydrocarbon waxes are shown in Figure 2. The thermograms are shown only to 120° C since no peaks were observed between 120 and 200° C. Paraffin, ceresin, and polyethylene waxes had definite premelting, or solid-solid transitions, and litene and barnsdahl had indefinite and no premelting transition, respectively. Paraffin and ceresin had two endotherms at 32 and 35° C, and 34 and 37° C, respectively. Thermograms of paraffin wax at a heating rate of 5° C/min. resulted in complete resolution of the two

---

* Ross Company, Jersey City, New Jersey.

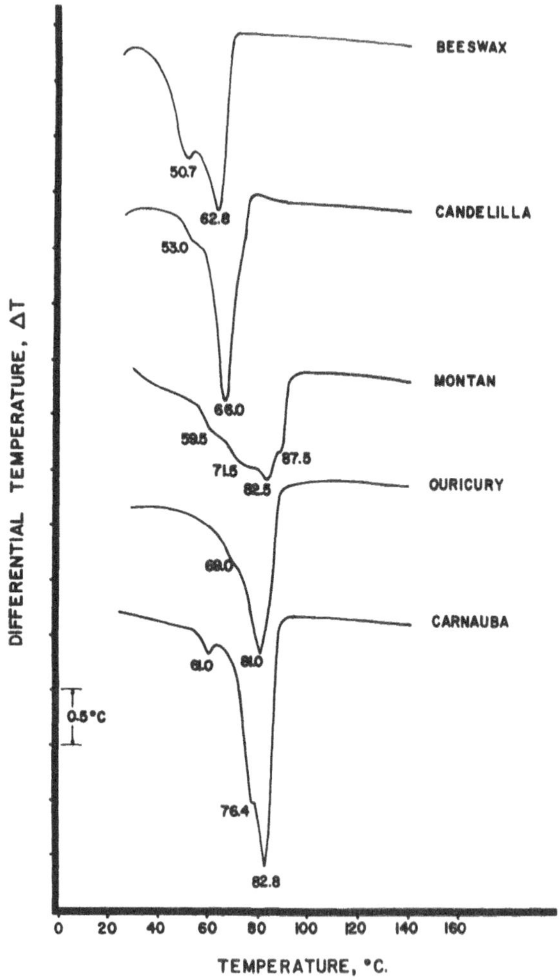

Figure 3 Thermograms of some natural waxes in nitrogen.

Table 1 - Transition Temperatures of Paraffin and Mixtures of Paraffin (P) and Carnauba (C) Wax

Transition	Composition and Transition Temperatures			
$Solid_1$ - $Solid_2$	25% P-75% C	50% P-50% C	75% P-25% C	100% P
	30.6° C	31.1° C	32.0° C	32.1° C
$Solid_2$ - $Solid_3$	50% P-50% C	25% P-75% C	75% P-25% C	100% P
	34.9° C	35.3° C	35.4° C	35.5° C
$Solid_3$ - Liquid	25% P-75% C	50% P-50% C	100% P	75% P-25% C
	51.1° C	51.8° C	52.0° C	52.3° C
ANOVA - ($F_{3,48}$ 31.1 > 4.9***)				

Table 2 - Transition Temperatures of Carnauba and Mixtures of Carnauba (C) and Paraffin (P) Wax

Transition	Composition and Transition Temperatures			
Solid - Liquid	25% C-75% P	75% C-25% P	50% C-50% P	100% C
	73.4° C	74.3° C	74.5° C	79.6°C
ANOVA - ($F_{3,32}$ = 115.5 > 5.2***)				

Table 3 - Total Heats of Transition of Various Batches of Waxes

Wax	Batch Number	ΔH_i	Cal g^{-1}
Paraffin	193	50.6	(2.5)[a]
	303	54.3	(3.4)
	274	55.1	(1.5)
	319	55.7	(3.2)
Carnauba	B 1845s	47.0	(1.9)
	B 2318	47.8	(3.3)
	B 1845c	51.2	(5.6)
	B 1956	51.5	(1.7)
Bee's	B 2183	40.3	(2.5)
	B 2170	41.9	(2.5)
	B 1788	43.7	(2.8)
	B 2247	45.3	(1.1)

[a] Numbers in parentheses are standard deviations determined from five individual determinations of ΔH_i.

Table 4 - Heats of Transition of Waxes

Wax	Transition	ΔH_i	Cals gm^{-1}
Montan	Total	38.4	(3.7)[b]
Candelilla	Total	40.7	(3.4)
Beeswax[a]	Total	42.8	(2.1)
Ouricury	Total	43.6	(3.6)
Ceresin	Premelting	6.6	(1.4)
	Melting	39.0	(3.6)
	Total	45.6	(5.0)
Carnauba[a]	Premelting	3.8	(1.1)
	Melting	45.5	(1.3)
	Total	49.4	(2.0)
Polyethylene	Premelting	5.3	(0.7)
	Melting	45.4	(3.9)
	Total	50.7	(4.6)
Barnsdahl	Total	53.1	(3.0)
Paraffin[a]	Premelting	11.4	(0.6)
	Melting	42.4	(1.9)
	Total	53.9	(2.5)
Litene	Total	55.6	(5.1)

[a] Means of 20 replications rather than 5 for other waxes.
[b] Numbers in parentheses are standard deviations.
H_o: rejected since $F_{9,76} = 14.2 > 3.0^{***}$ and accept that ΔH_i's are different.

endotherms at 32 and 35° C. In addition to the melting transitions of paraffin and ceresin of 52 and 53.5° C, ceresin had a broad transition at 73° C. Polyethylene wax had what appeared to be three melting transitions at 54, 77 and 90.5° C.

The thermograms of natural waxes, which have high concentrations of esters, as well as some hydrocarbons, are shown in Figure 3. Carnauba wax had the most distinct premelting transition at 61° C and melting transitions at 76.4 and 82.8° C. Montan wax was characterized by a broad melting transition consisting of a number of incompletely resolved peaks, while beeswax was typified two similar but incompletely resolved melting transitions at 50.7 and 62.8° C.

Since carnauba and beeswax are frequently added to paraffin wax in dental compounding to alter the properties of the latter, mixtures of each of these two waxes with paraffin were examined by DTA at 25-75, 50-50, and 75-25 weight percents. The transition temperatures of paraffin were found to be different in the presence of high concentrations of carnauba wax as shown by the F statistic and Duncan's ranking test as shown in Table 1. The temperatures connected by underscoring were not statistically different at the 95% confidence level. The addition of 50% carnauba decreased the solid-solid transitions and 75% carnauba decreased the solid-liquid transition of paraffin, although the decreases were small (0.4 to 1.2° C). The effect of the additions of paraffin on the transitions of carnauba wax is shown in Table 2, where 25% paraffin and more decreased the melting transition of carnauba wax about 5.5° C.

Addition of beeswax to paraffin and visa versa did not affect the corresponding transition temperatures with $F_{2,42}=0.37 < 3.23*$ and $F_{1,18}= 0.30 < 4.41*$. The ANOVA for additions of paraffin to beeswax is somewhat misleading since the transition at 62.8° C was not observed when 50% or more of paraffin was added to beeswax.

The waxes most used in compounding dental waxes are paraffin, carnauba, and beeswax; and therefore, the variation in heats of transition of various batches of wax of the same grade was determined. Four batches each of paraffin, carnauba, and beeswax were selected at random by the manufacturer and the total heats of premelting and melting transitions are listed in Table 3. Analysis of variance indicated the total heats of transitions were different and the results of ranking tests are shown in Table 3 by the vertical lines to the right of the values. The lines connect ΔH values that are not different at the 95% confidence level.

The heats of transition for the waxes are listed in Table 4. The two premelting transitions observed for both paraffin and ceresin were not sufficiently resolved in the calorimetric curves and thus the values of 11.4 and 6.6 cal g^{-1} represent the sum of the solid-solid heats of transitions. The heat of transition for the 38° C peak for polyethylene wax was 5.3 cal g^{-1}. Carnauba wax was the only ester or plant wax which showed a distinct premelting transition having a value of 3.8 cal g^{-1}. In general, the hydrocarbon waxes were characterized by higher total heats of transition, ΔH_T, (51 to 56 cal g^{-1}) and the ester containing waxes by lower ΔH_T values (38 to 44 cal g^{-1}). Exceptions to this generalization were ceresin wax in the hydrocarbon group (45.6 cal g^{-1}) and carnauba wax in the ester group (49.4 cal g^{-1}).

Statistical analysis and ranking also is shown in Table 4. Montan, candelilla and beeswax were different from ouricury and ceresin but beeswax and ouricury could not be distinguished. Carnauba, polyethylene, barnsdahl and paraffin could not be distinguished from each other on the basis of ΔH_T, nor could barnsdahl, paraffin, and litene.

The heats of transition of paraffin were determined in mixtures containing carnauba and beeswax. The solid-solid heats of transitions of paraffin, ΔH_a, per gram of mixture were plotted as a function of composition of paraffin-carnauba mixtures in Figure 4. Regression analysis was conducted and the curve showed regression and a high correlation with r equal to 0.956 (perfect correlation r=1).

Curve fitting established the equation of the line was best expressed (least error) by the quadratic equation listed below Figure 4. The heat

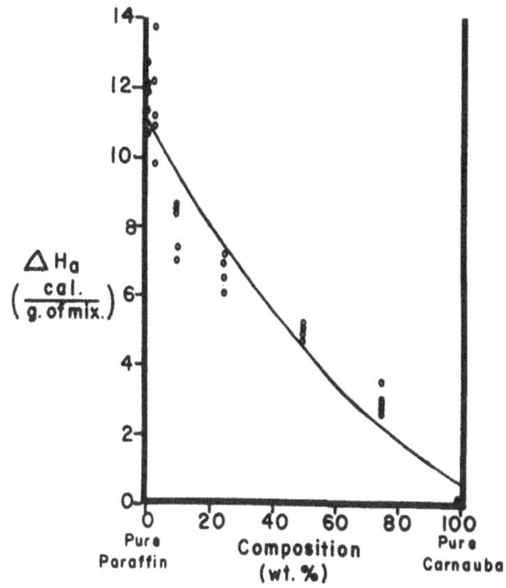

$y = 11.0 - 0.157x + 0.0005x^2$; $r = 0.956$

Fig. 4-ΔH_a of solid-solid transitions of paraffin in compositions containing carnauba wax.

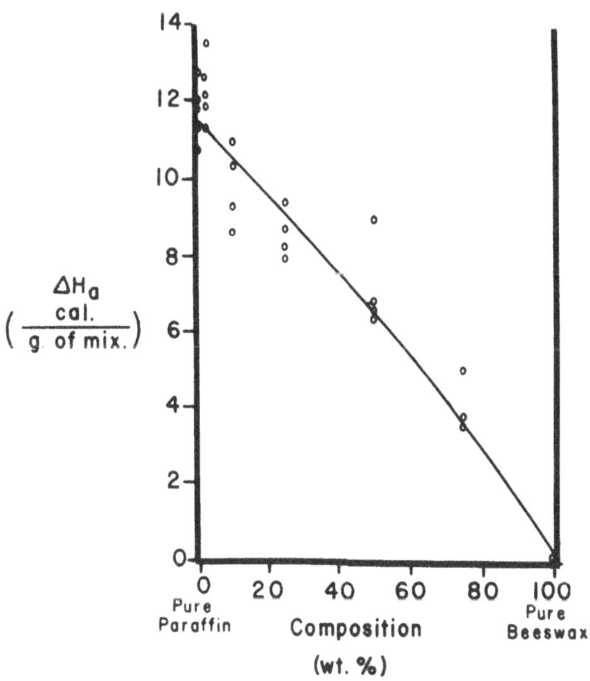

$y = 11.6 - 0.0931x - 0.0002x^2$; $r = 0.976$

Figure 5 - ΔH_a of solid-solid transitions of paraffin in compositions containing beeswax.

of transitions, ΔH_a, for mixtures of paraffin and beeswax are shown in Figure 5. Again regression and high correlation (r=0.929) was obtained and again the points were best fitted by a quadratic equation; the curve was concave rather than convex toward the composition axis as observed for the paraffin-carnauba compositions.

The heat of fusion of paraffin as a function of composition in mixtures containing carnauba wax is given in Figure 6. Regression and high correlation was obtained (r=0.968) and the quadratic equation, indicated below Figure 6, best fit the experimental points.

The total heats of premelting and melting of paraffin-carnauba and paraffin-beeswax mixtures as a function of composition are plotted in Figures 7 and 8. Low regression and correlation (r=0.096) was observed for paraffin-carnauba mixtures. Higher correlation(r=0.711) was found for the paraffin-beeswax mixtures.

Discussion

The two endothermic, first-order, transitions observed in the solid state for paraffin and ceresin wax are in contradiction to the thermograms reported by other investigators[7,9,10,11] studying commercial waxes. Barmby[11] did report one first-order and one second-order transition in low melting paraffin waxes although the premelting peaks for the low melting paraffins were broad suggesting the presence of two unresolved peaks. If two peaks were present it would present difficulties in interpretation of the x-ray data. In order to establish whether the two solid-solid transitions were a general phenomenen several commercial low-melting paraffin waxes* were tested and in all instances two endotherms were observed in the temperature range of 30 to 37° C. Possible explanations of these results are that the lower transition may be a mono-or triclinic transformation to orthorhombic lattice structure. Another possible explanation could be that the lower transition represents a lattice dimension change of the orthorhombic structure and the higher transition is a change from an orthorhombic to a hexagonal structure. The possibility of the two endotherms resulting from a mixture of two narrow fractions appears remote based on the lack of evidence of two melting endotherms or even a broad melting endotherm.

The decrease in the transition temperatures of paraffin by the addition of carnauba wax is in agreement with the observation[11] that impurities added to hydrocarbon waxes cause a decrease in the transition temperature. The lack of this effect with beeswax may result in the large hydrocarbon wax content of this wax as suggested by the large endotherm at 50.7° C. At a composition of 75% beeswax and 25% paraffin wax the endotherm at 32° C disappeared and only the 35° C endotherm remained; the thermogram of a 75% carnauba- 25% paraffin mixture showed both solid-solid endotherms. As expected for commercial paraffinic waxes the transition temperatures were much farther (15 to 20° C) from the melting transition than in pure hydrocarbons[15] (2 to 7° C).

The values for the heats of transition for different batches of paraffin, carnauba and beeswax indicate the variation in composition of waxes of the same grade. In general, the hydrocarbon waxes had higher total heats of transition (premelting and melting) than the ester-containing waxes; ceresin and carnauba wax were exceptions. The data did not indicate a relationship between melting transition temperature and heats. The calorimetric data showed litene, polyethylene, and ceresin were different but no distinction could be made between litene, paraffin, and barnsdahl or between paraffin, barnsdahl or polyethylene. The difficulty arises from rather high coefficients of variation (standard deviation x100/ΔH_T) of 4 to 12% and a mean coefficient of about 8%.

The premelting heats of transition of ceresin and paraffin of 6.6 and 11.4 cal g^{-1} may be a result of the paraffin being a narrower fraction and the influence of the impurity in ceresin suggested by the endotherm at 73° C.

---

* Texwax, Gulfwax.

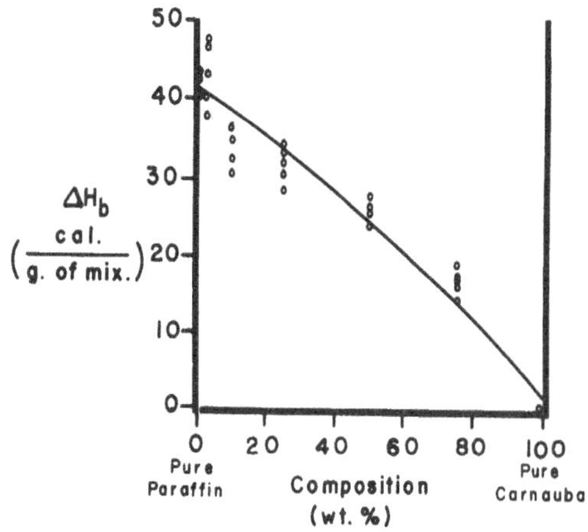

$y = 41.4 - 0.261x - 0.0014x^2$; $r = 0.968$

Figure 6 - ΔH_b of the melting transition of paraffin in compositions containing carnauba wax.

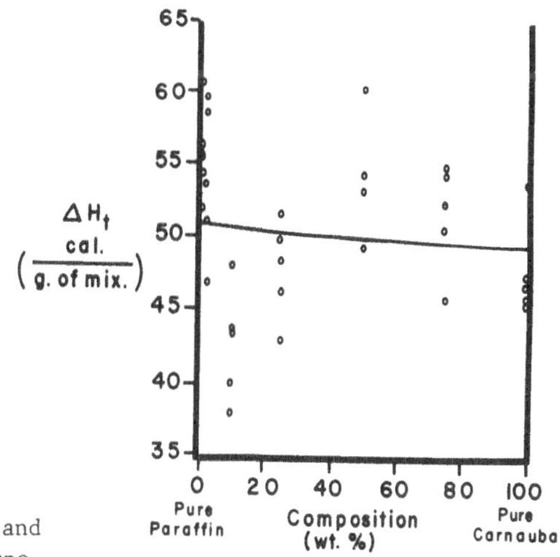

$y = 51.0 - 0.0234x + 0.0001x^2$; $r = 0.096$

Figure 7 - Total heat of premelting and melting of paraffin-carnauba wax compositions.

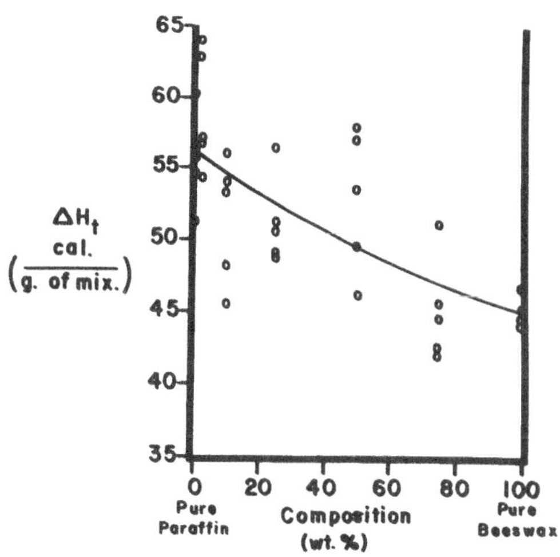

$y = 56.1 - 0.152x + 0.0004x^2$; $r = 0.711$

Figure 8 - Total heat of premelting and melting of paraffin-beeswax compositions.

These values are lower than those reported by Kawasaki, et al.[9] and Barmby, et al.[11] suggesting the waxes above were broader fractions. The solid-solid transitions percentage of the melting transitions were 27, 17, and 8.3 percent for paraffin, ceresin and carnauba wax which suggests that paraffin was the most and carnauba the least pure of the three waxes.

Carnauba and beeswax are added to paraffin wax to control the properties of a dental wax and there has been some question whether true mixtures are formed. The plots of the various heats of transition for mixtures of these waxes as a function of composition resulted in lines best represented by quadratic equations rather than linear as would be expected for mixtures. Therefore, interaction does occur as a result of mixing these waxes. It is interesting to note in Figures 4 and 6 that addition of carnauba to paraffin results in lower values of ΔH_a and higher values of ΔH_b than those expected for true mixtures. Figure 7 shows little change in ΔH_T for carnauba-paraffin mixtures as expected from the values in Table 4 where carnauba could not be distinguished from paraffin on the basis of ΔH_T. Inspection of Figure 5 shows the addition of beeswax to paraffin results in higher ΔH_a values than those expected for a mixture, while Figure 8 indicates lower ΔH_T values than those of a mixture with no interaction.

References

1. R.G. Craig, J.D. Eick, and F.A. Peyton, J. Dent. Res., 44, 1308 (1965).
2. M. Ohashi and G.C. Paffenbarger, J. Am. Dent. Assoc., 72, 1141 (1966).
3. D.C. Smith, R. Earnshaw,and J.W. McCrorie,Brit.Dent.J., 118, 437(1965).
4. R.G. Craig, J.M. Powers, and F.A. Peyton, J. Dent. Res., 46, 1090(1967).
5. J.M. Martin, Jr., R.W.B. Johnston, H.J. Cannon, and M.J. O'Neal, Anal. Chem., 30, 1005 (1958).
6. F.J. Ludwig, Anal. Chem., 37, 1737 (1965).
7. J. Lange and H. Jochinke, Fette, Seifen, Anstrichmittel, 67,89(1965).
8. B. Lorant, Selfen - Oele - Fette - Washse, 93, 57 (1966).
9. S. Kawasaki, H. Komizu, and T. Uchida,Kogyo Kagaku Zasshi,69,1315(1966).
10. B.R. Currell, and B. Robinson, Talanta, 14, 421 (1967).
11. D.S. Barmby, L.G. Bostwick, and J.A. Huston, Jr., Proc. 6th World Petrol. Cong., 6, 161 (1963).
12. W.M. Mazee, Rec. Tran. Chemique, Pays Bas, 67, 197 (1948).
13. H.H. Hatt and J.A. Lamberton, Res. Sci. Appl. Ind., 9, 138 (1956).
14. R.T. Edwards, Ind. Eng. Chem., 49, 750 (1957).
15. M.G. Broadhurst, J. Res. Nat. Bur. Stds., 66A, 241 (1962).
16. S. Kawasaki and T. Uchida, Kogyo Kagaku Zasshi, 70, 1281 (1967).
17. W.C. Guenther, Analysis of Variance, Prentice-Hall, Englewood Cliffs, New Jersey, 1964.
18. D.B. Duncan, Biometrics, 11, 1 (1955).
19. G.W. Snedecor and W.G. Cochran, Statistical Methods, 6th ed., Iowa State Univ. Press, Ames, Iowa, 1967, pp. 447-71.

Acknowledgement

This investigation was supported in whole by USPHS Research Grant, DE-01234 from the National Institute for Dental Research, NIH, Bethesda, Md.

CALORIMETRIC STUDIES OF MILK FAT BY DIFFERENTIAL THERMAL ANALYSIS

Robert A. Yoncoskie
Dairy Products Laboratory
Eastern Utilization Research and Development Division
U. S. Department of Agriculture
Washington, D. C. 20250

Synopsis

The heat of melting and heat capacities of milk fat were determined with a differential thermal analysis calorimeter. Heat of melting values varied from 20.4 cal/g (average value) to 25.1 cal/g (average value) depending on tempering conditions. Heat capacity values were .423 cal/g/°C (average value) at -43C and .521 cal/g/°C (average value) at 70C.

Introduction

The use of conventional DTA in heats of fusion studies has met with a number of difficultues which makes its application in quantitive work very restrictive. A cell design, that permits quantitative treatment of thermogram areas was illustrated by Barrall (1) and David (2). A comparison of the results obtained from conventional DTA and calorimetric DTA was demonstrated by the former author (1).

A further interesting application for the DTA calorimeter is in heat capacity measurements. The determination of this useful thermodynamic property by DTA calorimetry has been described in other publications (2,3, 4).

The development of suitable methods in DTA calorimetry for obtaining much needed data on heats of melting and heat capacities of milk fat was a prime objective of the work reported in this paper. Suitable procedures developed for milk fat would appear to have applicability for determining thermodynamic properties of other fats and oils.

Experimental[1]

Apparatus

The Du Pont Calorimeter, a plug-in accessory for the Du Pont 900 Differential Thermal Analyzer, was used to obtain all calorimetric data.

Calibration

The calibration of the calorimeter for heats of fusion and melting was performed using a programmed heating rate of 5C/min. Occasional checks, made up to heating rates of 10C/min., were found, within experimental error, to agree with the calibration established at 5C/min. Standards employed for the calibration were the following high purity metals: mercury, gallium, indium, tin and zinc. Calibration coefficients, used to establish a calibration curve as a function of peak temperatures, were calculated from the determined peak areas and known heats of fusion by substitution in the following formula:

$$\Delta H\ (\text{mcal/mg}) = E\frac{A\Delta T_s}{MC}$$

where E = calibration coefficient, mcal/°C-min.; A = peak area, sq. in.; ΔT_s = Y axis sensitivity, °C/in.; M = sample mass, mg; and C = chart speed, in./min.

[1] Reference to certain products or companies does not imply an endorsement by the Department over others not mentioned.

For highest accuracy in heat capacity determinations a second calibration curve based on known heat capacities should be established. In the work reported here, a second calibration curve was constructed using the following high purity organic materials: palmitic acid, benzoic acid, napthalene and anthracene. These calibrating standards permitted the calculation of heat capacities from -50 to 100C. The calculation of heat capacities with the DTA calorimeter is based on a heat lag curve which is generated when a sample is heated. An accurate measurement of the difference in the heat lag curve under no sample and sample conditions permits the calculation of heat capacities. The calculation of calibration coefficients for constructing a heat capacity calibration curve was performed through the use of the following formula:

$$(C_p)_T \frac{(\text{millicalories})}{\text{mg-}^\circ\text{C}} = \frac{(\Delta T_x - \Delta T_{ns})\, E_T}{Ma}$$

where $(C_p)_T$ = heat capacity at temperature T; ΔT_x = absolute differential temperature at sample conditions, °C; T_{ns} = absolute differential temperature at no-sample conditions, °C; E_T = calibration coefficient at temperature T, mcal/°C-min; M = mass, mg; and a = heating rate, °C/min.

Procedure

For heats of melting and heats of fusion all samples except gallium were weighed into seamless aluminum liners; platinum liners were used for gallium. An empty liner similar to that used for the samples was employed in the reference holder for the above determinations. For heat capacity measurements a set of about four aluminum liners were employed repeatedly. A similar liner, weighing slightly less than the sample liners, was used in the reference holder. Once a heat lag curve was established for the above designated sample liner, this curve was used in all subsequent determinations.

The calculation of heat capacities at temperatures as low as -50C made it necessary to modify the procedure which is normally used for specific heat determinations above ambient temperatures. In order to establish an accurate heat lag curve at low temperatures, it was necessary to employ a simultaneous heating and cooling operation to bring sample and reference thermocouples to the same temperature. An external recorder was used in this modified procedure.

All heats of fusion and heats of melting data reported in this study, as well as heat capacity values, were calculated with the formulas previously defined.

Results and Discussions

Because of the lack of a true fusion temperature in milk fat, the previously defined method for calculating heats of fusion was not directly applicable. Two separate approaches were considered for calculating this calorific value(5). A first approach consisted of dividing the total peak area into small increments and using E calibration coefficients corresponding to each increment area. The heat of melting was calculated as the summation of these calorific values. A second approach consisted in employing an average E calibration coefficient corresponding to the temperature at 50% of the thermogram area. For a milk fat sample tempered at 22C for 47 hours, values of 21.36 cal/gm and 21.40 cal/gm were obtained from the first and second methods respectively.

Heats of fusion values obtained for representative pure compounds, as well as for heats of melting of milk fat samples tempered at different temperatures, are given below in Tables I and II.

TABLE I

Results of ΔH_f Determinations

Sample	Sample size mg	Heating Rate °C/min	Cal. 15°/gram determined ΔH_f(5)	Cal. 15°/gram known ΔH_f
Trimyristin (β)	2.97	5	49.7	50.3[a]
Tripalmitin (β)	2.45	5	53.4	53.1[a]
Tristearin (β)	2.16	10	54.9	54.6[a]
Benzoic Acid	2.01	10	34.1	33.9[b]
Benzoic Acid	4.03	5	33.6	33.9
Palmitic Acid	3.05	10	51.0	50.6[a]
Naphthalene	2.34	10	35.8	35.6[b]
Silver Nitrate	8.46	10	16.4	16.2[c]

a Reference 6
b Reference 7
c Reference 8

TABLE II

Heats of Melting of Milk Fat

Tempering Conditions	Sample Wt. mg	Heating Rate °C/min	Heat of Melting (5) cal/gm
no tempering	15.21	5	20.4
"	11.70	10	20.6
Tempered at 22C for 43 hr	11.51	5	21.7
Tempered at 22C for 44 hr	11.28	5	21.6
Tempered at 0C for 18 hr	14.20	5	24.9
Tempered at 0C for 25 hr	8.37	5	25.6

NOTE: All samples were cooled from either the conditioning or tempering temperature to -90C at about 5C/min.

The determination of heat capacities of milk fat and other materials at low temperatures presented no problems once the difficulty of defining the zero differential temperature point at low temperatures was resolved. Heat capacities of organic materials and milk fat are given in Tables III and IV.

TABLE III

Results of Heat Capacity Determinations

Sample	Sample size mg	Heating Rate °C/min	Cal. 15°/gm/°C C_p, determined		Cal.15°/gm°C C_p, known
Anthracene	11.53	10	50°C 100°C	.302 .351	.308[b]
Anthracene	12.49	5	50°C 100°C	.314 .342	
Benzoic Acid	14.01	5	20°C	.292	.287[b]
Dextrose (anhydrous)	10.10	5	0°C 20°C	.269 .274	.277[b] .275
Naphthalene	11.99	5	0°C	.312	.313[b]
Naphthalene	9.06	10	0°C	.326	
Tristearin (β Form)	7.68	10	-60°C -30°C	.300 .338	.277[a] .310
Palmitic Acid	3.36	10	-50°C	.304	.306[b]
Palmitic Acid	4.74	10	0°C	.387	.382
Lauric Acid	15.14	5	-30°C	.437	.430

a Reference 6
b Reference 7

TABLE IV

Heat Capacities of Milk Fat (9)

Sample Size mg	Heating Rate °C/min	Cal. 15°/gm/°C C_p, determined	
10.00	5	-43°C	.422
5.81	10	"	.419
8.25	10	"	.425
10.00	5	70°C	.522
8.25	10	"	.520
8.49	5	"	.521

Prior work on the heat of melting of milk fat between -10C and 60C gave a value of 19.5 cal/gm (10). A more recent paper (11) gave values of 24.52 and 19.13 cal/gm based on two different thermal treatments of cream from which the calculations were made. Previous work on heat capacities of milk fat gave a value of .524 cal/gm/°C at 74C (12).

Conclusions

As contrasted to standard calorimetry, this method requires less time to perform, gives fairly accurate data and needs only milligrams amounts of sample. Because of its applicability to milk fat, one of the more complex fat systems known, DTA calorimetry would appear to be a useful technique for other fats and oils.

References

1. Barrall, E. M., R. S. Porter, and J. F. Johnson, Anal. Chem. 36 2172 (1964).

2. David, D. J., Anal. Chem. 36 2162 (1964).

3. Wendlandt, W. W., Delta Therma, Application Note No. 58, Technical Equipment Corporation, 917 Acoma Street, Denver, Colorado.

4. Thermal Analysis Newsletter No. 4., Perkin-Elmer Corporation, Norwalk, Connecticut.

5. Yoncoskie, R. A., JAOCS 44 446 (1967).

6. Mattil, K. F., F. A. Norris, A. J. Stirton and D. Swern, "Bailey's Industrial Oil and Fat Products," ed. D. Swern, Interscience Publishers, Inc., Division of John Wiley and Sons, Inc., New York, N. Y. 1964.

7. Hadgman, C. D., "Handbook of Chemistry and Physics," Chemical Rubber Publishing Co., Cleveland, 1960.

8. Lange, H. A., "Handbook of Chemistry," McGraw-Hill Book Company, Inc., New York, 1961.

9. Yoncoskie, R. A., unpublished data.

10. Jack, E. L., and J. R. Brunner, J. Dairy Sci. 26 169 (1943).

11. Sherbon, J. W., and S. T. Coulter, J. Dairy Sci. 49 1376 (1966).

12. Mikhaylov, L., Mol. Prom. 10 (1) 25 (1949).

THERMAL STUDIES OF MILK FAT

by
J. W. Sherbon
Department of Food Science
Cornell University, Ithaca, New York 14850

The study of the solidification and melting of fats by thermal methods can be fascinating. It seems there is always one more factor that should have been considered after the experiment is finished. If you added an antioxidant to the fat to preserve it, did you determine the effect the added compound had on the crystallization of the fat? How do you know that the solid phase on which you just finished composition studies was completely free from adhering or entrapped liquid phase?

Besides being interesting, the study of the thermal behavior of fat is also useful. Knowledge of fat crystallization and melting patterns is necessary if we are to improve the food we eat. Butter straight from the refrigerator has the consistency of a brick. Some margarines, on the other hand, have to be packaged in a cup. I am sure most of you can cite better examples of your own.

My purpose is to demonstrate to you some of the possibilities and limitations associated with the study of fat crystallization using thermal methods. I cannot avoid reference to other techniques, nor can I be complete in the material covered. Instead, the material will reflect my own interests and prejudices, and probably my own ignorance.

Fats have long been known to exhibit both monotropic polymorphism and the ability to form solid solutions. In general, those working with simple triglycerides and fractions of natural fats are more interested in polymorphism while those working with intact natural fats are more interested in solid solution formation. Both groups, however, need to be aware of both phenomena. Both groups also use thermal methods extensively.

Most of the evidence for the polymorphism of fats has been obtained on samples of pure fatty acids and triglycerides, or simple mixtures of them. Some fractions of natural fat have also been shown to exhibit polymorphism (2, 11), but the behavior of a fraction may have little relation to the melting behavior of the fat from which it was derived (8). At least three crystal forms may be exhibited by simple fats. The nomenclature used to describe these forms can be confusing unless one is aware of the scheme being used, the one proposed by Malkin or the one proposed by Lutton. This paper will follow Chapman's suggestion (6) of using a subscript L to signify that the Lutton scheme is being used. The names of the various forms of crystalline fat and the allowed transitions between them are shown in Figure 1. The Δ symbolizes the absorption of heat by the transition and the arrow indicates the direction. The particular transition occurring in a sample depends partly upon the rate of addition or removal of heat, the greater the rate the longer the jump between forms. The chemical species modifies the pattern in that even-chain, monoacid triglycerides rarely show an α_L to β'_L transition under any conditions. The best available evidence indicates that the stable β_L form occurs as a triclinic crystal and the β'_L form occurs as a monoclinic crystal. By extrapolation, then, the α_L form may be tetragonal.

Several factors influence the formation of metastable polymorphic forms of fats. Fast cooling leads to supercooling of the liquid, and crystallization from a supercooled liquid is more likely to be in a metastable form. Impurities often promote crystallization in a metastable form. Of course, the composition of the fatty mixture is also important, the simpler the mixture the more likely the formation of metastable forms.

It should be recognized that it is possible for more than one crystal form to exist simultaneously in the same sample. Figure 2 shows the thermograms obtained for a sample of tristearin cooled rapidly from the melt (by two different means) compared to the powder form received from the manufacturer. The original sample melted at 71-73 C with a heat of fusion (ΔH_f) of 54.9 cal/g. Although glad that these values agree with those in the literature (2), I prefer to call your attention to the endotherm at 54 C and the exotherm between the two peaks. The 54 C endotherm corresponds to melting of the α_L form of tristearin, but the ΔH_f is 3/8 of previously reported values. The ΔH_f of the exotherm is -9.5 cal/g, lower than the expected -11.4 cal/g for a β'_L to β_L transition, and the ΔH_f of the endotherm corresponding to the melting of the β_L form is 29.4 cal/g. The next step is to do enthalpy balances using proposed melting schemes to see if one actually fits the observed net values. A simple transition of α_L to β_L to liquid is eliminated quickly. Neither is the lower endotherm a transition of a part of the α_L to liquid and the rest to β_L. Only by assuming the simultaneous existence of more than one crystal form within the sample can the observed transition energies be calculated. A similar situation exists with tripalmitin (11). The calorimeter used for these studies is rather large (10-15 g of sample), and the sample is partially divided by heater surfaces. I should add that I have never seen the same phenomenon using spectroscopy or a different type of calorimeter in which there are only a few milligrams of sample.

Bailey (2) has cautioned that although polymorphism is pronounced in a binary mixture of homologues, the more complex the fat the less evidence there is for the existence of polymorphism. Further, it is doubtful whether polymorphism *per se* affects the plasticity of fatty mixtures (8). Many natural fats are good examples of systems capable of forming mixed crystals. Mulder (9) has pointed out that: a) the lower the temperature at which milk fat is solidified the lower the melting point of the solid formed; b) stepwise cooling results in the solidification of a smaller amount of fat than direct cooling; c) recrystallization can occur during tempering; and d) if the fat is rapidly cooled to below the solidus line, the effect of solidification temperature will be removed. In a fat containing as many species of triglycerides as milk fat, solid solution formation and entrapment of the liquid phase probably occur simultaneously.

A major task confronting the researcher interested in the solid phase of natural and modified fats is to recognize which phenomenon is occurring, polymorphism or solid solution formation. It is especially difficult to distinguish the two phenomena because a metastable crystal may simply melt, showing an endotherm, instead of transforming to a more stable crystal form before melting. The thermogram in the latter case would show an exotherm followed by an endotherm. Both polymorphism and solid solution formation can result in a lowering of the melting point. The best course is to confirm the results postulated from the thermal data using spectroscopy or x-ray diffraction. Even though the nature of a transition may not be clearly defined on the basis of thermal data alone, the transition energies and heat capacities obtained are still valuable for application to processing problems.

Although a lowering of the melting point can be caused by either polymorphism or solid solution formation, thermal data may be used to make a good guess if the composition of the fat can be varied. The problem is presented by the thermogram shown in Figure 3, wherein a quickly cooled mixture of 3 parts of tristearin to 1 part of triolein is heated at 1 deg/min. The exotherm at 30 C represents a solid-solid transition, although it would appear that not all of the sample underwent the transition. We are still left with the dilemma of whether the endotherm at 67 C is the melting of a metastable form or a solid solution. Thermograms of mixtures having different tristearin-triolein ratios were obtained. The temperature at which the

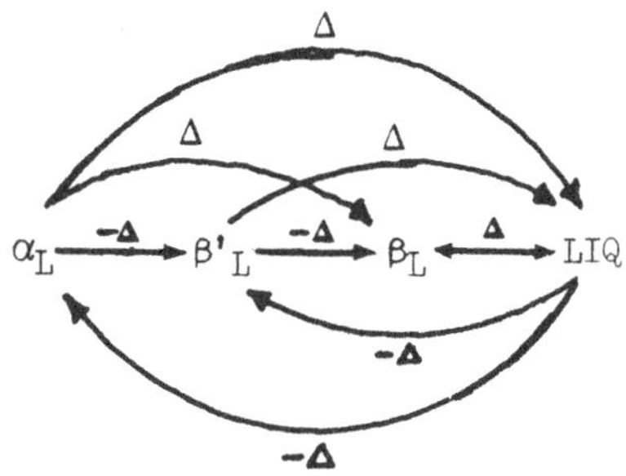

Figure 1. Transitions between the various forms of triglycerides.

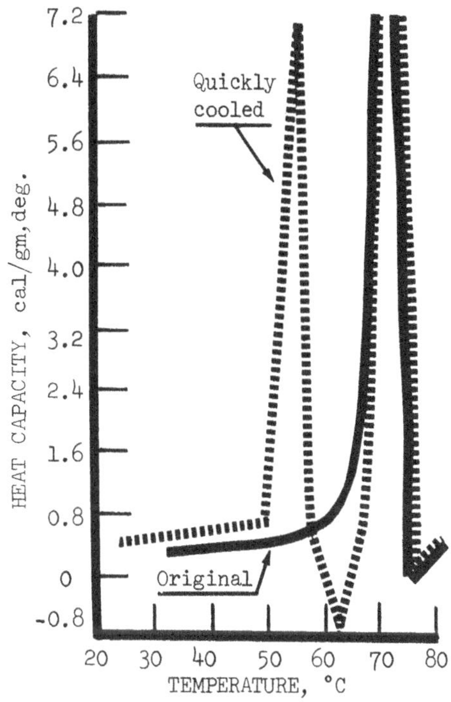

Figure 2. Thermograms of tristearin after different cooling treatments.

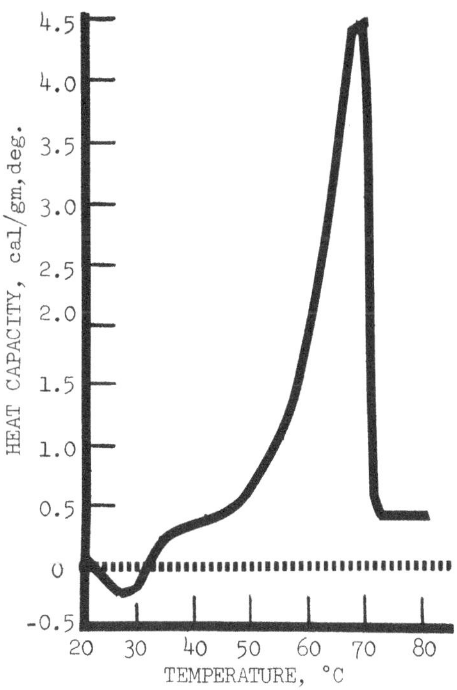

Figure 3. Thermogram of a quickly cooled mixture of 3 parts tristearin to 1 part triolein.

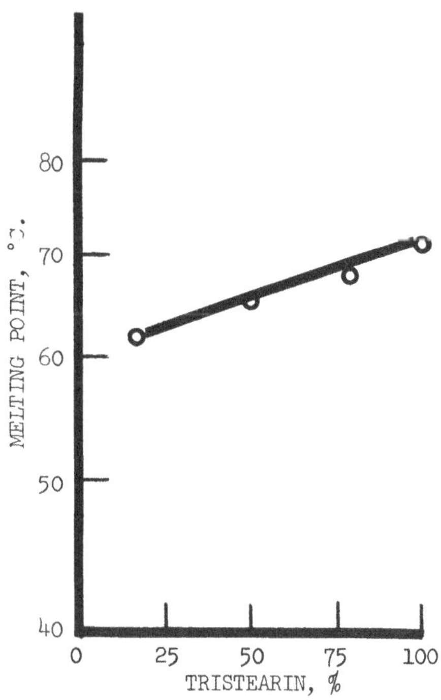

Figure 4. The melting point of tristearin-triolein mixtures.

last endotherm occurs gradually decreases to 61 C for a 1:3 mixture. If the transition had been the melting of a metastable form of tristearin, the transition temperature would not have decreased with the addition of triolein. Infrared spectra of these samples show a minor absorption band at 890 cm^{-1} and a stronger band at 720 cm^{-1}. Chapman (3,4,5) has shown that this pattern is unique to the β_L form.

Let us examine binary mixtures of tristearin and triolein more closely. Figure 4 shows the relationship between the melting point of the stable form and the composition of the mixture. Between 25 and 100% tristearin, the logarithm of the melting point is proportional to tristearin content. The log relationship may be fortuitious; the important point is that when the line is extrapolated to pure triolein, one obtains a melting point of 59 C. This is much too high. There must be a sharp break somewhere below 25% tristearin. Rossell (10) has summarized data reported by Lutton, Ward, et al., Kung, and McGowan that shows a large change in curvature at about 15% triolein. The continuous nature of this solidus curve is good evidence for solid solution formation. The radical change in curvature at 15% tristearin might indicate that there is a limit to the amount of triolein in the solid solution. In any event, solid phases formed by slow cooling of 5-25% solutions of tristearin in triolein have infrared patterns typical of β_L crystals.

The thermograms (1) of a 1:1 mixture of tristearin and triolein in different states of dispersion are presented in Figure 5. The data for the two emulsions were corrected for the aqueous phase, thus the thermograms are for the fat alone. The fat dispersed as a coarse emulsion has two endotherms. The one occurring at 50 C has the same melting point as the solid phase of a 15% tristearin in triolein mixture. The larger endotherm, occurring at 62 C, could represent the melting of the bulk of the fat mixture. From the melting point, however, one concludes that the composition of this phase is more like 35% tristearin than 50% tristearin. The thermogram of the 1:1 mixture dispersed in a fine emulsion shows only a hint of the 50 C component found in the coarse emulsion, but it shows a substantial fraction of the fat to be melting at 74 C. The fat in the fine emulsion probably exists in two crystalline phases, pure tristearin and the solid solution. This alternative, rather than polymorphism, is chosen as the explanation simply because the solid melting at about 62 C shows some tendency to change as the composition of the fat is changed. As noted above, the same fats in bulk were found to be in the β_L form by infrared spectroscopy.

The extension of the behavior of a fat in continuous dispersion to the behavior of the fat when it is dispersed in an emulsion is not perfect. Albright (1) was able to detect transitions of metastable forms to more stable forms in quickly cooled emulsions of a 1:1 mixture of tristearin and triolein but was unable to detect the same transitions if the fat was in the continuous phase. He concluded that the coarse dispersion (3.2 μ globules) allowed more supercooling than the fine dispersion (1.3 μ globules) as the finely dispersed fat was more like bulk fat in solidification characteristics than the coarsely dispersed fat. The critical particle size has not been determined.

We theorize that the effect of dispersion on fat solidification is an effect of specific surface. If the globule surface can function as a nucleation site, the removal of the mote effect by subdividing the fat will at some point be compensated by the additional surface created.

This brings up some interesting points concerning the thermal instrumentation used to study fats. Albright's data were collected in a calorimeter using a sample size around 15 g and a heating rate of about 0.5 - 1.0 deg/min. I have been unable to produce metastable crystalline forms of milk fat in this calorimeter (11), even at cooling rates of -40 C/min. However, DeMan has evidence for such forms. These were produced by rapid cooling in an x-ray diffractometer mount (7) and on a NaCl flat (14). These devices

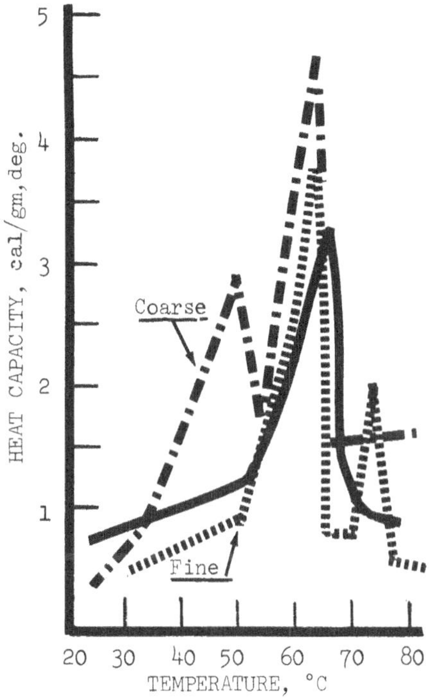

Figure 5. Effect of dispersion on the melting of a 1:1 mixture of tristearin and triolein.

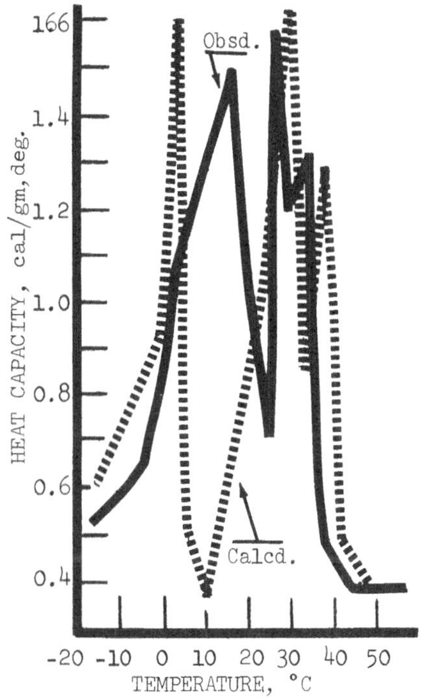

Figure 6. Solid solution formation in a mixture of high- and low-melting milk fat.

TABLE 1. The thermodynamic properties of a fall milk fat (12).

Cooling rate	Hours held at less than -40 C	Melting range limits lower C	Melting range limits upper C	Heat of fusion cal/g	Average heat capacity of melt cal/g, deg
Fast	0	1.00	35.00	23.05*	0.43
Intermediate	0	-18.39	42.82	26.27	0.47
Slow	0	-18.39	36.21	29.02	0.46
Fast	24	0.02	35.45	24.80*	0.41
Fast	24	-20.45	35.76	29.11	0.42
Slow	20	-21.82	37.12	32.73	0.42

*Value includes contribution of premelting from -17 C.

have much higher specific surfaces than our calorimeter. We also use a Perkin-Elmer DSC-1 calorimeter for some of our work. This instrument requires only a few milligrams of material to be spread on the bottom of a pan less than 0.5 cm in diameter. Preliminary runs, however, indicate that rapid cooling in this instrument does not produce metastable forms of milk fat.

Solid solution formation has a direct effect on the processing of fats. Compare, as in Figure 6, the thermogram obtained (12) for a mixture containing equal parts of high- and low-melting fractions of milk fat mixed after cooling the components to the thermogram calculated on the basis of no interaction of the two components. The low melting fraction melted at a higher temperature in the mixture than it did alone, whereas the high melting fraction melted at a lower temperature in the mixture than it did alone. Fatty acid analysis of each phase showed that there was a considerable interchange of fats between the liquid and solid phases, even when the liquid fat was blended with the solid fat. The general tendency is for the two phases to move towards a common composition, indicative of the recrystallization predicted by Mulder. If the fractions were mixed before solidification, the difference between the compositions of the liquid and solid phases was even smaller. This indicates that in a plastic fat, co-crystallization from the melt affects phase composition more than recrystallization. Practically speaking, a liquid fat added to partially solidified butterfat would have more effect on the spreading properties of the butter than it would if it were added to liquid butterfat.

A comparison of the thermograms of milk fat (13) after cooling at different rates shows that fast cooling increases and sharpens the lower melting limit. This results in a reduction of the melting range of the fat. Tempering the fat after cooling did not materially affect the thermograms obtained, indicating that the initial crystallization was complete and the phase compositions were near the equilibrium point. The various properties of the milk fat calculated from these thermograms are listed in Table 1. Note the gradual reduction of the heat of fusion as the cooling rate of the fat is increased. The melting range is also affected, as mentioned above. These data are indicative of solid solution formation.

The amount of fat present in the liquid state in a plastic fat is a useful value to calculate, as it is extremely important in determining the spreading properties of the fat. This can be done by taking the fraction of the total heat of fusion absorbed by warming the fat to the temperature of interest. With milk fats, I assume the heat of fusion to be constant over the entire melting range. Using this technique, it has been found that the increase in liquid fat content of butter caused by stepwise cooling of the cream from which it is made (the "winter" method) is more marked at slower cooling rates. The difference is also more pronounced if the butter is at 20 C rather than at 10 C (13).

Summarization of the study of fat crystallization by thermal methods is not difficult. Calorimetry is especially useful, not only in describing the melting behavior of the fat and in determining apparent heat capacities, but also for determining the amount of liquid phase present. The position of endotherms obtained by melting solid solutions is affected by the composition of the mixture and by the cooling rate used during solidification. Thermal methods are also useful in following the recrystallization phenomenon. The coexistence of crystal forms is detected by thermal methods if done quantitatively. Thermal methods alone, however, are insufficient. Final proof of a particular crystal form produced in a fatty sample must be based on other techniques.

References

(1) Albright, L. D. 1965. The effect of dispersion on the physical state of a simple fat system. M.S. Thesis, Cornell University.

(2) Bailey, A. E. 1950. Melting and Solidification of Fats. Interscience Publishers, Inc., New York.

(3) Chapman, D. 1956. Infrared spectra and the polymorphism of glycerides. J. Chem. Soc. 1956:55.

(4) Chapman, D. 1957. Infrared spectra and the polymorphism of glycerides. III. Palmitodistearins and dipalmitins. J. Chem. Soc. 1957:2715.

(5) Chapman, D. 1958. Infrared spectra and the polymorphism of glycerides. IV. Myristopalmitins and myristostearins. J. Chem. Soc. 1958:3186.

(6) Chapman, D. 1962. The polymorphism of glycerides. Chem. Rev. 62:433.

(7) DeMan, J. M. 1961. Physical properties of milk fat. II. Some factors influencing crystallization. J. Dairy Res. 28:117.

(8) DeMan, J. M. 1963. Polymorphism in milk fat. Dairy Sci. Abst. 25: 219. (Review article).

(9) Mulder, H. 1953. Melting and solidification of milk fat. Netherlands Milk & Dairy J. 7:149.

(10) Rossell, J. B. 1967. Phase diagrams of triglyceride systems. Ch. 9 in Adv. in Lipid Res., Vol. 5. Paoletti and Kritchevsky, Ed. Academic Press, New York.

(11) Sherbon. J. W. 1963. The physical state and thermodynamic properties of selected milk fat systems. Ph.D. Thesis, Univ. of Minnesota.

(12) Sherbon, J. W. and Coulter, S. T. 1966. Solid solutions and the hardness of fatty mixtures. J. Dairy Sci. 49:1126.

(13) Sherbon, J. W. and Coulter, S. T. 1966. Relation between thermal properties of butter and its hardness. J. Dairy Sci. 49:1376.

(14) Woodrow, I. L. and DeMan, J. M. 1967. Polymorphism in milk fat. J. Dairy Sci. 50:962. Abstr. paper read at 62nd. Annual Mtg., Amer. Dairy Sci. Assn., Ithaca, New York.

MEASUREMENT OF THE KINETICS OF DISSOCIATION OF DICYCLOPENTADIENE DERIVATIVES WITH THE DIFFERENTIAL SCANNING CALORIMETER

by

W. E. Franklin, C. H. Mack, and S. P. Rowland
Southern Regional Research Laboratory*, New Orleans, La. 70119

INTRODUCTION

In the course of studies of potential thermally reversible crosslinking agents for cotton cellulose, it became necessary to develop a method of measuring the kinetic parameters for the dissociation of dimers of methyl cyclopentadienecarboxylate. It has previously been reported that tetracyanoethylene and N-phenylmaleimide are suitable trapping agents for the products of these dissociations (1). A method capable of measuring the kinetics of such reactions, both in solution and in the cellulose fiber matrix, was required. The use of the differential scanning calorimeter for measuring the kinetics of these reactions in solution is discussed here.

Thermal analytical methods for measurement of reaction kinetics have been discussed by Wendlandt (2) and by Reed (3). Differential enthalpic analysis, also known as differential scanning calorimetry, measures the difference in heat flow to the sample and to an inert reference as they are being held isothermal with respect to each other and as the temperature of both is being raised at a constant rate (4). The small amount of sample required and the good thermal contact between the sample and the heating and sensing elements in commercial differential scanning calorimeters minimize the usual instrumental difficulties in the measurement of reaction kinetics by thermal methods. The fact that the differential scanning calorimeter gives data in the form of rates of heat evolution or absorption greatly simplifies the calculation of kinetic parameters from data obtained with this instrument. The kinetics of a number of reactions (5,6,7,8) and polymer transitions (9,10) have been measured with the differential scanning calorimeter. The availability of hermetically sealed sample containers makes possible the extension of the use of the differential scanning calorimeter to the measurement of kinetics of reactions in relatively volatile solvents.

EXPERIMENTAL

Apparatus and Reagents

The differential enthalpic analysis curves were obtained with the Perkin-Elmer DSC-1 differential scanning calorimeter. The samples and reference solvents were contained in aluminum capsules and hermetically sealed with the Perkin-Elmer Volatile Sample Sealer Assembly. Reagent grade tetracyanoethylene was used and commercial N-phenylmaleimide was recrystallized to a constant melting point of 87.5-89° before use. Practical grade diglyme and tetraglyme were distilled from sodium under nitrogen and stored under nitrogen before use. Practical grade dicyclopentadiene was recrystallized before use. The preparation of methyl dicyclopentadienecarboxylate (4-carbomethoxytricyclo[5,2,1,$0^{2,6}$]deca-3,8-diene) (11) and dimethyl dicyclopentadienedicarboxylate (4,9-dicarbomethoxytricyclo[5,2,1,$0^{2,6}$]deca-3,8-diene) (1,11) have already been described.

Method

Solutions of 0.2 mmole of the dicyclopentadiene derivative and 0.6 or 0.7 mmole of the trapping agent (tetracyanoethylene or N-phenylmaleimide; 2.0 or 2.5 molar excess) in 350 µl. of tetraglyme or diglyme were prepared. In one set of experiments, the concentrations of dimethyl dicyclopentadienedicarboxylate and tetracyanoethylene were one-quarter of the usual values in order to provide a concentration identical to that previously studied (1).

---

*A laboratory of the Southern Utilization Research and Development Division, Agricultural Research Service, United States Department of Agriculture.

Weighed samples of 13 to 15 µl. of the reaction solution were hermetically sealed in aluminum capsules. An equal volume of the solvent in a sealed capsule was used as the inert reference. Pan support covers were used on the sample and reference pans to make the emissivities of the two more nearly equal and, thus, to give a flatter baseline.

The thermograms were obtained at the following instrument settings: chart speed, 0.75 in./hr.; range, 2 mcal./sec. at half-scale deflection; scan, 10°/min. or 20°/min.; and slope, 510 to 620, adjusted as required to give the best baseline. The thermograms were measured in the temperature range 350-525° K. (350-475° K. for reactions involving tetracyanoethylene because of the exothermic decomposition of the adducts above 460° K.). To measure the reaction of dimethyl dicyclopentadienedicarboxylate with N-phenylmaleimide, it was necessary to use the following settings: chart speed, 3 in./min.; range, 8 mcal./sec. at half-scale deflection; and scan, 80°/min. At the high scan rate, the sample passed rapidly through the temperature range (below 420° K.) at which two isomeric adducts were formed (1) and consequently minimized the distortion of the reaction curve from the rearrangement of the adduct formed only at lower temperatures. The sample was allowed to remain undisturbed as it cooled, and the run was repeated immediately to obtain the baseline.

TREATMENT OF DATA

Figure 1 shows a typical thermogram obtained from the reaction of methyl dicyclopentadienecarboxylate with N-phenylmaleimide. The figure shows the baseline superimposed on the thermogram. The curvature in the baseline arises from the difference in emissivity between the sample pans and from other instrumental factors, such as the position of the sample pan covers. Repeating the run on the reacted, undisturbed sample gives a baseline for which the instrumental variables are identical with those for the reaction run. A baseline superimposable on both ends of the thermogram indicates that no significant change in the heat capacity of the sample has occurred during the reaction.

The ordinate of the thermogram provides a measure of the difference between the rates of heat flow into the sample and into the reference, while they are being held isothermal with respect to each other as the temperature of both is being increased at the selected scan rate. The abscissa provides a measure of both time and temperature. As a reaction occurs in the sample, a lower (Figure 1) or higher rate of heat flow to the sample is required to maintain it isothermal with respect to the reference. The total area, A_t, between the thermogram line and the baseline is proportional to the total heat of reaction, Q_t, of the sample (4):

$$Q_t = K_2 A_t,$$

where K_2 is the proportionality constant. The distance, d (in Figure 1), is proportional to the rate of heat evolution, dQ/dt, at that point of the reaction:

$$dQ/dt = K_3 d,$$

where K_3 is the proportionality constant.

At any specific time during the run, the area, A (in Figure 1), between the baseline and the thermogram line to the right of that time and temperature is related to the heat of reaction, Q, of the reagent remaining in the sample at that time:

$$Q = K_2 A.$$

The number of moles, n, of the reagent in the sample is related to Q and to ΔH, the heat of reaction per mole of reagent, by the following equation:

$$n = Q/\Delta H.$$

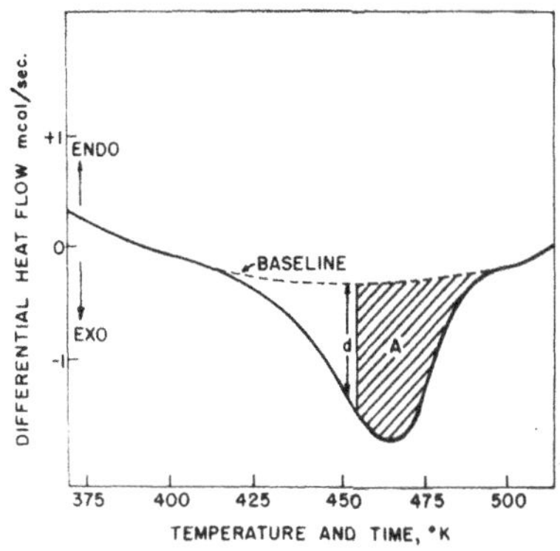

Fig. 1. Differential Scanning Calorimeter Thermogram for the reaction of methyl dicyclopentadienecarboxylate with N-phenylmaleimide.

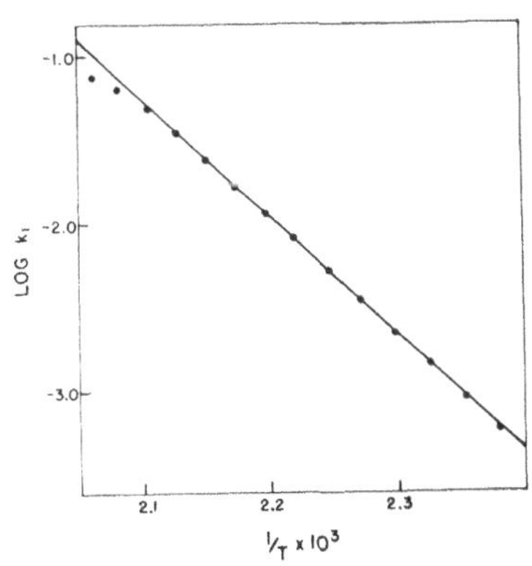

Fig. 2. Arrhenius plot for the reaction of methyl dicyclopentadienecarboxylate with N-phenylmaleimide.

Table I. Calculations of Rate Constants for the Reaction of Methyl Dicyclopentadienecarboxylate with N-phenylmaleimide

T °K	d in.	A $in.^2$	d/A $in.^{-1}$	$k_1 x 10^3$ $sec.^{-1}$
420	0.10	2.027	0.0493	0.602
425	0.15	2.004	0.0749	0.913
430	0.24	1.967	0.122	1.49
435	0.36	1.911	0.188	2.30
440	0.53	1.828	0.290	3.53
445	0.75	1.708	0.439	5.36
450	1.01	1.543	0.655	7.98
455	1.27	1.329	0.956	11.66
460	1.50	1.069	1.403	17.11
465	1.57	0.779	2.015	24.57
470	1.43	0.498	2.873	35.04
475	1.05	0.265	3.958	48.27
480	0.57	0.113	5.026	61.29
485	0.23	0.038	5.990	73.05

Mention of a company and/or product by the U. S. Department of Agriculture does not imply approval or recommendation of the product to the exclusion of others which may also be suitable.

Since the dependence of ΔH on temperature for most reactions is small, it can be assumed that ΔH remains constant within the limits of accuracy of the measurements over the temperature range (50-75°) of the reaction peak. The rate of change of moles of reagent in the sample may therefore be written as follows:

$$-\left(\frac{dn}{dt}\right) = \frac{1}{\Delta H}\left(\frac{dQ}{dt}\right).$$

The value of ΔH may be found from Q_t and the original value of n.

Since the reactions reported here follow first order kinetics (1), the first order rate constant, k_1, was calculated at each temperature of measurement from the following relationship:

$$k_1 = \frac{-(dn/dt)}{n} = \frac{\Delta HK_3d}{\Delta HK_2A} = \left(\frac{K_3}{K_2}\right)\left(\frac{d}{A}\right).$$

The values of d and A were measured at 5° temperature intervals throughout the reaction peaks. A rate constant was calculated from each pair of values of d and A. The values of K_2 and K_3 were obtained from the calibration of the instrument and the settings used in the run. The following values were used in these calculations: at range = 2 and chart speed = 0.75, $K_2 = 7.29X10^{-2}$ cal. in.$^{-2}$ and $K_3 = 8.89X10^{-4}$ cal. sec.$^{-1}$ in.$^{-1}$; at range = 8 and chart speed = 3.0, $K_2 = 7.29X10^{-2}$ cal. in.$^{-2}$ and $K_3 = 3.56X10^{-3}$ cal. sec.$^{-1}$ in.$^{-1}$.

Table I shows the calculations from a typical run of the reaction of methyl dicyclopentadienecarboxylate with N-phenylmaleimide.

The activation energies, E_a, and the pre-exponential factor terms, log A, were calculated in the usual manner from plots of log k_1 against the reciprocal absolute temperature. Figure 2 shows the plot from the data in Table I.

RESULTS AND DISCUSSION

The kinetics of the dissociations of the following compounds were measured: dicyclopentadiene (I), methyl dicyclopentadienecarboxylate (II), and dimethyl dicyclopentadienedicarboxylate (III).

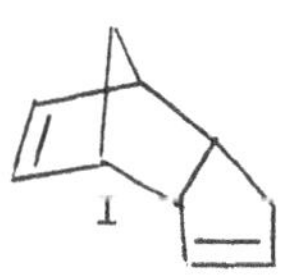

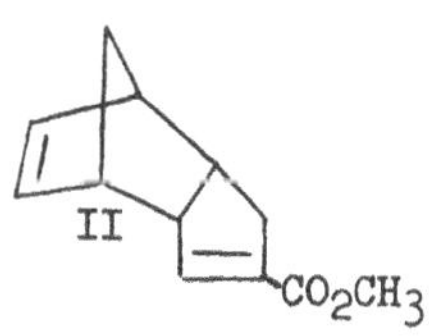

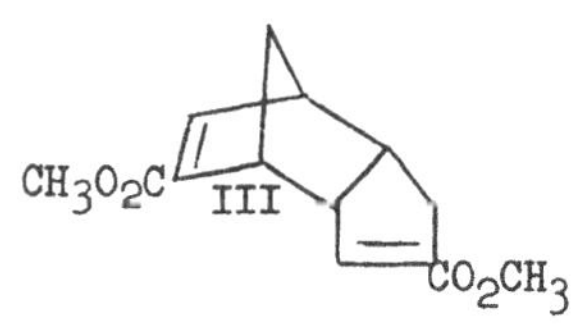

N-phenylmaleimide (IV) and tetracyanoethylene (V) were used as active dienophiles to trap the cyclopentadiene or methyl cyclopentadienecarboxylate produced in the dissociations. Table II summarizes the reactions investigated and the Arrhenius parameters obtained. The heats of reaction, calculated from the total areas of the reaction peaks on the thermograms, are also given in the table. The standard deviations are also given in the table.

The reactions studied here are assumed to be first order with respect to the dicyclopentadiene derivative and independent of the trapping agent. Both the dissociation of the dicyclopentadiene derivative and the subsequent reaction of the dissociation products with the trapping agent contribute to ΔH, but, since the reaction of the dissociation products with the trapping agent is much faster (12) than the dissociation reaction, the observed rate of reaction depends only on the rate of dissociation. This was previously found to be true for the reaction of III with V in diglyme (1). The fact that the same kinetics were observed for the dissociation of III in the

Table II. Arrhenius Parameters and Heats of Reaction Calculated from Differential Scanning Calorimeter Data

Reaction[a]	Solvent[b]	E_a Kcal./mole	log A	ΔH of reaction Kcal./mole.
I + IV	T	35.1 ± .2	13.7 ± .1	-31.2 ± 3.9
II + IV	T	32.2 ± .2	13.5 ± .1	-26.9 ± 1.1
III + IV	T	30.5 ± .5	13.1 ± .3	-27.7 ± 2.4
III + V[c]	T	30.0 ± .2	13.3 ± .1	-22.5 ± 1.9
III + V	D	31.1 ± 1.5	13.8 ± .7	-16.3 ± 1.2
III + V[d]	D	30.6 ± 2.1	13.4 ± 1.1	-20.8 ± 4.2

a/ I = Dicyclopentadiene, II = Methyl Dicyclopentadienecarboxylate, III = Dimethyl Dicyclopentadienedicarboxylate, IV = *N*-phenylmaleimide, V = Tetracyanoethylene.

b/ T = Tetraglyme, D = Diglyme.

c/ Average from three runs. All other values are averages from four or more runs.

d/ Dilute solution, concentrations of reagents are 0.25 of those in other reactions.

A report of a similar method of measuring the kinetics of reactions in nonvolatile solvents by K. E. J. Barrett appeared in *J. Appl. Polymer Sci.*, *11*, 1617 (1967) after this abstract was complete.

presence of IV and in the presence of V, while different heats of reaction were observed, supports this assumption.

The kinetics of two of these dissociations have been described in the literature. The dissociation of endo-dicyclopentadiene (I) as the pure liquid has been reported (13) to have the following Arrhenius parameters: $E_a = 35.3 \pm 0.6$ Kcal./mole, and log $A = 13.6 \pm 0.4$. These values are in excellent agreement with our values. Other kinetic values for the dissociation of I in paraffin solution (13) and in the gas phase (14) are slightly lower. The solvent effects in the pure liquid are expected to be the most similar to the solvent effects in our system. The agreement of the present values for E_a and log A with those in the literature tends to confirm the assumption that ΔH is constant over the temperature range of the reaction.

A least squares analysis of the data previously given for the reaction of III with V in diglyme (1) gives the following Arrhenius parameters: $E_a = 25.23 \pm .09$ Kcal/mole and log $A = 10.5 \pm .5$, which are in poor agreement with the present measurements. An estimate of the maximum possible errors in the previous data by the method of Peterson (15), based on the least squares standard deviations of the individual rate constants, gave the following Arrhenius parameters: $E_a = 25.2 \pm 4.8$ Kcal/mole and log $A = 10.5 \pm 2.7$. Although the agreement is poor, the range from the previous data overlaps the range of the present measurements, which are considered more reliable.

As a further check on our method, the Arrhenius parameters for each run were calculated from the values of d and A at the peaks of the DSC curves, using equations derived for the DSC analogously to equations derived for TGA data by Fuoss (16). The values obtained with these equations were similar to those in Table II, except that the standard deviations were generally one order of magnitude greater. It is felt that our method of calculating rate constants at arbitrary temperature intervals and finding the best slope from the series of constants is inherently superior in accuracy and precision to the second method, which depends on finding accurately the peak of a somewhat flattened curve.

The use of the differential scanning calorimeter for measuring reaction kinetics provides a convenient and rapid method for studying many reactions in solution. The technique requires only milligrams of sample to measure rate constants over temperature ranges of up to 70°. The precision and accuracy of this technique are similar to those of other standard kinetic techniques. This technique does not suffer from the well-known deficiencies of other thermal methods of measuring reaction rates, such as DTA and TGA.

In addition to the usual limitations on the kinetic measurement of solution reaction systems, several other limitations apply to reactions to be measured by the differential scanning calorimeter. The reaction must have an appreciable endothermic or exothermic heat of reaction. A total heat change of approximately 0.1 cal. from the 15 mg. of solution is ordinarily required for accurate measurement of rate constants. The relatively high standard deviation in the Arrhenius parameters measured for reaction of the dilute solution of III and V arises from the fact that the total heat evolved from the reaction of the sample was only 0.05 cal.

Side reactions occurring at the high temperature end of the thermogram make it difficult or impossible to measure accurately the heat of reaction and, consequently, the concentration of the reagents during the reaction. Nevertheless, it is sometimes possible to separate reactions on the differential scanning calorimeter by adjusting the scan rate. Because of the strongly exothermic decomposition above 460° K. of the adduct of cyclopentadiene with V, the kinetics of reactions of I and II with V could not be measured on the differential scanning calorimeter.

If there is an extraneous peak in the thermogram immediately preceding the reaction peak, it is sometimes possible to cool the sample after the first reaction is complete and rerun the complete thermogram to obtain a good peak for the reaction to be measured. Although this procedure may sacrifice accuracy in the measurement of the enthalpy of reaction, it has no effect on the measurement of the kinetics of first order reactions.

The choice of solvents is usually not critical. The use of hermetically sealed sample containers extends the useful range of a solvent to about 50° above its boiling point. The use of unsealed sample containers severely limits the choice of solvents and reactants. The only solvent that we have found suitable for the measurement of reactions by the method of Rodgers (6) was polyphenyl ether, and even with this solvent such reagents as tetracyanoethylene and dibenzoylethylene were unsuitable because the reaction peaks were masked by the endothermic distillation of the reagents from the unsealed sample containers.

Acknowledgment.—The authors are grateful to the Cotton Producers Institute for partial support of this work.

LITERATURE CITED

(1) W. E. Franklin, S. P. Rowland, and C. H. Mack, J. Org. Chem., in press.

(2) W. W. Wendlandt, "Thermal Methods of Analysis," John Wiley & Sons, Inc., New York, 1964, p. 174.

(3) R. L. Reed, L. Weber, and B. S. Gottfried, Ind. Eng. Chem., Fundamentals, 4, 38 (1965).

(4) E. S. Watson, M. J. O'Neill, J. Justin, and N. Brenner, Anal. Chem., 36, 1233 (1964).

(5) C. H. Mack and S. R. Hobart, Applied Polymer Symposia, 2, 133 (1966).

(6) R. N. Rogers and E. D. Morris, Anal. Chem., 38, 412 (1966).

(7) M. J. Uricheck, Instrument News (Perkin-Elmer Corp.), Vol. 17, No. 2 (1966).

(8) C. H. Mack and D. J. Donaldson, Textile Res. J., in press.

(9) S. M. Ellerstein, J. Phys. Chem., 69, 2471 (1965).

(10) S. M. Ellerstein, Applied Polymer Symposia, 2, 111 (1966).

(11) K. Alder, F. H. Flock, A. Hausweiler, and R. Reeber, Ber., 87, 1752 (1954).

(12) J. Sauer, H. Wiest, and A. Mielert, Ber., 97, 3183 (1964).

(13) B. S. Khambata and A. Wasserman, J. Chem. Soc., 375 (1939).

(14) W. C. Herndon, C. R. Grayson, and J. M. Manion, J. Org. Chem., 32, 526 (1967).

(15) R. C. Peterson, J. H. Markgraff, and S. D. Ross, J. Am. Chem. Soc., 83, 3819 (1961).

(16) R. M. Fuoss, I. O. Salyer, and H. S. Wilson, J. Polymer Sci., Part A, 2, 3147 (1964).

THERMAL ANALYSIS OF KAOLIN CLAYS: CORRELATION WITH X-RAY CRYSTALLINITY DATA

by

Donald E. Koopman
Materials Research
Central Research Division
The National Cash Register Company
Dayton, Ohio 45409

Thermal analysis and x-ray diffraction data have been obtained on five samples of selected kaolinite. These samples are essentially mono-mineralic and differ in crystallinity from well-crystallized to poorly crystallized.

EXPERIMENTAL METHODS AND APPARATUS

An empirical x-ray crystallinity index for each clay sample was determined using a modified method of Hinckley (1). The crystallinity index is expressed as the ratio of the sum of the peak amplitudes of the $1\bar{1}0$ and $11\bar{1}$ reflections above diffraction background to the total peak height of the $1\bar{1}0$ reflection above general background.

Diffraction measurements were made using a Norelco high angle x-ray unit and strip chart goniometer fitted with a rotating sample spinner. A specially designed powder mount (2) was used to minimize any preferred orientation that might be introduced into the sample during preparation. CuKα radiation and a scan speed of 0.5° 2θ/minute were used for all samples.

Differential thermal analysis data were obtained on DuPont 900 DTA equipped with the DuPont High Temperature Cell. Instrument parameters and sample preparation suggested by Langer and Kerr (3) in their study of variables influencing kaolinite DTA curves were used. Heating rates were 10° and 30°C./minute from ambient to 1200°C. For each run, 28 milligrams of sample was packed into the sample holder with a sample jig to insure uniform packing. The reference material was calcined Al_2O_3 ground to clay size.

Thermal gravimetric analyses, using a DuPont 950 TGA, were performed on samples weighing 22 to 25 milligrams at a 10°C/minute heating rate from ambient to 1200°C. A nitrogen gas flow of < 1 ft.^{3}/hour was employed.

Derivative thermal gravimetric analysis was carried out by interfacing a Cahn 2030 Time Derivative Computer with the DuPont 950 TGA. Sample weights of 28 milligrams, a heating rate of 15°C./minute to 1200°C and a nitrogen gas flow of < 1 ft.^{3}/hour were used. The resulting derivative traces were resolved into components on a DuPont 310 Curve Analyzer.

The variation of crystallinity in kaolinites has been thoroughly discussed by several investigators (4-6) and needs no further comment. In order to show that the variation in crystallinity was due to the structural arrangement of the kaolinite lattice and not to a wide variation in particle size, electron micrographs were prepared and the surface areas determined for each sample.

TABLE I

Kaolinite	Crystallinity Index	Surface Area (4 pt. BET) M^2/gram	% Wt. Loss by TGA
OG 6978, Mine 11A	1.15 ± .03	10.2	13.81
OG 6979, Mine 23	0.78 ± .02	13.8	13.05
OG 6980, Mine 15	0.75 ± .02	19.0	13.64
OG 6982, Montezuma	0.55 ± .02	22.9	14.03
OG 6981, Renfrose	0.34 ± .05	23.4	13.56

In general, the electron micrographs show a loss in euhedral character and a slight decrease in particle diameter with decreasing crystallinity. Approximately 90% of the particles viewed in all samples appeared to be between .1 and 1μ in diameter. No attempt was made to obtain particle size distributions from the micrographs. Only a two fold increase in surface area (Table I) was measured over the range of crystallinity, supporting the evidence presented in the electron micrographs showing a slight decrease in particle size with decreasing crystallinity. It is concluded that the crystallinity is of a structural nature.

The differential thermograms of kaolinite show a well-defined but broad endotherm due to dehydroxylation that begins about 350°C, peaks around 590°C and recovers to the baseline in the neighborhood of 800°C. A sharp exothermic reaction occurs around 1000°C which is thought to be the reaction $SiO_2 + Al_2O_3 \rightarrow \gamma$ mullite. In the well-crystallized kaolins several investigators have seen a small but discernible endotherm just before the exotherm. This endotherm was also observed in this study. Grim and Bradley concluded that with dehydroxylation of poorly crystallized kaolinites, the crystalline structure is completely lost, but in well crystallized kaolinites some of the lattice structure remains along with some water. This structural remnant (and water) is finally lost at the time of the exothermic reaction (7). In this study no evidence of any water left in the system at these temperatures was observed. However, we agree that some remnants of structure persists in the dehydrated well crystallized kaolinites and accounts for the higher temperatures necessary to achieve the exothermic reaction. The small endotherm immediately preceeding the strong exotherm is probably a minor glass transition occurring in an essentially amorphous material that does contain some skeletal remains. Enough energy must be "pumped" into the system to overcome this residual structure necessitating higher temperatures. Figure 1 shows the dependence of peak temperature of the exothermic reaction on the initial crystallinity of the kaolinite samples.

Three basic assumptions were made for the interpretation of the derivative thermal gravimetric analysis data: 1) the kaolinite lattice has three different kinds of hydroxyls, 2) with increasing temperature the hydroxyls are driven off in a Gaussian distribution, 3) the component areas resolved under the derivative traces are proportional to the per cent of total hydroxyl of each type of hydroxyls.

In an infinite lattice of kaolinite of perfect crystallinity only two kinds of hydroxyls are present, those occupying a sheet-like position parallel to the basal plane and those in the lattice proper itself occupying positions common to both the tetrahedral and octahedral layers of the structure. The per cent hydroxyl distribution of the former is 75% and the latter 25%. In a finite crystal of kaolinite,

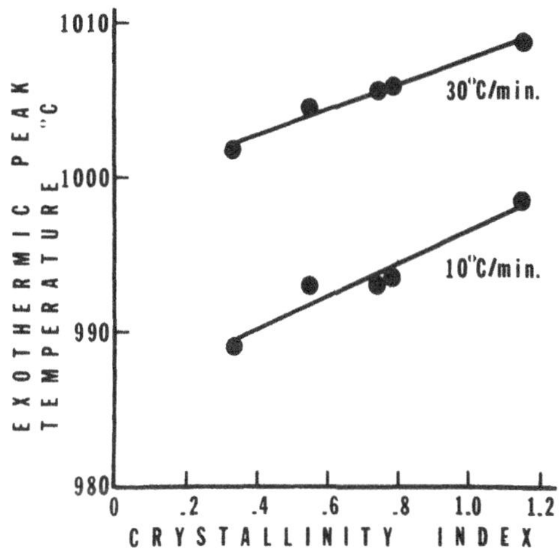

Figure 1

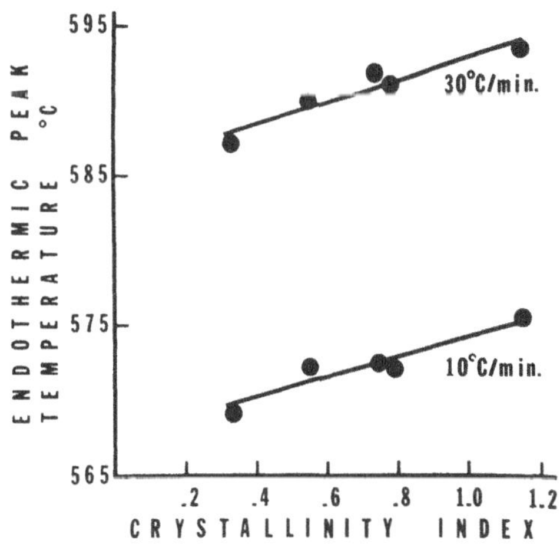

Figure 2

exposed hydroxyl positions exist around the broken edges and terminal basal planes. A value of 1% has been approximated for the "edge" hydroxyl on a 1μ diameter particle of kaolinite having perfect crystallinity.

In nearly perfect and perfectly crystallized kaolins the above three hydroxyl models of kaolinite can be used or at least approximated. In real kaolinites having order-disorder phenomenon, limited cation substitution, etc., this "sheet-lattice-edge hydroxyl" model probably has limited value. These positions are of course occupied in real kaolinites, but instead of considering a hydroxyl position model, a model can be postulated in which the hydroxyls are bonded to the lattice by three different energies which vary with the crystallinity of the sample.

Figure 3 shows the resolved derivative traces of a well crystallized and a poorly crystallized kaolinite. The outer trace is the derivative curve. The component peaks are labled E, S, and L with the lowest temperature peak being E. It is apparent that the magnitude of E is much smaller than L in the well crystallized sample, the converse being true in the poorly crystallized one.

Figure 4 is a composite of the data from the resolved derivative traces. A linear relationship is evident and one can extrapolate to some interesting conclusions. If line E is extrapolated to zero per cent area (intercept on the crystallinity index at 1.25) which would represent zero edge hydroxyl in a perfect kaolinite structure, and lines L and S are extended to the ordinate at this intercept, values of 70% and 30% respectively are obtained. This is reasonably close to the 3 to 1 ratio of the sheet to lattice hydroxyl in a perfect kaolin structure. It becomes more reasonable when one considers that in the dehydroxylation reaction a volume decrease is observed in the sample. It is speculated that some of the S hydroxyls are trapped by this collapse of the structure and are finally driven off at a higher temperature with the L hydroxyls. As this would only amount to about 5% of the total hydroxyls in a well crystallized sample, this seems to be a reasonable speculation.

In the region of low and zero crystallinity, L approaches zero and S and E converge. We believe that it is reasonable to project that the mineral allophane, an amorphous hydrated alumino-silicate, would yield data of this type. The hydroxyls in allophane would be weakly bonded to the structure and would be driven off at a relatively low temperature. These conclusions agree with the data reported by Nutting (8) and Grim and Rowland (9).

Figure 2 shows an apparent linear relationship of peak temperature of the dehydroxylation with crystallinity. In light of the derivative data, as the crystallinity increases the magnitude of the L hydroxyl increases and the E hydroxyl decreases, shifting the entire thermal spectrum to higher temperatures.

The data obtained from thermal gravimetric analysis (Table I) does not correlate with any parameters determined in this study. The per cent weight loss is reasonably close to the accepted theoretical value of 13.96%.

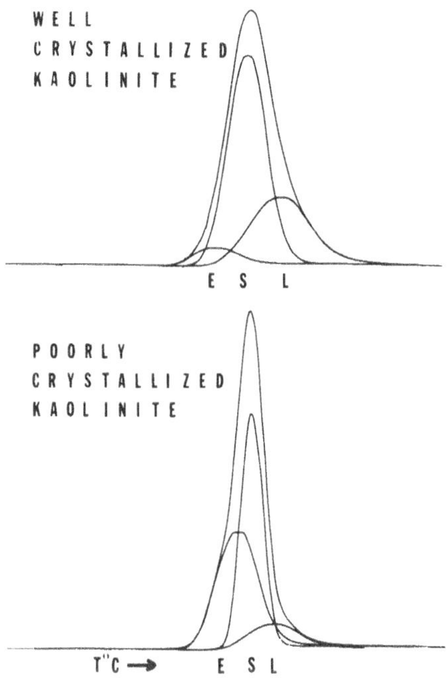

Figure 3

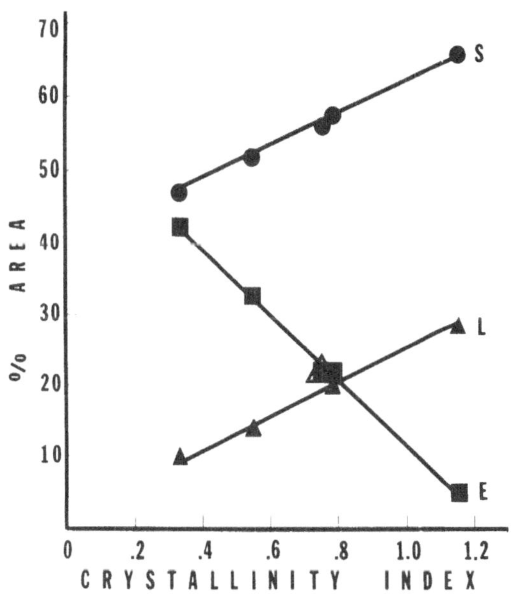

Figure 4

ACKNOWLEDGMENT

The author would like to thank the Georgia Kaolin Company for supplying the kaolinite samples and the Proctor and Gamble Company for allowing him access to their DuPont 310 Curve Resolver. Acknowledgment is also given to Dr. R. Tettenhorst, The Ohio State University and Dr. P. Bouffard, The National Cash Register Company for their valuable aid and critical review throughout the course of this investigation.

BIBLIOGRAPHY

1. Hinckley, D. N., Proc. 11th Nat. Conf. on Clays and Clay Minerals, pp. 229.

2. Morton, V., The NCR Co., personal communication.

3. Langer, A. M. and Kerr, P. F., DuPont Thermogram 3, No. 1, pp. 1 (1966).

4. Brindley, G. W. and Robinson, K., Trans. Faraday Soc., 42B, 198-205 (1946).

5. Grimshaw, R. W., Heaton, E., and Robert, A. L., Trans. Brit. Ceram. Soc. 44, 69-92, (1945).

6. Grim, R. E., Am. Mineral., 32, 493-501 (1947).

7. Grim, R. E. and Bradley, W. F., Am. Mineral., 33, pp 50-59 (1948).

8. Nutting, P. G., U. S. Geol. Survey Profess. Paper, 197E, pp. 197-216 (1943).

9. Grim, R. E. and Rowland, R. A., Am. Mineral., 27, pp. 746-761, 801-818 (1942).

CALORIMETRIC MEASUREMENTS ON METAL SULFATES AND THEIR HYDRATES: ELECTRODE POTENTIALS AND THERMODYNAMIC DATA FOR AQUEOUS IONS OF TRANSITION ELEMENTS*

John W. Larson and Loren G. Hepler

(Contribution from Departments of Chemistry, Carnegie-Mellon University and University of Louisville, Pittsburgh, Pa., and Louisville, Ky.)

Knowledge of reversible electrode potentials is important for many aspects of analytical electrochemistry, and also as a source of thermodynamic data that are finding increasing applications in analytical chemistry. Unfortunately, there are substantial experimental difficulties associated with determination of reversible electrode potentials for metals that are very reactive and for ill-behaving metals that often yield "irreversible" electrodes. Many years ago, Lewis developed and applied indirect electrochemical methods for determination of the potentials for such active metals as sodium in aqueous systems. Subsequently, Latimer and others have applied thermodynamic data to calculation of aqueous potentials for aluminum, fluorine, etc.

Although the thermodynamic relations that are used in calculating potentials from thermal data are certainly sound, there are legitimate questions about the reliability of the thermal data for several of the transition elements and thence the derived potentials that presently interest us. We have therefore chosen to investigate copper, zinc, and cadmium because the electrochemical potentials are well known and provide a useful check on the overall reliability of potentials calculated from thermal data. Further, these measurements also yield useful thermodynamic data for the aqueous ions of these elements. The excellent agreement of the calculated potentials with the electrochemical potentials for these metals gives us confidence in the potentials that we are able to calculate from thermal data for iron, cobalt, and nickel. We also report useful thermodynamic data for aqueous ions of these latter elements.

EXPERIMENTAL

The calorimetric apparatus used for these measurements was patterned after one previously described (ref. 1). The calorimeter consists of a Dewar vessel suspended in a brass can submerged in a water bath maintained at constant temperature. Temperatures in the calorimeter were determined with a nickel resistance thermometer connected to a Leeds and Northrup Mueller G-2 Bridge with a Liston-Becker Model 14 Breaker Amplifier. The resistance thermometer and a manganin heater (300 ohms) were contained in a glass spiral filled with mineral oil. All of the calorimetric data were obtained at $25.00 \pm 0.15°$.

The metal sulfates used were recrystallized from distilled water to obtain the hydrates in equilibrium with saturated solution. These samples were removed from solution by filtration and pressed on the filter. Excess water was then removed by storing in vacuum desiccators with $CaCl_2$. When analysis showed that a sample was approaching the desired composition, it was gently ground and stored for two weeks in a closed container. After final analysis, calorimetric runs were made. Further dehydration with $CaCl_2$ and heating were used to obtain lower hydrates and anhydrous sulfates.

Special care was taken in all of the work with ferrous sulfate to exclude oxygen. All calorimetric measurements were made with water that had been boiled and stored to exclude oxygen or through which nitrogen had been bubbled for at least an hour. On the basis of analysis with thiocyanate, it was found that solid ferrous sulfates contained less than 0.1% Fe^{+3} and that less than 0.5% of the total iron was present as Fe^{+3} in the calorimetric solution *after* each calorimetric measurement was complete.

*Parts of this work are also reported in the following: J. W. Larson, Ph.D. Thesis, Carnegie-Mellon University, 1968; J. W. Larson, P. Cerutti, H. K. Garber, and L. G. Hepler, J. Phys. Chem., in press.

RESULTS AND CALCULATIONS

Results of our heat of solution measurements have been combined with heat of dilution data (refs. 2 and 3) to obtain the standard heats of solution represented by ΔH° in Table I. Total uncertainties in these values range from about 0.03 to 0.1 kcal/mole.

Table I

Standard Heats of Solution

Compound	ΔH°(kcal/mole)
$CuSO_4(c)$	-17.56
$CuSO_4 \cdot 4H_2O(c)$	- 1.31
$CuSO_4 \cdot 5H_2O(c)$	+ 1.43
$ZnSO_4(c)$	-19.9
$ZnSO_4 \cdot 6H_2O(c)$	- 0.15
$ZnSO_4 \cdot 7H_2O(c)$	+ 3.18
$CdSO_4(c)$	-12.41
$CdSO_4 \cdot H_2O(c)$	- 7.66
$CdSO_4 \cdot \frac{8}{3}H_2O(c)$	- 4.32
$FeSO_4 \cdot H_2O(c)$	-10.6
$FeSO_4 \cdot 4H_2O(c)$	- 3.3
$FeSO_4 \cdot 7H_2O(c)$	+ 2.82

Copper

Adami and King (ref. 4) have reported $\Delta H_f^\circ = -183.4$ kcal/mole for $CuSO_4(c)$. Using the best data for sulfuric acid (ref. 5) and CuO (ref. 6), we recalculate their results to obtain $\Delta H_f^\circ = -184.0$ kcal/mole for $CuSO_4(c)$. Combination of this value with our ΔH° of solution from Table I and ΔH_f° for $SO_4^{-2}(aq)$ (ref. 5) gives $\Delta H_f^\circ = 15.7$ kcal/mole for $Cu^{+2}(aq)$.

For the reaction

$$CuSO_4 \cdot 5H_2O(c) = Cu^{+2}(aq) + SO_4^{-2}(aq) + 5\ H_2O(liq) \qquad (1)$$

we have $\Delta G^\circ = -RT \ln s^2\gamma^2 a_w^5$, where s is the molal solubility, γ is the mean activity coefficient in saturated solution, and a_w is the activity of water in the saturated solution. The solubility is taken from Seidell (ref. 7) and activity data from Robinson and Stokes (ref. 8), leading to ΔG° = 3.58 kcal/mole.

Combination of this ΔG° with our ΔH° from Table I gives ΔS° = -7.2 gibbs/mole for reaction (1). We use the Third Law entropy of $CuSO_4 \cdot 5H_2O(c)$ ($S_{298}^\circ = 72.0$ gibbs/mole) (ref. 9) with this ΔS° and entropies (ref. 5) of $SO_4^{-2}(aq)$ and $H_2O(liq)$ to calculate $\bar{S}_2^\circ = -23.6$ gibbs/mole for $Cu^{+2}(aq)$.

Combination of ΔH_f° and $\bar{S}_2^\circ$ values for $Cu^{+2}(aq)$ given above with entropies (refs. 5 and 10) for Cu(c) and $H_2(g)$ gives $\Delta G_f^\circ = 15.9$ kcal/mole for $Cu^{+2}(aq)$. Using ΔG° =-nFE°, we calculate $E^\circ = -0.34_5$ v as the oxidation potential (ref. 11) for $Cu(c) = Cu^{+2}(aq) + 2\ e^-$. We estimate that total uncertainties in our ΔG_f° and E° values are about ± 0.4 kcal/mole and ±0.009 v.

Electrochemical data for the $Cu/Cu^{+2}(aq)$ electrode have been reviewed by deBethune, Licht, and Swendeman (ref. 12). Potentials ranging from 0.337 to 0.3448 v have been reported. We therefore see that the thermal and electrochemical potentials are in satisfactory agreement for this system.

On the basis of our estimates of the uncertainties in the data and our weighted averaging procedure we adopt E° =-0.34$_0$ v for the Cu/Cu^{+2} potential and the corresponding "best" ΔG_f° = 15.7 kcal/mole for $Cu^{+2}(aq)$. Of several possible values for ΔH_f° of $Cu^{+2}(aq)$, we believe ΔH_f° = 15.6 kcal/mole derived by combination of ΔG_f° = 15.7 kcal/mole with $\bar{S}_2^\circ$ = -23.6 gibbs/mole to be the "best" value.

Hugus (ref. 13) has analyzed earlier equilibrium data on the $Cu/Cu^+/Cu^{+2}$ system. We combine his results with thermodynamic properties of $Cu^{+2}(aq)$ cited above to obtain ΔG_f° = 12.0 kcal/mole, ΔH_f° = 17.2 kcal/mole, and $\bar{S}_2^\circ$ = 9.8 gibbs/mole for $Cu^+(aq)$. These values are in satisfactory agreement with the measurements and calculations reported by Tourky and Khairy (ref. 14). The free energies lead to E° = -0.52 v for the Cu/Cu^+ oxidation potential and E° = -0.16 v for the Cu^+/Cu^{+2} oxidation potential.

The data in Table I permit calculation of heats of dehydration of $CuSO_4 \cdot 5H_2O(c)$ and $CuSO_4 \cdot 4H_2O(c)$. There is some disagreement in the literature as to whether $CuSO_4 \cdot 4H_2O(c)$ exists as a definite hydrate or only as a mixture of the pentahydrate and the trihydrate. This disagreement is not resolved by our data.

Zinc

We have used the new (ref. 5) ΔH_f° values for sulfuric acid and ZnO in recalculating the results of Adami and King (ref. 4) to obtain ΔH_f° = -233.7 kcal/mole for $ZnSO_4(c)$. Our own analysis of the available thermochemical data for ZnO suggests that the "best" ΔH_f° is about 1/3 kcal/mole more negative than the value (ref. 5) used in this calculation, which would lead to a correspondingly more negative ΔH_f° for $ZnSO_4(c)$. Combination of the quoted ΔH_f° for $ZnSO_4(c)$ with our standard heat of solution from Table I and the ΔH_f° for $SO_4^{-2}(aq)$ (ref. 5) leads to ΔH_f° = -36.3 kcal/mole for $Zn^{+2}(aq)$. Adoption of more negative ΔH_f° values for ZnO and $ZnSO_4$ would make this ΔH_f° of $Zn^{+2}(aq)$ correspondingly more negative.

Using solubility (ref. 7) and activity (ref. 8) data, we have calculated ΔG° = 2.62 kcal/mole for the standard free energy of solution of $ZnSO_4 \cdot 7H_2O(c)$ as described by an equation similar to that given earlier for $CuSO_4 \cdot 5H_2O(c)$. We combine this ΔG° with our ΔH° of solution of $ZnSO_4 \cdot 7H_2O(c)$ from Table I to obtain ΔS° = 1.9 gibbs/mole for the solution reaction. Further combination of this ΔS° with entropies (ref. 5) for $ZnSO_4 \cdot 7H_2O(c)$, $H_2O(liq)$, and $SO_4^{-2}(aq)$ gives $\bar{S}_2^\circ$ = -26.9 gibbs/mole for $Zn^{+2}(aq)$.

Combination of ΔH_f° and $\bar{S}_2^\circ$ values for $Zn^{+2}(aq)$ given above with entropies (refs. 5 and 10) for Zn(c) and $H_2(g)$ gives ΔG_f° = -34.6 kcal/mole for $Zn^{+2}(aq)$ and thence E° = +0.75$_1$ v for the standard oxidation potential associated with $Zn(c) = Zn^{+2}(aq) + 2e^-$. Adoption of the more negative ΔH_f° for $Zn^{+2}(aq)$ mentioned above would make ΔG_f° correspondingly more negative and E° more positive.

Another path, which does not depend on data for ZnO, is also available for determination of the thermodynamic properties of $Zn^{+2}(aq)$. We combine data from Giauque, Barieau, and Kunzler (ref. 15) with data from Covington, Dobson, and Wynne-Jones (ref. 16) to obtain ΔG_f° = -612.59 kcal/mole for $ZnSO_4 \cdot 7H_2O(c)$, which is the same as the value given by Wagman (ref. 5). Combination of this ΔG_f° with ΔG_f° values (ref. 5) for $SO_4^{-2}(aq)$ and $H_2O(liq)$ and our ΔG° of solution calculated above leads to ΔG_f° = -35.2 kcal/mole for $Zn^{+2}(aq)$ and thence E° = +0.76$_4$ v for the standard oxidation potential. These values support our suggestion that ΔH_f° values quoted earlier for ZnO, $ZnSO_4$, and $Zn^{+2}(aq)$ should be revised to more negative values.

The ΔH_f° of $ZnSO_4 \cdot 7H_2O(c)$ is calculated (ref. 5) from the above ΔG_f° and the Third Law entropy, and is therefore independent of heat of solution data. We combine this ΔH_f° with our standard heat of solution from Table I and ΔH_f° values (ref. 5) for $SO_4^{-2}(aq)$ and $H_2O(liq)$ to obtain ΔH_f° = -36.9

kcal/mole for $Zn^{+2}(aq)$. We regard this value, which is consistent with the ΔG_f° in the paragraph above and our $\bar{S}_2^\circ$, as the best ΔH_f° for $Zn^{+2}(aq)$.

Electrochemical data (refs. 12 and 17) give E° = 0.763 v for the zinc oxidation potential and thence ΔG_f° = -35.2 kcal/mole for $Zn^{+2}(aq)$, in excellent agreement with our value calculated from data for $ZnSO_4 \cdot 7H_2O$ and in fair agreement with our value calculated from data for $ZnSO_4$.

Data from Table I permit calculation of heats of dehydration of $ZnSO_4 \cdot 7H_2O(c)$ and $ZnSO_4 \cdot 6H_2O(c)$.

Cadmium

Combination of the new (ref. 5) ΔH_f° values for sulfuric acid and CdO with results of Adami and King (ref. 4) gives us ΔH_f° = -223.0 kcal/mole for $CdSO_4(c)$. We use this value with our heat of solution and ΔH_f° for $SO_4^{-2}(aq)$ to obtain ΔH_f° = -18.1 kcal/mole for $Cd^{+2}(aq)$.

With solubility (ref. 7) and activity (ref. 8) data we calculate ΔG° = 2.58 kcal/mole for solution of $CdSO_4 \cdot \frac{8}{3}H_2O(c)$ and combine with our ΔH° of solution from Table I to obtain ΔS° = -23.0 gibbs/mole for the standard entropy of solution. Further combination of this last quantity with entropies (refs. 5 and 10) of $CdSO_4 \cdot \frac{8}{3}H_2O(c)$, $SO_4^{-2}(aq)$, and $H_2O(liq)$ gives $\bar{S}_2^\circ$ = -17.4 gibbs/mole for $Cd^{+2}(aq)$.

Combination of ΔH_f° and $\bar{S}_2^\circ$ values for $Cd^{+2}(aq)$ with entropies (refs. 5 and 10) of Cd(c) and $H_2(g)$ gives ΔG_f° = -18.6 kcal/mole for $Cd^{+2}(aq)$ and thence E° = 0.40_3 v for the standard oxidation potential to be associated with $Cd(c) = Cd^{+2}(aq) + 2\ e^-$. We estimate total uncertainties to be about ±0.4 kcal/mole and ±0.009 v. Latimer (ref. 17) has cited 0.402 and 0.403 v for this potential.

Data from Table I permit calculation of heats of dehydration of $CdSO_4 \cdot \frac{8}{3}H_2O(c)$ and $CdSO_4 \cdot H_2O(c)$.

Iron

Now consider thermodynamics of the reaction represented by

$$FeSO_4 \cdot 7H_2O(c) = Fe^{+2}(aq) + SO_4^{-2}(aq) + 7\ H_2O(liq) \qquad (2)$$

We calculate ΔG° = 3.3 kcal/mole for this reaction from the solubility (ref. 7) of the heptahydrate at 25°C and estimated activity and osmotic coefficients for the saturated solution based on experimental values (ref. 8) for these quantities for $MgSO_4$, $MnSO_4$, $ZnSO_4$, and especially $NiSO_4$, at the same concentration. Combination of this ΔG° with ΔH° from Table I gives ΔS° = -1.6 gibbs/mole for reaction (2). We combine this ΔS° with entropies (refs. 5 and 18) of $SO_4^{-2}(aq)$, $H_2O(liq)$, and $FeSO_4 \cdot 7H_2O(c)$ to obtain $\bar{S}_2^\circ$ = -25.6 gibbs/mole for $Fe^{+2}(aq)$.

Adami and Kelley (ref. 19) have reported ΔH_f° = -719.98 kcal/mole for $FeSO_4 \cdot 7H_2O(c)$. We have recalculated their results in terms of the best data (ref. 5) for sulfuric acid to obtain ΔH_f° = -720.48 kcal/mole for this compound. Combination of this value with our standard heat of solution from Table I and ΔH_f° values (ref. 5) for $H_2O(liq)$ and $SO_4^{-2}(aq)$ leads to ΔH_f° = -22.1 kcal/mole for $Fe^{+2}(aq)$. Since this value differs significantly from the ΔH_f° = -21.0 kcal/mole listed in NBS Circ. 500 (ref. 20), it is worth considering other related data.

Koehler and Coughlin (ref. 21) have reported ΔH_f° = -81.86 kcal/mole for $FeCl_2(c)$. We combine this value with the heat of solution from Li and Gregory (ref. 22) to obtain ΔH_f = -21.5 kcal/mole for $Fe^{+2}(aq)$. We may also estimate the heat of dilution of ferrous chloride to obtain ΔH° of solution from the above ΔH and thence ΔH_f° somewhat more negative than -21.5 kcal/mole for $Fe^{+2}(aq)$. Similar treatment of the heat of solution of $FeBr_2(c)$ from Li and Gregory (ref. 22) and its heat of formation (ref. 20) leads to ΔH_f° = -22.5 kcal/mole for $Fe^{+2}(aq)$.

The results cited above strongly support a ΔH_f° more negative than -21.0 kcal/mole for $Fe^{+2}(aq)$ and we adopt our ΔH_f° = -22.1 kcal/mole for furthur calculations. Combination of this ΔH_f° with our entropy of $Fe^{+2}(aq)$ and entropies (refs. 5 and 10) of Fe(c) and $H_2(g)$ gives ΔG_f° = -21.8 kcal/

mole for $Fe^{+2}(aq)$. This free energy leads to the following oxidation potential:

$$Fe(c) = Fe^{+2}(aq) + 2\,e^- \qquad E° = 0.47_3\ v$$

We believe that the total uncertainty in our ΔG_f° for $Fe^{+2}(aq)$ is no more than ±0.5 kcal/mole, which corresponds to ±0.01 v.

Combination of our recalculated ΔH_f° of $FeSO_4 \cdot 7H_2O(c)$ from Adami and Kelley (ref. 19) with entropies of this compound (ref. 18) and the elements (refs. 5 and 10) gives $\Delta G_f^\circ = -599.94$ kcal/mole for $FeSO_4 \cdot 7H_2O(c)$. Further combination of this ΔG_f° with the $\Delta G°$ of solution given earlier and ΔG_f° values (ref. 5) for $SO_4^{-2}(aq)$ and $H_2O(liq)$ leads directly to $\Delta G_f^\circ = -21.8$ kcal/mole for $Fe^{+2}(aq)$ and $E° = 0.47_3$ v for the Fe/Fe^{+2} standard potential. It should be recognized that this ΔG_f° for $Fe^{+2}(aq)$ that is identical with the value given above is not independent, but does demonstrate the internal consistency of our calculations.

On the basis of earlier electrochemical measurements, Latimer (ref. 17) adopted $E° = 0.440$ v for $Fe(c) = Fe^{+2}(aq) + 2\,e^-$. More recently, Patrick and Thompson (ref. 23) have reported $E° = 0.409$ v for this potential. Still more recently, Hurlen (ref. 24) has made measurements that lead to $E° = 0.467$ v at 20°. Since combination of thermal data with the Gibbs-Helmholtz equation shows that $(dE°/dT)(5°)$ is negligible, we can take Hurlen's E° as also applying at 25°. Each of these different E° values is based on results of careful experiments, and none can be rejected without further experimental evidence. In addition to the electrochemical significance of this disagreement, it should be noted that a spread of 0.058 v in the standard potential corresponds to 2.7 kcal/mole in $\Delta G°$ or $\Delta H°$ or to 9 gibbs/mole in $\Delta S°$. Our calculated potential clearly supports the potential reported by Hurlen (ref. 24) rather than either of the electrochemical potentials reported earlier (refs. 17 and 23).

The best available data (ref. 17) lead to $E° = -0.771$ v for the Fe^{+2}/Fe^{+3} potential, which we combine with our ΔG_f° of $Fe^{+2}(aq)$ to obtain $\Delta G_f^\circ = -4.0$ kcal/mole for $Fe^{+3}(aq)$.

We have several sources of $\Delta H°$ values for the reaction

$$Fe^{+2}(aq) + H^+(aq) = Fe^{+3}(aq) + \tfrac{1}{2}\,H_2(g) \qquad (3)$$

The ΔH_f° values cited in NBS Circ. 500 (ref. 20) correspond to $\Delta H° =$ 9.8 kcal/mole for this reaction. The entropies in an earlier review (ref. 25) correspond to $\Delta H° = 12.0$ kcal/mole. Connick and McVey (ref. 26) calculated $\Delta H = 10.0$ kcal/mole from equilibrium measurements at different temperatures with solutions of $\mu = 0.5$ and quoted a calorimetric $\Delta H =$ 9.5 kcal/mole for similar solutions. Other equilibrium measurements (ref. 27) in 1 M $HClO_4$ led to $\Delta H = 9.7$ kcal/mole. Calorimetric measurements with hydrogen peroxide (ref. 28) and with bromine (ref. 29) as oxidizing agents for $Fe^{+2}(aq)$ lead to $\Delta H°$ values of 9.8 ± 1.2 and 10.6 ± 0.9 kcal/mole for reaction (3). Because of serious difficulties in correcting the experimental data for dilution, complexing, and hydrolysis effects, the best we can reasonably do is adopt $\Delta H° = 10$ kcal/mole for reaction (3).

Combination of the "best" $\Delta H° = 10$ kcal/mole for reaction (3) with our data for $Fe^{+2}(aq)$ and our ΔG_f° for $Fe^{+3}(aq)$ above gives $\Delta H_f^\circ = -12$ kcal/mole and $\bar{S}_2^\circ = -67$ gibbs/mole for $Fe^{+3}(aq)$. Although the uncertainties in these latter quantities are somewhat greater than ± 1 kcal/mole and ± 3 gibbs/mole, they are the best values presently available.

Heats of hydration calculated from our results in Table I are in good agreement with results of earlier workers (ref. 19).

Cobalt

In a previous paper (ref. 30) it was reported that $\Delta S° = -1.9$ gibbs/mole for solution of $CoSO_4 \cdot 7H_2O(c)$, determined by a procedure analogous to that described above for $FeSO_4 \cdot 7H_2O(c)$. Now we repeat the earlier calculation leading to the entropy of $Co^{+2}(aq)$, this time using the most recent entropies (ref. 5) of $SO_4^{-2}(aq)$ and $H_2O(liq)$, to obtain $\bar{S}_2^\circ = -26.6$ gibbs/mole for $Co^{+2}(aq)$.

Using the most recent data (ref. 5) for sulfuric acid, we recalculate the results of Adami and King (ref. 4) to obtain $\Delta H^\circ = -212.55$ kcal/mole for $CoSO_4(c)$. Combination of this value with the previously determined (ref. 30) heat of solution of $CoSO_4(c)$ and ΔH_f° for $SO_4^{-2}(aq)$ leads to $\Delta H_f^\circ = -14.0$ kcal/mole for $Co^{+2}(aq)$.

Combination of ΔH_f° and $\bar{S}_2^\circ$ values for $Co^{+2}(aq)$ with entropies (refs. 5 and 10) of Co(c) and $H_2(g)$ leads to $\Delta G_f^\circ = -13.2$ kcal/mole for $Co^{+2}(aq)$. This ΔG_f° gives $E^\circ = 0.28_7$ v for $Co(c) = Co^{+2}(aq) + 2\ e^-$. We believe that the total uncertainty in these values is less than ±0.5 kcal/mole or ±0.01 v and see that our calculated E° is in satisfactory agreement with the three higher potentials cited by Latimer (ref. 17), but not in good agreement with the lower potentials.

Nickel

Very careful high temperature equilibrium measurements by Busey and Giauque (ref. 31) lead to $\Delta H_f^\circ = -73.0$ kcal/mole for $NiCl_2(c)$. This value is in agreement with the results of other less precise investigations (refs. 32 and 33). Combination of $\Delta H_f^\circ = -73.0$ kcal/mole with $\Delta H^\circ = -19.7$ kcal/mole for the standard heat of solution (ref. 34) of $NiCl_2(c)$ and ΔH_f° for $Cl^-(aq)$ from (ref. 5) leads to $\Delta H_f^\circ = -12.8$ kcal/mole for $Ni^{+2}(aq)$.

Stout, Archibald, Brodale, and Giauque (ref. 35) have reported $\Delta H^\circ = -1.836$ kcal/mole for

$$\alpha\text{-}NiSO_4\cdot 6H_2O(c) + H_2O(liq) = NiSO_4\cdot 7H_2O(c) \qquad (4)$$

We combine this result with $\Delta H^\circ = 1.15$ kcal/mole for the standard heat of solution (ref. 30) of α-$NiSO_4\cdot 6H_2O(c)$ to obtain $\Delta H^\circ = 2.89$ kcal/mole for the standard heat of solution of $NiSO_4\cdot 7H_2O(c)$ at 298°K. Using solubility data from Seidell (ref. 7) and activity and osmotic coefficients from Robinson and Stokes (ref. 8), we calculate $\Delta G^\circ = 3.09$ kcal/mole for the standard free energy of solution of $NiSO_4\cdot 7H_2O(c)$ at 298°K. These values of ΔG° and ΔH° of solution lead to $\Delta S^\circ = -0.67$ gibbs/mole and then by combination with entropies (refs. 5 and 35) of $NiSO_4\cdot 7H_2O(c)$, $SO_4^{-2}(aq)$ and $H_2O(liq)$, we calculate $\bar{S}_2^\circ = -31.6$ gibbs/mole for $Ni^{+2}(aq)$.

Combination of our ΔH_f° and $\bar{S}_2^\circ$ values for $Ni^{+2}(aq)$ with entropies (refs. 5 and 10) for Ni(c) and $H_2(g)$ gives $\Delta G_f^\circ = -10.5$ kcal/mole for $Ni^{+2}(aq)$. This free energy leads to $E^\circ = 0.22_8$ v for $Ni(c) = Ni^{+2}(aq) + 2\ e^-$. Again, we believe that our total uncertainties are no more than ±0.5 kcal/mole and ±0.01 v. Thus our thermal potential is in excellent agreement with E° = 0.232 v from Carr and Bonilla (ref. 36), but not with the higher value (E° = 0.250 v) adopted by Latimer (ref. 17). Taking the "best" E° values for Co and Ni to be 0.29 and 0.23 v, we see that the difference is significantly greater than the difference derived from the equilibrium investigation cited by Latimer (ref. 17).

ACKNOWLEDGMENT

We are grateful to the National Science Foundation for support of this research, to Paul Cerutti and Harry Garber for help with some of the experimental work, and to Lewis Gedansky for helpful discussions concerning copper.

REFERENCES

1. W. F. O'Hara, C. H. Wu, and L. G. Hepler, J. Chem. Educ., 38, 512 (1961).
2. E. Lange in "The Structure of Electrolyte Solutions," W. J. Hamer, Ed., John Wiley and Sons, Inc., New York, N. Y., 1959.
3. E. Lange and W. Miederer, Z. Electrochem., 60, 34 (1956).
4. L. H. Adami and E. G. King, U. S. Bureau of Mines Report of Investigations NO. 6617, Director, Mines Bureau, Pittsburgh, Pa., (1965).

5. D. D. Wagman, W. H. Evans, V. B. Parker, I. Halow, S. M. Bailey, and R. H. Schumm, Technical Note 270-3, U. S. Natl. Bur. Stds., Washington, D. C., (1968).
6. A. D. Mah, L. B. Pankratz, W. W. Weller, and E. G. King, U. S. Bureau of Mines Report of Investigation No. 7026, Director, Mines Bureau, Pittsburgh, Pa., (1967).
7. A. Seidell, Solubilities of Inorganic and Metal Organic Compounds, 3rd ed., D. Van Nostrand Co., Inc., New York, (1940).
8. R. A. Robinson and R. H. Stokes, Electrolyte Solutions, 2nd ed., Revised, Butterworths, London, (1965).
9. J. W. Stout, personal communication, (1967).
10. K. K. Kelley and E. G. King, U. S. Bureau of Mines Bulletin 592, U. S. Government Printing Office, Washington, D. C., (1961).
11. In spite of considerable discussion in recent years, general agreement is still lacking on "sign conventions" for potentials. Much of the confusion arises because "sign" can be either electrical or algebraic, with the two kinds of signs not necessarily interchangeable. All potentials cited here are oxidation half reaction potentials with algebraic signs. This question has been discussed in more detail elsewhere: R. N. Goldberg and L. G. Hepler, Chem. Rev., 68, 229 (1968).
12. A. J. deBethune, T. S. Licht, and N. Swendeman, J. Electrochem. Soc., 106, 616 (1959).
13. Z Z. Hugus, Jr., J. Am. Chem. Soc., 73, 5459 (1951).
14. A. R. Tourky and M. M. Khairy, J. Chem. Soc., 2626 (1952).
15. W. F. Giauque, R. F. Barieau, and J. E. Kunzler, J. Am. Chem. Soc., 72, 5685 (1950).
16. A. K. Covington, J. V. Dodson, and W. F. K. Wynne-Jones, Trans. Faraday Soc., 61, 2050 (1964).
17. W. M. Latimer, Oxidation Potentials, 2nd ed., Prentice-Hall, Inc., New York, N. Y., (1952).
18. D. N. Lyon and W. F. Giauque, J. Am. Chem. Soc., 71, 1647 (1949).
19. L. H. Adami and K. K. Kelley, U. S. Bureau of Mines Report of Investigation No. 6260, Director, Mines Bureau, Pittsburgh, Pa., (1963).
20. "Selected Values of Chemical Thermodynamic Properties," National Bureau of Standards Circular 500, U. S. Government Printing Office, Washington, D. C., (1952).
21. M. F. Koehler and J. P. Coughlin, J. Phys. Chem., 63, 605 (1959).
22. J. C. M. Li and N. W. Gregory, J. Am. Chem. Soc., 74, 4670 (1952).
23. W. A. Patrick and W. E. Thompson, J. Am. Chem. Soc., 75, 1184 (1953).
24. T. Hurlen, Acta Chem. Scand., 14, 1533 (1960).
25. W. M. Latimer, K. S. Pitzer, and W. V. Smith, J. Am. Chem. Soc., 60, 1829 (1938).
26. R. E. Connick and W. H. McVey, J. Am. Chem. Soc., 73, 1798 (1951).
27. L. B. Magnusson and J. R. Huizenga, J. Am. Chem. Soc., 75, 2242 (1953).
28. D. K. Bewley, Trans. Faraday Soc., 56, 1629 (1960).
29. M. M. Birky and L. G. Hepler, unpublished data.
30. R. N. Goldberg, R. G. Riddell, M. R. Wingard, H. P. Hopkins, C. A. Wulff, and L. G. Hepler, J. Phys. Chem., 70, 706 (1966).
31. R. H. Busey and W. F. Giauque, J. Am. Chem. Soc., 75, 1791 (1953).
32. K. Sano, F. Ishikawa Anniversary Vol., Science Reports Tohuko Univ., 37, No. 1, 1 (1953).
33. S. N. Flengas and T. R. Ingraham, Can. J. Chem., 35, 1254 (1957).
34. C. N. Muldrow, J. S. Sweet, and L. G. Hepler, unpublished data.
35. J. W. Stout, R. C. Archibald, G. E. Brodale, and W. F. Giauque, J. Chem. Phys., 44, 405 (1966).
36. D. S. Carr and C. F. Bonilla, J. Electrochem. Soc., 99, 425 (1952).

ENTHALPIMETRIC ANALYSIS

by
Joseph Jordan and Peter W. Carr
Department of Chemistry
The Pennsylvania State University
University Park, Pennsylvania 16802

There are two principal calorimetric methods which may be conveniently applied to problems of analytical chemistry in solution. Thermometric Enthalpy Titration (TET) is in essence an end point determinative technique,[1] whereas Direct Injection Enthalpimetry (DIE) is the complementary concentration determinative procedure.[2]

These two techniques have become feasible by the advent of the thermistor, a semiconductor material with a large negative temperature coefficient of resistance. The thermistor allows for virtually instantaneous measurement of very small changes in temperature (±0.001°C).

Representative THERMOMETRIC ENTHALPY TITRATION CURVES ("ENTHALPOGRAMS") are illustrated in Fig. 1. Experimentally TET is implemented by placing the sample in an insulated container (e.g. a dewar flask). The solution is stirred and allowed to come to thermal equilibrium, as indicated by the line $\overline{AB}$ in Fig. 1a. A standardized titrant is then added via a constant rate buret. Subsequently a temperature-time curve corresponding to region $\overline{BC}$ of the figure is recorded as the off-balance potential of a Wheatstone bridge. The end point is taken as the intersection of lines $\overline{BC}$ and $\overline{CD}$: The corresponding time on the abscissa is used as the measure of the total number of moles of sample (="the titrate") present.

DIRECT INJECTION ENTHALPIMETRY is a quasi-instantaneous thermometric titration, performed by injecting an excess of unstandardized reagent. The signal obtained is shown in Fig. 2. The amplitude of the "temperature jump" represents a direct measure of concentration.

Both TET and DIE are ideally suited to the needs of organic functional group analysis for the following reasons.

1. The monitored variable (temperature change engendered by the heat of reaction) is perhaps the most general property of chemical reactions.

2. The completeness of the "titration reaction" at the equivalence point is far less critical than, for instance, in potentiometric titrations.[3] This is inherent in the fact that TET is a "linear titration" method, i.e. the measured variable (T) is a linear - rather than logarithmic - function of concentration.

3. Intrinsic compensation of the integral heat evolved and heat capacity (both of which are proportional to sample volume) allow for easy adaptation to microanalysis.

Several general limitations are imposed by TET. The heat of reaction must be large enough to produce an accurately measurable temperature change (0.01°C) in as short a time as possible, at most three minutes. The heat of dilution of the titrant must be small as compared to the heat of reaction. The titration curve should be controlled solely by equilibria and not by the rates of any slow chemical reaction. Virtually correct stoichiometric end points can be obtained even where equilibria are appreciably incomplete at the equivalence point. This is illustrated in Fig. 1b. Kinetic complications are not as readily corrected and simple linear extrapolation will fail as shown in Fig. 1c. The first two of the above limitations apply to DIE as well, but the kinetic limitation is not as severe since a large excess of reagent may be added to increase the overall reaction rate.

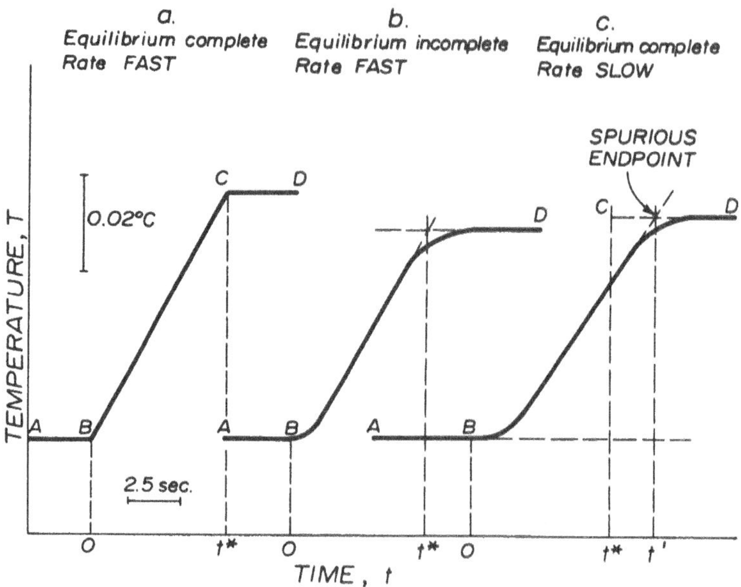

FIG. I Idealized TET Enthalpograms. a. $HCl + NaOH$; b. $H_3BO_3 + NaOH$; c. $R\text{-}NO_2 + NaOH$. $\overline{AB}$ - base line; B - start of titration; C - theoretical equivalence point; $\overline{CD}$ - excess reagent line.

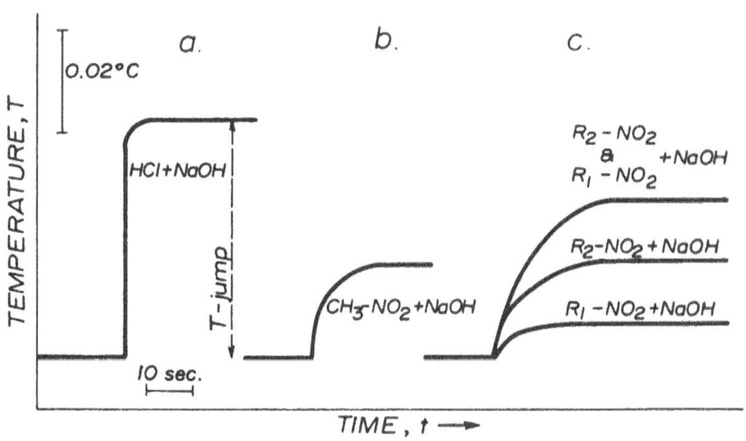

FIG. 2 Idealized DIE Enthalpogram.

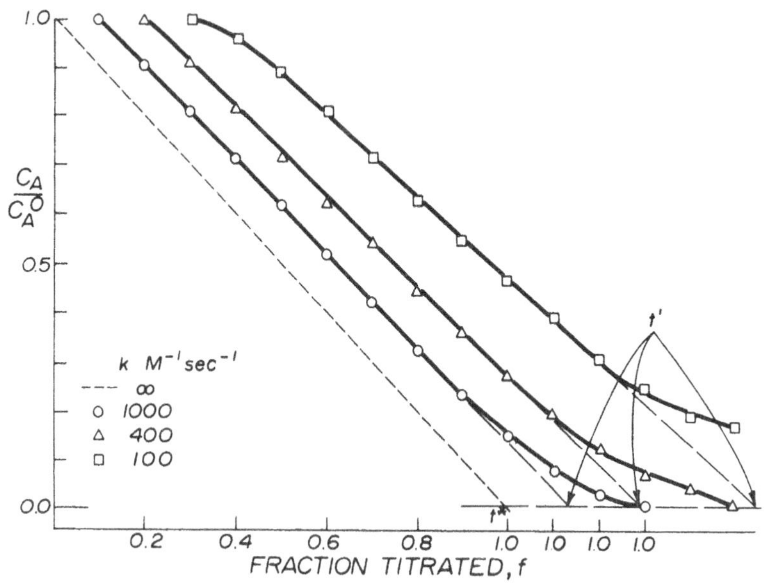

FIG. 3 Normalized Titration Curves Computed from Equation 6 for $C_A^0 = 10^{-3}$M, $\rho = 10^{-5}$M/sec, $t^* = 100$sec.

At best TET is capable of an accuracy and precision of 0.2-0.5% whereas DIE is best utilized when a precision of 3-5% is acceptable at millimolar concentration levels. This difference in attainable accuracy is normal for an end point determinative method (TET) versus a direct measurement (DIE). The precision and accuracy of the two methods depends upon both heat of reaction and concentration. Quite generally the larger the heat of reaction or concentration the better the precision and accuracy. At the lowest concentration levels amenable to measurement by either of these techniques, DIE is usually more precise than TET, due to the imperfect adiabaticity of the reaction vessel.

A particularly interesting application of DIE lies in the area of differential kinetic analysis.[4] In this approach one adds a known excess of reagent to a mixture of two compounds with the same functional groups. The experimental enthalpogram (Fig. 2c, top curve) may be resolved by any of the standard mathematical procedures used in kinetic analysis.[5] Table I gives a listing of some of the applications of thermochemical methods to functional group analysis.

Since most reaction heats are large enough to allow for precise titrations, the major limitation of TET in the analysis of functional groups is the absence of titrants which react rapidly. This rather severe limitation of the methodology is related to the rate of the titration reaction:

$$A + R \xrightarrow{k_A} \text{Products} \tag{1}$$

where $\underline{A}$ and $\underline{R}$ denote the "titrate" and the titrant respectively. The relevant kinetics are described by the simultaneous differential equations (2) and (3) and the applicable initial conditions (4).

$$\frac{dC_A}{dt} = -k_A C_A C_R \tag{2}$$

$$\frac{dC_R}{dt} = \rho - k_A C_A C_R \tag{3}$$

$$C_A(0) = C_A^\circ; \quad C_R(0) = 0 \tag{4}$$

where ρ denotes the rate of titrant addition expressed in appropriate units [e.g. moles/lsec if the concentrations ($\underline{C}$) are given in moles/l]. Solving (by standard mathematical procedures) for

$$\overline{C_A} \equiv \frac{C_A}{C_A^\circ} \tag{5}$$

one obtains

$$\overline{C_A} = \frac{C_A}{C_A^\circ} = \frac{\exp[k_A t(C_A^\circ - \frac{\rho t}{2})]}{1 + C_A^\circ k_A \int_0^t \exp[k_A t(C_A^\circ - \frac{\rho t}{2})]dt} \tag{6}$$

A family of "normalized titration curves" (plots of $\overline{C_A}$ versus fraction titrated, $\underline{f}$) was computed from eq. (6) for various assignments of k_A and is shown in Fig. 3. The "genuine equivalence point time" ($t^* = C_A^\circ/\rho$, see Fig. 1) was arbitrarily fixed at 100 sec. in all instances. The zero ordinate intercepts in Fig. 3 correspond to the type of "spurious extrapolated end point times" ($\underline{t'}$) exemplified in Fig. 1c. Titration errors calculated on this basis are listed in Table II. It is apparent that even when rates are quite high ($1000M^{-1}sec^{-1}$)serious analytical errors will occur. We have developed a method of obtaining useful analytical results in situations where the titration curve is significantly distorted by kinetics. The initial concentration $\underline{C_A^\circ}$ (="the unknown") which is proportional to $\underline{t^*}$ can be

TABLE I

FUNCTIONAL GROUP ANALYSIS BY THERMOCHEMICAL METHODS

FUNCTIONAL GROUP	REAGENT	REACTIONS	PRECISION AND ACCURACY ATTAINABLE	METHOD
-COOH	Strong Base	Acid-base Neutralization	1%	TET
$-COO^-$ or $-NH_2$	Strong Acid	Acid-base Neutralization	1%	TET
-C=O or -CHO	$NH_2OH \cdot HCl$	Oximation	0.6%	TET
-CHO	$H_2SO_4 - Na_2SO_3$	Bisulfite Addition	0.6%	TET
$Ar-NH_2$	$NaNO_2$	Diazotization	0.2%	TET
$Ar-SO_2NH_2$	NaOCl	Oxidation	3-5%	TET
Ar-OH	$Br_2 + HCl$	Bromination	3-5%	DIE
R_1-NO_2 plus R_2-NO_2	Strong Base	Proton Abstraction	7%	Thermokinetic Analysis

TABLE II

TITRATION ERRORS

k_A $(M^{-1}sec^{-1})$	% Error $(\frac{t'-t^*}{t^*} \cdot 100)$
∞	0
1000	4
400	8
100	28

evaluated from $C_A(t)$ and t, provided both k_A and ρ are known from independent studies. The mathematical operation involves taking the anti-function of eq. (6), which is quite complicated. To solve this problem we have developed simpler graphical procedures with the aid of a high speed digital computer. The fundamental equation was simplified by noting that

$$f \equiv t/t^* \tag{7}$$

and introducing the dimensionless parameter

$$\beta \equiv k_A C_A^{\circ 2}/\rho \tag{8}$$

Substitution into eq. (6) and rearrangement yielded

$$\overline{C_A} = \frac{\exp[\beta/2]\ \exp[-\beta/2(f-1)^2]}{1 + \sqrt{\pi\beta/2}\ \exp[\beta/2](\mathrm{erf}\sqrt{\beta/2}(f-1) + \mathrm{erf}\sqrt{\beta/2})} \tag{9}$$

Kinetic titration errors (Table II) were corrected for and valid equivalence point assignments for t^* were obtained as follows. First a tentative assignment of β was made from a preliminary value of either C_A° or t^*; then f was calculated from an appropriate graphical plot based on eq. (9) which in turn yielded a new value of t^* hence β; iteration was continued until t^* no longer varied; finally a series of t^* assignments were obtained from an entire titration curve and averaged.

The utilization of thermochemical titrations in the area of functional group analysis is almost entirely contingent on the availability of suitable titrants. The main difficulty which has hitherto handicapped their use was sluggish reactivity. The treatment described in this paper has minimized this impediment. We hope that its application will now make the field of organic functional group analysis accessible to automation, which is long overdue in this day and age.

Acknowledgments

The investigation described in this paper was supported in part by the United States Atomic Energy Commission under Contract AT(30-1)-2133 with the Pennsylvania State University (Report No. NYO-2133-48). The junior author (P.W.C.) holds a National Defense Education Act (Title IV) Graduate Fellowship.

References Cited

1. For a review see e.g.: J. Jordan and G. J. Ewing, "Thermometric Titrations" in Handbook of Analytical Chemistry, L. Meites, Ed., McGraw-Hill, New York, 1963, Sect. 8, pp. 3-7.

2. J. C. Wasilewski, P. T-S Pei and J. Jordan, Anal. Chem., 36, 2131 (1964).

3. J. Jordan, Chimia, 17, 101 (1963).

4. R. A. Henry, Thesis, The Pennsylvania State University, 1967.

5. S. Siggia, Quantitative Organic Analysis via Functional Groups, John Wiley and Sons, Inc., New York, 1963, pp. 655-682.

A SIMPLE GENERALIZED THEORY FOR THE ANALYSIS OF DYNAMIC THERMAL MEASUREMENT

Allan P. Gray
Perkin-Elmer Corporation
Norwalk, Conn.

Virtually all of the theoretical discussions of differential thermal analysis which have appeared in the literature begin with two simple relationships, one arising from the conservation of energy and the other from the linear dependence of heat flow on temperature differential known as Newton's Law. Usually these equations have been stated and further developed for the purpose of describing the performance of some particular apparatus or to provide a basis for the analysis of data obtained in the investigation of some particular class of problem. As a result, mathematical complexities which may arise in the analysis of a particular apparatus or problem are incorporated in the theoretical development so that the conclusions lose their generality and are often of little value to the thermal analyst employing other equipment or investigating different areas of application. It is the object of this paper to show that the important performance properties of scanning thermal analysis systems, including DTA, differential scanning calorimetry (DSC) and thermogravimetric analysis (TGA), can be derived from the simplest possible forms of the energy conservation and heat transfer equations. The results can be interpreted as representing the performance of an ideal scanning thermal device and therefore provide useful criteria for evaluating instrumentation and data quality. Further, since several modern apparatus designs actually approach "ideal" behavior quite closely, the development is not entirely an academic exercise; the equations lead to graphical methods for correcting thermal analysis data for instrumental effects which are simple and quite useful in practice.

It is not claimed that there is anything fundamentally novel in the treatment to follow as the initial equations and assumptions are essentially the same as those employed by Vold (1), Boersma (2), Borchardt and Daniels (3), Reed, Weber and Gottfried (4), and many others. It does appear, however, that the implications of these equations have not been fully exploited or developed in a way which clearly brings out the important instrumental design factors which influence the quality of thermal analysis data. Also, because of the variety of cell designs which are currently used in the practice of thermal analysis, some simple theoretical basis for understanding their various performance characteristics is considered to be necessary.

The essential parts of a thermal analysis cell are shown schematically in Figure 1. They consist of the sample and its container at a temperature, T_s; a source of thermal energy, at temperature T_p, and a path having a certain thermal resistance, R, through which the thermal energy flows to or from the sample at a rate, dq/dt. We assume that the sample temperature, T_s, is uniform and equal to that of the container; that the total heat capacity of the sample plus container, C_s, and the controlling thermal resistance, R, and constant over the range of interest and adopt the convention that heat generated by the sample per unit time, dh/dt, is positive and heat absorbed, negative. Heat flow from the thermal energy source to the sample, dq/dt, is taken as positive. Although we will be considering both exothermic and endothermic effects, we will develop the equations initially for the case where the sample is generating energy due to an exothermic reaction.

At any instant of time the sample is generating energy at a rate, dh/dt. Energy generated by the sample can either increase the sample temperature or be lost to the surroundings. As energy must be conserved, the sume of these two effects must equal dh/dt. Therefore,

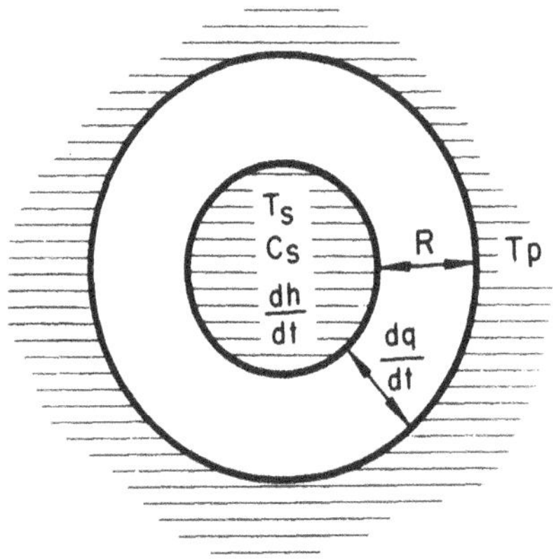

Fig. 1. Thermal analysis cell schematic.

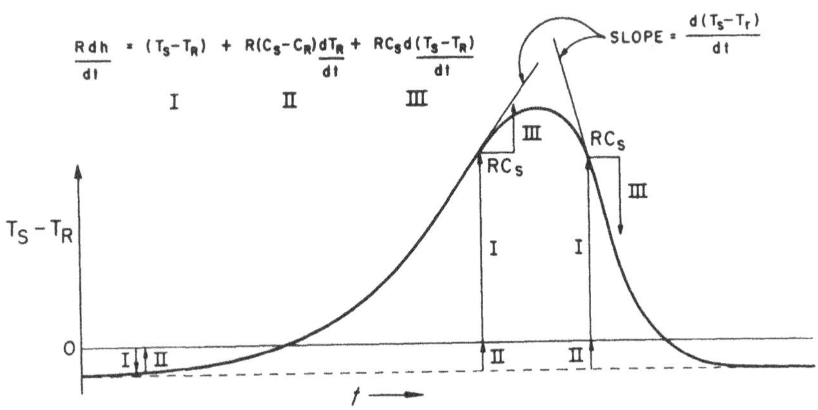

Fig. 2. Graphical determination of Rdh/dt from experimental DTA curve.

$$\frac{dh}{dt} = C_s \frac{dT_s}{dt} - \frac{dq}{dt} \tag{1}$$

The rate of heat loss to the surroundings is controlled by the thermal resistance and the temperature difference between sample and surroundings in the same way as electrical current through a resistance is controlled by the potential difference across the resistance. The thermal analog of Ohm's Law, which is referred to as Newton's Law, is simply

$$\frac{dq}{dt} = \frac{T_p - T_s}{R} \tag{2}$$

Now we can substitute for dq/dt in equation (1) to obtain

$$\frac{dh}{dt} = C_s \frac{dT_s}{dt} + \frac{T_s - T_p}{R} \tag{3}$$

Differential Thermal Analysis

A differential thermal analysis apparatus is one which measures a difference in temperature between a sample cell and a thermally inert reference cell and records this difference versus time or temperature. By writing an equation equivalent to (3) for the reference cell, where dh/dt = 0, subtracting this equation from (3) and rearranging, we obtain an expression for the instantaneous rate of heat generation by the sample which is the unknown quantity we are trying to determine.

$$\frac{Rdh}{dt} = (T_s - T_r) + R\,(C_s - C_r)\frac{dT_r}{dt} + RC_s \frac{d\,(T_s - T_r)}{dt} \tag{4}$$

Here we have assumed that R for the reference cell is the same as for the sample cell since this is reasonable, and the effect of inequality makes no significant difference to the analysis in any case. C_r, however, will not in general be equal to C_s, and the inequality is therefore explicitly considered. Note that $dT_r/dt = dT_p/dt$, the scanning rate, and is therefore a constant.

We see from equation (4) that, at any time, Rdh/dt can be considered as the sum of three terms in units of temperature. These we will consider as representing certain distances, I, II and III respectively on a DTA recording. The distance I = $T_s - T_r$ is always the differential temperature continuously recorded by the instrument. In the steady state after the instrument has reached scanning equilibrium but while dh/dt is effectively zero, the third term is zero and the recorded signal, I = $T_s - T_r$, is equal to but opposite in sign to the constant second term, so that I = - II. The distance, II, therefore represents the baseline displacement from the zero signal level. The third term is the slope of the curve at any point of interest multiplied by a constant, RC_s. This constant has the units of time; it is in fact the thermal time constant of the system and has a most important influence on its performance. If we assume for the moment that we know the value of RC_s, we can determine the magnitude of the third term by a simple graphical construction. A tangent is drawn to the curve at any point of interest. From the point a horizontal distance, RC_s, is laid off in the direction of increasing time and a vertical line drawn through the terminus. As can be seen in the example of Figure 2, this vertical line will cut the tangent at some point; and we have constructed a right triangle whose hypotenuse is the tangent of slope, $d(T_s - T_r)/dt$ and whose base is RC_s. The height of the triangle is then $RC_s d(T_s - T_r)/dt$ which is the distance III.

At any point on the curve, Rdh/dt is equal to the sum I + II + III, or if the tangent has a negative slope, to I + II - III. Thus, knowing RC_s, we can graphically construct a curve which directly reflects the instantaneous thermal behavior of the sample. We can then interpret this corrected curve very simply in kinetic terms. The displacement from the baseline at any time (or temperature) is proportional to the reaction rate at that time, and the fraction of the total area swept out up to that time will be the fractional amount of the sample which has reacted. Note that if R is known and constant, dh/dt and the total area can be obtained in calorimetric units.

It is obviously advantageous in any instrumental design to make RC_s as small as possible and ideally so small that the distance, III, is negligible compared to I + II. In such a case, no allowance for instrumental thermal lag, which this correction represents, would be necessary. The smaller and more constant RC_s, the more accurately will the instrument record the instantaneous thermal behavior of the sample. The degree to which RC_s is small and constant can be considered a measure of accuracy, speed of response, or resolution, whichever term is appropriate to the problem under consideration.

It should be noted from equation (4) that the recorded signal, $T_s - T_r$, for a given sample size, is proportional to R. In fact, in a DTA apparatus if R is zero, there is no signal at all. The requirement of high inherent sensitivity, which requires a large R, is therefore incompatible with the requirement for fast response or high resolution which demands that RC_s be small. In practice, however, sensitivity loss due to a small R can often be receovered by electronic signal amplification. The degree to which this can be done successfully depends upon the quality of the overall system design, particularly that of the programmer and differential temperature amplifier. For this reason the quality of a DTA apparatus with respect to sensitivity cannot be judged without a specification of the instrumental noise level.

Up to this point we have considered thermal behavior in the sample which is both temperature and time dependent. Another important class of thermal phenomena consists of those which can be considered temperature dependent only. For example, the melting of an absolutely pure material takes place at constant temperature, the melting point; and the temperature will not increase further until the melting is completed. In this case the endothermic energy required to melt the sample must be provided from the source of thermal energy and dh/dt in equation (4), which will now represent the rate of melting, is limited by the rate at which this energy can be supplied through the thermal resistance.

As we scan the sample up to and through its melting point, the instrumental readout can be divided into three successive stages; before, during, and after the melting. Before the melting point, just as discussed above, $T_s - T_r$ equals the second term of equation (4) with reversed sign so that,

$$T_s = T_r - R(C_s - C_r)\frac{dT_r}{dt} \tag{5}$$

At the instant when T_s becomes equal to T_m, $T_s = T_m$ and $dT_s/dt = 0$ until the melting is completed so that during the melting equation (4) becomes,

$$\frac{Rdh}{dt} = T_m - T_r - RC_r\frac{dT_r}{dt} \tag{6}$$

Since T_r is a linear function of time we can write for $T_s - T_r$ during the melt,

$$T_s - T_r = T_m - T_r = T_m - (T_r^\circ + \frac{dT_r}{dt} t) \tag{7}$$

where T_r° is the reference temperature at the instant that melting begins. This is obtained from equation (5) by setting $T_s = T_m$. Substituing in (7),

$$T_m - T_r = -R(C_s - C_r)\frac{dT_r}{dt} - \frac{dT_r}{dt}t \quad (8)$$

Thus during the melting the differential temperature versus time is a straight line of slope $-dT_r/dt$, the negative of the scanning rate. This straight line will continue until the sample is completely melted; that is, until the integral of dh/dt equals Δh, the total heat of fusion of the sample. Substituting for $T_m - T_r$ from equation (8) into equation (6) we find

$$\frac{Rdh}{dt} = -\left(RC_s\frac{dT_r}{dt} + \frac{dT_r}{dt}t\right) \quad (9)$$

The time to the maximum of the peak can be obtained from equation (9) by integrating with respect to t from 0 to t_{max}, setting the integral of dh/dt equal to Δh and solving for t_{max}. The result is,

$$t_{max} = RC_s\left[\left(1 - \frac{2\Delta h}{RC_s^2\frac{dT_r}{dt}}\right)^{1/2} - 1\right] \quad (10)$$

Note that by our convention the sign of an endothermic Δh will be negative so that the term inside the square root bracket will be positive. Similarly one can obtain the result that the peak height, ΔT_{max}, <u>measured from the baseline</u> is

$$\Delta T_{max} = -t_{max}\frac{dT_r}{dt} \quad (11)$$

At the instant that melting is completed, at the peak maximum, dh/dt once again becomes zero. To determine what happens next, we solve equation (4) with dh/dt = 0 and the initial condition that $T_s - T_r = \Delta T_{max}$. It is then found that ΔT measured from the baseline is

$$\Delta T = \Delta T_{max}\, e^{-t/RC_s} \quad (12)$$

This equation shows that the signal decays exponentially from the peak to the baseline with the time constant, RC_s. The appearance of a DTA peak for a sharp transition is shown in Figure 3. Note the rectangular area "C" in the figure. This is the contribution of the first term of equation (9) to the integral and it represents part of the fusion energy provided to the sample during the melting. In this interval the sample temperature does not change so that the heat capacity energy provided before the melting is no longer required. The melting is complete at the peak maximum, and the sample must now catch up to the program temperature. The heat capacity energy must now be provided, and it appears in the exponential "catch-up" tail. The area "B" equals "C" so that the total peak area correctly represents the heat of fusion even though melting is actually complete at the peak.

Other important conclusions to be drawn from the equations above include the fact that the total area under the peak, including the exponential tail, in units of degree-seconds, is $R\Delta h$ and that two closely spaced transitions of the same magnitude will be resolved with at least a 50% valley between them if they are at least $(0.693RC_s + 1/2\, t_{max})\frac{dT_r}{dt}$ degrees apart. For a DTA apparatus R is most conveniently determined by simply measuring the area of a peak in degree seconds and dividing by the transition energy of the sample. C_s can often be calculated with sufficient accuracy, especially when its major component is the heat capacity of the container. The specific heats of metals and ceramics commonly used as container materials can be found in standard tables. If the sample heat capacity is a small fraction of that of

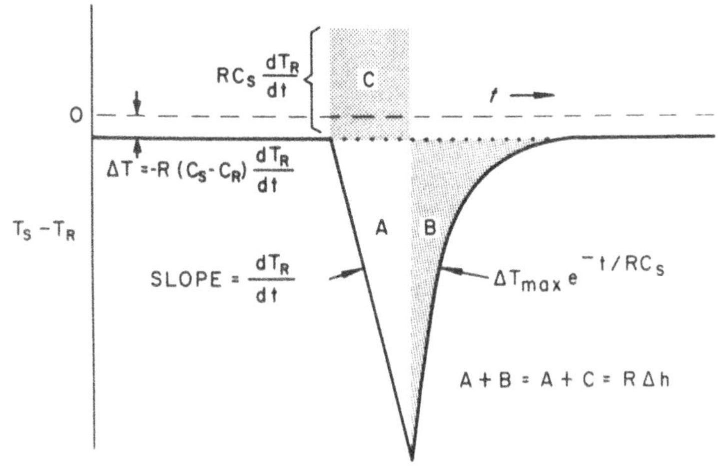

Fig. 3. DTA curve for a sharp transition.

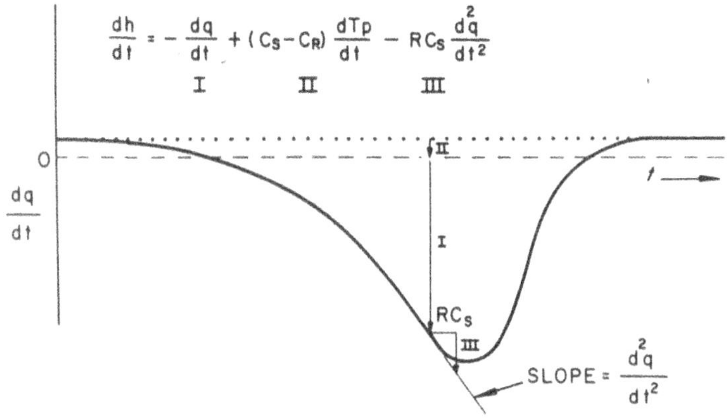

Fig. 4. Graphical determination of dh/dt from experimental DSC curve.

the container, its contribution to the total can be neglected or roughly approximated.

The relationships given above permit the calculation of peak height, peak width, peak area and resolving power as a function of sample size, transition energy, scanning rate and the instrumental parameters which have been discussed. Of course, these calculations are based upon the validity of the assumptions which were made initially. However, if these assumptions do not apply to the particular apparatus under consideration, it is easy to see in what respect the data will be of poor quality. For example, in an instrument where the sample itself makes a major contribution to the thermal resistance, R, so that R is not constant during the transition, the data will not be quantitative in the sense that the peak area will not be simply proportional to the transition energy. A DTA apparatus designed specifically for quantitative measurements is made to minimize variations in R from sample to sample. Although this can be achieved to a high degree in a given narrow temperature interval, R is invariably temperature dependent and calibration is required at numerous points over the range.

Differential Scanning Calorimetry

A Differential Scanning Calorimeter* is a device consisting of sample and reference holders which are insulated from each other but always maintained at the same temperature by closed loop control of the electrical power provided to heaters in each of the holders. Since, to maintain this condition of temperature equality, the energy per unit time which the sample absorbs from or provides to the sample holder must be exactly compensated by a corresponding variation in the differential electrical power provided to the heaters, measurement of this electric power is equivalent to measuring dq/dt as defined above. The theory and operating principle of such an instrument have been given by Watson, O'Neill, Justin and Brenner (5), and by O'Neill (6). The latter author discussed the response of a DSC apparatus to a sharp transition in some detail, including the case, "the sample limited case," where the thermal resistance of the sample itself is significant compared to the resistance between the sample pan and the sample holder. In actual practice, sample size and thickness are so small for sharp transitions that the thermal resistance of the sample is entirely negligible so that it is the "instrument limited case" which is further elaborated in the following. We therefore take the controlling thermal resistance, R, to be that between the sample container and the sample holder.

From equation (1) and the first time derivative of (2) we can obtain a basic equation relating dh/dt to the measured quantities just as was done for DTA in equation (4). We find that for DSC,

$$\frac{dh}{dt} = -\frac{dq}{dt} + (C_s - C_r)\frac{dT_p}{dt} - RC_s\frac{d^2q}{dt^2} \tag{13}$$

Here again dh/dt involves the sum of three terms, the first being the signal measured from the zero signal line, the second being the baseline displacement due to heat capacity mismatch between sample and reference, and the third being the slope of the recorded curve multiplied by the constant, RC_s. A DSC curve can be corrected for thermal lag represented by the third term by means of the graphical construction described previously and illustrated in Figure 4. Although a DSC record is analagous in many respects to those obtained in DTA, there are two important differences which should be noted. The first arises from the fact that the thermal resistance, R, occurs only in the third term of equation (13). It is therefore possible to reduce the thermal time

*The principle of Differential Scanning Calorimetry described herein and its embodiment in thermoanalytical instrumentation is proprietary to the Perkin-Elmer Corporation under U.S. Patent No. 3263484.

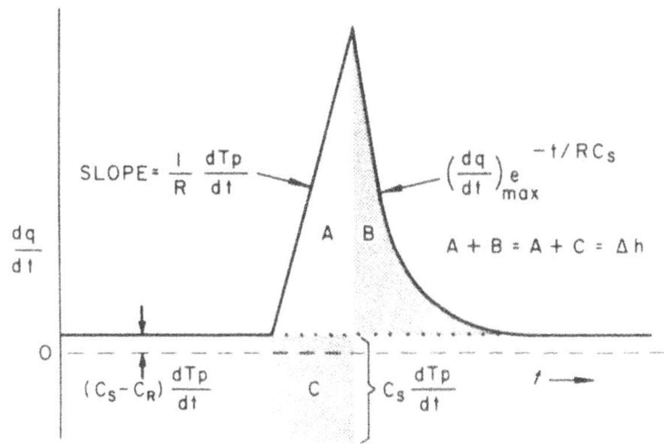

Fig. 5. DSC curve for a sharp transition.

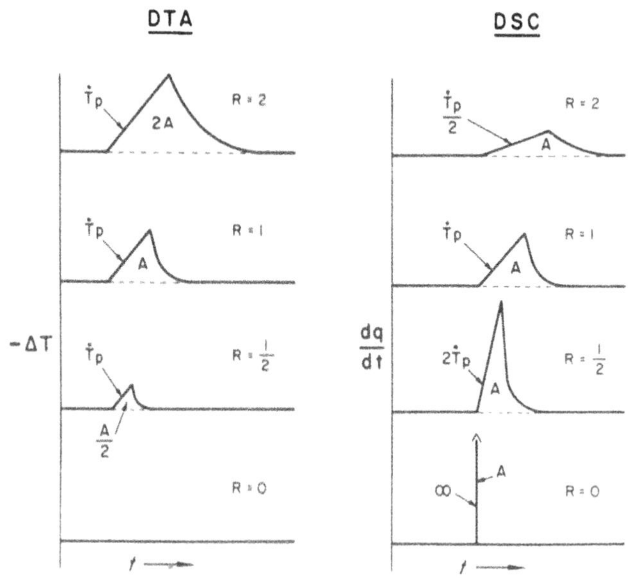

Fig. 6. Effect of thermal resistance on peak shape and area.

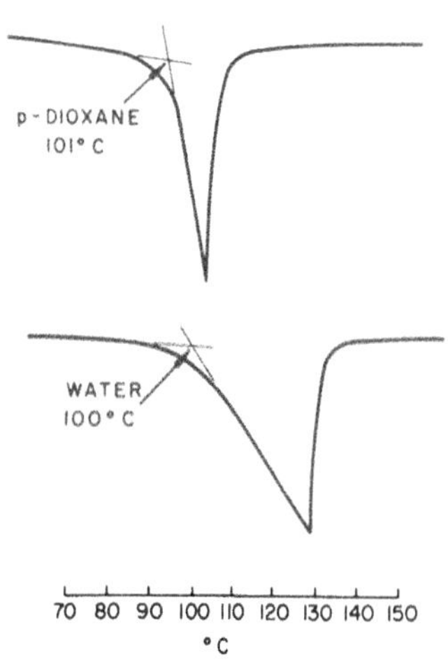

Fig. 7. Temperature calibration with boiling points.

constant, RC_s, by reducing R without sacrificing sensitivity--in fact, sensitivity increases as the thermal resistance decreases. The second major difference is that the area under a DSC peak directly equals $\Delta q = -\Delta h$. Insofar as calorimetric measurements are concerned, the value of R need not be known nor need it be constant. The calibration of a DSC is therefore not dependent on the temperature, and a single point calibration applies over the whole range. The "calibration constant" required to convert area to calories in DSC is a constant, electrical conversion factor, whereas in quantitative DTA it is a thermal "constant" which in practice varies with the temperature.

The fundamental reason for these differences is that a DTA apparatus is basically a temperature measuring device and energy information is obtained through secondary relationships. A DSC apparatus, on the other hand, is basically an energy measuring device and temperature information is obtained through secondary relationships, In DSC the sample holder temperature is monitored and maintained equal to the program temperature, T_p; the sample temperature, T_s, is not measured directly but is computed from equation (2). Whether or not T_s differs significantly from T_p depends upon the magnitude of R. Thermal lag is minimized and temperature accuracy and sensitivity maximized by reducing R to the smallest possible value. This is the principal design objective for a differential scanning calorimeter.

The equations which describe the shape of a DSC peak during a sharp time independent transition are obtained in the same way as was described for DTA. The analog of equation (8) for the straight line recorded during the melting is

$$\frac{dq}{dt} = (C_s - C_r)\frac{dT_p}{dt} + \frac{1}{R}\frac{dT_p}{dt}t \qquad (14)$$

In DSC the slope of this line is seen to be the scanning rate divided by the thermal resistance. This provides a convenient means of determining R for a DSC apparatus. The smaller R, the steeper the line and hence the sharper the peak. Equation (10) giving the time to the peak maximum is the same for DSC, and the analogs of equations (11) and (12) which give the peak height and the equation of the peak "tail" respectively are:

$$\frac{dq}{dt}_{max} = (t_{max})\frac{1}{R}\frac{dt_p}{dt} \qquad (15)$$

and

$$\frac{dq}{dt} = \frac{dq}{dt}_{max} e^{-t/RC_s} \qquad (16)$$

The appearance and characteristics of a DSC peak for a sharp transition are shown in Figure 4. A distinctive difference between DSC and DTA is illustrated in Figure 5 which shows the effect of a change in R on the peak area and shape of a sharp transition. The DSC area is invariant with R and resolution improves as R decreases. In DTA, resolution also improves as R decreases, but the peak area decreases proportionately and goes to zero as R goes to zero.

Thermogravimetric Analysis

In TGA we are not measuring a thermal property of the sample, but this does not mean that the thermal resistances and capacitances of the system do not influence the quality of the data. The thermal resistance between the sample and the furnace determines the rate at which heat can be provided to the sample, which for an endothermic reaction determines the rate of the reaction, which in turn determines the rate of weight loss. In an exothermic reaction the thermal resistance determines how rapidly the heat generated by the reaction is dissipated to the surroundings and therefore the degree of control of the sample temperature. Since R in a TGA apparatus is often several orders of magnitude greater than that in any DTA or DSC system, such effects can be extremely important.

Since in TGA we are instantaneously recording the weight of the sample versus time or temperature, the rate of weight loss is directly proportional to dh/dt. In fact,

$$\frac{dm}{dt} = \frac{\frac{dh}{dt}}{\frac{dh}{dm}} = \frac{1}{\Delta H}\frac{dh}{dt} \tag{17}$$

Now we can substitute for dh/dt in equation (3) and obtain a relationship between the rate of weight loss, the sample and furnace temperatures, the instrumental R and C_s, and the specific reaction energy, ΔH, of the reaction under consideration.

$$\frac{dm}{dt} = \frac{C_s}{\Delta H}\frac{dT_s}{dt} + \frac{T_s - T_p}{R\Delta H} \tag{18}$$

This expression can be handled in the same way as those dealt with previously. An example of the kind of important conclusion that can be drawn from this relationship is the fact that if a TGA apparatus is operated in the derivative mode and used to examine endothermic transitions, such as boiling points, during the vaporization one should obtain a straight line just as in DTA and DSC. The slope of this straight line when dm/dt is plotted against program temperature is $1/R\Delta H$. Thus, the resolution of a TGA measurement depends not only upon R but upon the heat of the particular reaction being studied. This prediction is verified in practice as shown in Figure 6 where derivative recordings of the boiling of water and of p-dioxane are compared. Measurement of the slope of such a curve for a material of known heat of vaporization provides a means of evaluating the effective R for a TGA apparatus. Knowing R and C_s, it is then possible to calculate most of the important performance characteristics of the instrument.

To summarize, the three principal thermoanalytical techniques, DTA, DSC, and TGA have been shown to have much in common and to be interrelated through the simple equations (1) and (2). A graphical method for correcting DTA and DSC reaction curves for instrumental thermal lag has been given, and the factors which control the performance of thermoanalytical instrumentation as well as means for their evaluation have been described.

References

(1) Vold, M. J., Anal. Chem., 21, 683 (1949).
(2) Boersma, S. L., J. Am. Ceramic Soc. 38, 281 (1955).
(3) Borschardt, H. J., and Daniels, F. J., J. Am. Chem. Soc. 79, 41 (1957).
(4) Reed, R. L., Weber, L., and Gottfried, B. S., Ind. & Eng. Chem., Fundamentals 4, 38 (1965).
(5) Watson, E.S., O'Neill, M. J., Justin, J., and Brenner, N., Anal. Chem. 36, 1233 (1964).
(6) O'Neill, M. J., Anal. Chem. 36, 1238 (1964).

THERMAL ANALYSIS OF MESOPHASES

Reinhard D. Ennulat
U. S. Army Electronics Command
Night Vision Laboratory, Fort Belvoir, Virginia

Introduction

Liquid crystals and mesomorphic materials are generally described as states of matter, which exhibit spontaneous anisotropic features usually associated with solid crystals as well as flow properties commonly related to ordinary liquids. Depending on the particular compound such states can be caused by fusion or solvation. In the former case, the various thermotropic phases are obtained in reproducible temperature ranges, while in the latter, the lyotropic phases are realised in certain concentration intervals. This distinction is not always sharp, since certain lyotropic phases show a definite thermotropic behavior (1).

The scheme for the classification of the phases occuring in these spontaneously ordered materials is based mainly on optical evidence. For each type of phase the characteristic anisotropic appearances, the so-called textures, can be assigned as described by Gray (2) for the thermotropic and by Rosevear (3) for the lyotropic case. From this optical evidence and other physical information, in part x-ray diffraction data, structural building principles are derived for the molecular arrangement of the various phases. Thus the existence of certain characteristic textures not only classifies the type of the particular phase, but also specifies its molecular building principle. This empirical approach is due to the lack of a deeper theoretical foundation.

Mabis (4) recently found a lyotropic phase, for which the optically derived molecular arrangement differs from the actual structure as determined by x-ray diffraction. A certain lyotropic phase of the smectic kind did not have the anticipated molecular layer arrangement, but exhibited a three dimensional lattice, which was not even of the layer type. The discrepancy can be resolved, if one distinguishes between liquid crystalline and mesomorphic states and excludes the former from the optical phase identification scheme. This requirement was recognized by Friedel (5) long before the discovery of liquid crystals. According to Friedel liquid crystals are defined as three dimensional lattice structures differing from solid crystals only in other quantitative features, which are not important in this context. On the other hand, mesomorphic states are characterized by a different type of molecular arrangement, which at best can be described as extremely degenerated lattices of less than three dimensions.

To date no liquid crystalline states were found in purely thermotropic phases. In addition, Friedel's phase classification was derived from empirical evidence predominantly obtained from such phases. Therefore, it is reasonable to adopt as a working hypothesis that the phase classification scheme applies to all thermotropic mesophases. The feedback obtained from new experimental evidence may either lead to a refinement of this classification or may force us to abandon it. Obviously the same hypothesis is not applicable for the classification of lyotropic phases. However, this deficiency is made up by the more readily available x-ray diffraction data, which permits a finer distinction between the various lyotropic phases.

The following will be exclusively devoted to thermotropic mesophases. The important aspect of the contribution of temperature scanning thermal analysis to the phase identification of thermotropic mesophases will be discussed. In particular, the detection of mesomorphic phases, the determination of transition temperatures, and the appraisal of the impurity content of the sample will be investigated. These subjects have been discussed previously by the author in a more general context (6). It is the purpose of this paper to extend the thermodynamic considerations and to present some pertinent experimental examples.

Phases and Textures

Smectic, nematic and cholesteric are the three types of thermotropic mesophases. In the smectic phase the molecules are arranged in parallel layers with the molecular axis approximately parallel to the layer normal.

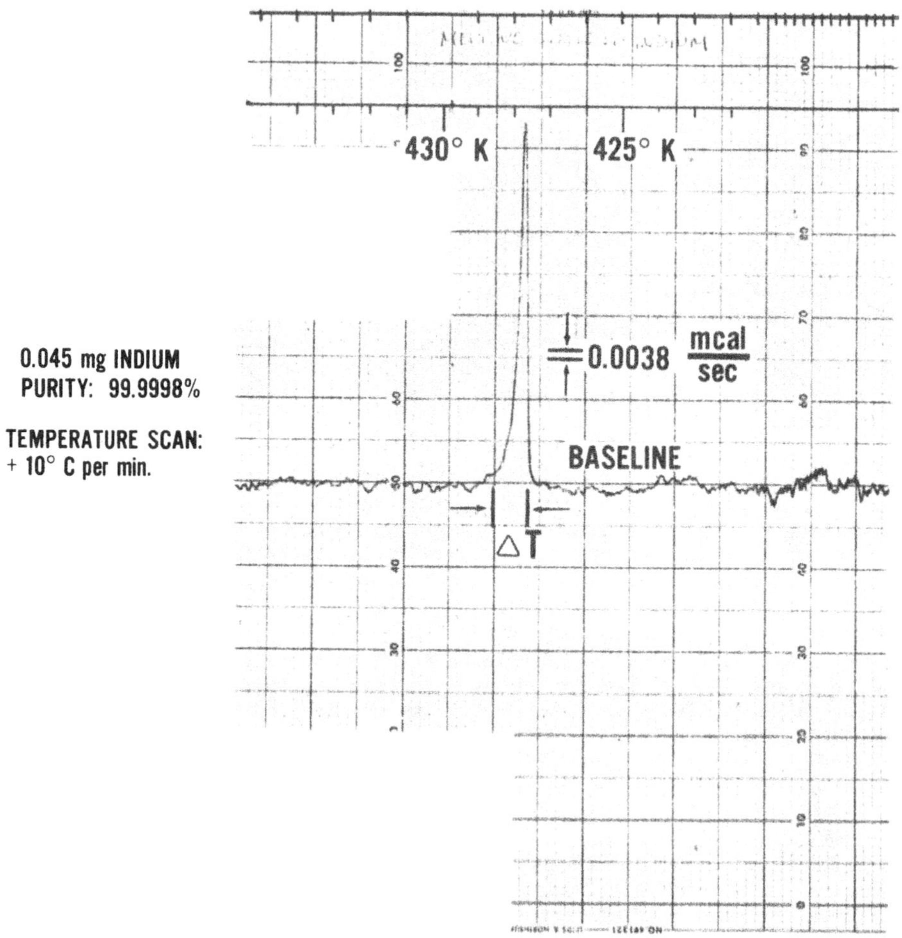

Fig. 1. Melting curve of indium.

For the nematic phase experimental and theoretical evidence indicate an orientation of the molecules about an average direction within so-called swarms, consisting of about 10^6 molecules (7). These building principles are compatible with the textures observed in the respective phases. Such a conclusion cannot be derived for cholesteric mesophases. The proposed molecular arrangements are not supported by sufficient experimental and theoretical evidence to explain the existence of the purely birefringent textures and the optical active texture, which also exhibits the colorful interference effects of light. This classification of mesophases is complicated by the occurence of successive smectic phases because general conclusions cannot be derived about the associated changes of the molecular order.

Mesophases are defined by characteristic optical features of certain textures. This is in contradiction to the phase definition accepted in thermodynamics, where phases are considered as homogeneous regions of matter, characterized by continuous, homogeneous entropy functions, which depend exclusively on thermodynamic variables (8). In this sense the optically distinguishable left and right hand quartz constitute the same phase. According to Tisza's extended definition of the distinction of phases, they are phase doublets (8), since both manifestations are associated with the same entropy function and differ only in symmetry. However, the occurance of mesomorphic phase transitions, as determined by temperature reversible optical effects, was confirmed by thermal analysis whenever applied. In all cases, definite enthalpy changes were observed. This fact justifies the application of optical criteria for mesophase transitions on a tentative basis. Thermodynamic data are required to establish the existence of phases, while optical data should only be used to identify the type of the mesophase.

Several distinct textures are usually observed within each of the thermally established mesophases. This raises the question whether such textures really represent different phases or optically distinguishable phase doublets or only diverse manifestations of the same phase. According to Gibb's phase rule, distinct textures coexisting in equilibrium within a finite temperature interval, constitute the same phase for one-component mesomorphic materials. This reasoning can be extended to mesomorphic compounds of 99 percent purity or better, in which minor and major components have about the same molecular weight. If the textures were caused by phase separation, they should occupy about one percent of the sample and not, as commonly observed, substantially larger regions. In other instances certain textures of a mesophase exist only in definite temperature intervals. It can be argued that the latent heats, which could indicate a phase transition, may be too small to be measurable with existing equipment. Such phase changes are unlikely because the occurance of these textures is not temperature reversible (9) in spite of the small relaxation time as indicated by the low viscosity. In principle, textures may only differ in their symmetry and thus represent phase doublets. Because of the nature of mesomorphic structures and of the poor definition of the samples, it is difficult to determine symmetry elements (such as mirror-symmetry and inversion) by optical or x-ray diffraction analysis. Even if the structures of various textures are known, it may be a problem to uniquely define such differences because of statistical features (10). Although the possibility of phase doublets cannot be discounted in special cases, it has to be rejected for most textures because of their drastic differences in appearance. All this evidence leads to the conclusion that for sufficiently pure compounds the equilibrium textures of a mesophase do not indicate sub-phases but are different manifestations of the same building principle. In this sense textures are analogous to the habits of solid crystals.

However, it should be pointed out that none of the previous arguments deny the existence of thermally established polymesomorphism.

Detection of Mesophases

The three types of mesophases can only be distinguished according to optical features of characteristic textures. Therefore customary polarizing microscopy is required to identify such phases. Unfortunately this approach has certain shortcomings (6). The most significant one is the difficulty to observe phase transitions and thus to determine transition temperatures.

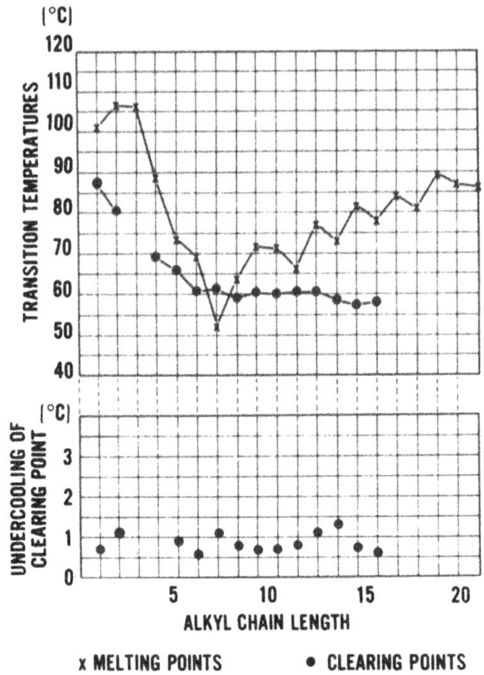

Fig. 2. Transition temperatures of cholestanyl n-alkyl carbonates.

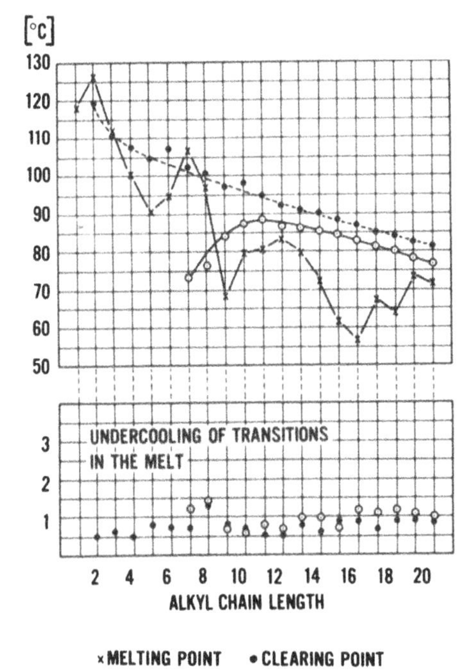

Fig. 3. Thiocholesteryl esters transition temperatures.

In most cases this problem can be expediently eliminated by the application of temperature scanning thermal analysis.

An absolute detection of mesophases cannot be guaranteed by this method, since neither the smallest enthalpy change of mesomorphic transitions nor the smallest temperature interval separating consecutive transitions are known. A certain type of temperature scanning calorimeter (11) appears to be best suited for the detection of mesophases. In this instrument a time linear temperature variation is imposed on the sample and the heating rate required for the phase transition is recorded as a function of time. The area enclosed by this transition curve and the extrapolated base line (see Fig 1) is proportional to the enthalpy change. For extremely pure sample materials the heating rate $\dot{q}$ increases with time t as

$$\dot{q} = \frac{\dot{T}_S}{R} t$$

where $\dot{T}_S$, is the constant temperature rate of change and R represents the thermal source resistance of the instrument. At the end of the transition the heating rate drops back to the baseline closely following an e-function with the time constant of about a second. This instrument limited condition is obtained for sample weights smaller than 0.2 $R\rho k$*, where ρ is the density and k the thermal conductivity of the sample. For larger samples the heating rate levels off to a certain degree before returning to the base line. The width ΔT at the base of a transition curve is a convenient measure for the resolution of consecutive transitions. Two transitions separated by the width ΔT of the first one can always be distinguished, while for smaller separations this is possible only under certain conditions. Of course, in all cases the peak of the transition curve has to be well above the noise level and the spurious fluctuations of the base line. Therefore an optimum performance can be realized, when the sample weight is large enough to obtain a high peak and yet small enough to achieve an almost instrument limited condition.

Figure 1 demonstrates the capability of the temperature scanning calorimeter DSC1 (Perkin Elmer Corp). This instrument was modified to obtain a tenfold increase of sensitivity, amounting to 0.0038 mcal per unit deflection. The melting process of 0.045 mg indium with a purity of 99.9998% is shown for a 10°C per minute temperature scan. The resolution ΔT of the indium sample is about 1°C. The actual phase transition, i.e. the leading edge of the curve, accounts for only 0.2°C of ΔT, while the instrument time constant is responsible for the remaining 0.8°C. The area under the transition curve corresponds to a heat of fusion of 0.3 mcal. Obviously it is possible to detect about one quarter of this value, i.e. 0.075 mcal. If this heat would have been generated by a pure sample of 5 mg during a mesophase transition, 0.015 cal/g would represent the minimum detectable enthalpy change. However, impurities in the available mesomorphic materials cause the heating rate to increase slower than in the instrument limited case. Even for the purest mesomorphic materials this results in a transition curve of about 2°C width and a corresponding reduction of the peak value by a factor of approximately two. Therefore, 0.03 cal/g can be estimated for the minimum detectable enthalpy change of mesophase transitions. This compares favorably with the smallest known transition heat of 0.034 cal/g or 24 cal/mol (12). (See also Figure 6). Furthermore successive mesophase transitions of similar peak height and width can still be distinguished when separated by about 1°C. This is explained by the sharp drop of the heating rate at the end of the first transition and its slow rise for the beginning second transition.

Mesophase Transitions

Once the melting and the freezing point of a compound are known, the presence of a mesophase can be usually determined from thermal analysis data. They are simply the phases exhibited by the melted sample above the solidification point. However, this approach can yield erroneous results,

*This expression follows directly from equation 19 of (11).

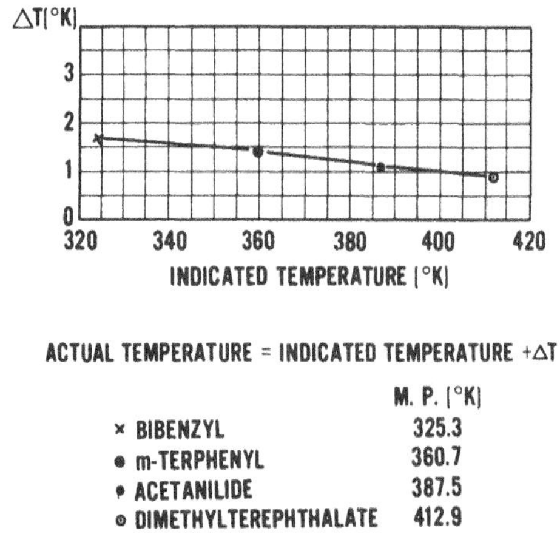

Fig. 4. Calibration curve of DSC-1.

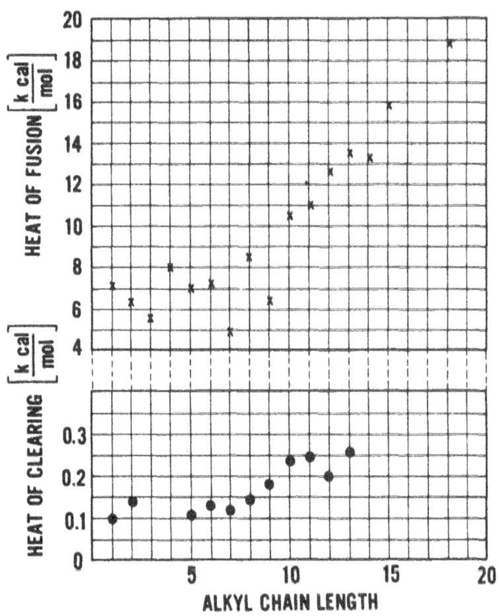

Fig. 5. Transition heats of cholestanyl n-alkyl carbonates.

when the melting point is strongly dependent on the thermal history. For example, methyl cholestanyl carbonate exhibits two melting points: one at 100.3°C after slow freezing (2.5°C per minute or slower) and the other one at 89.7°C after fast freezing (10°C per minute) and short subsequent storage. If the sample is frozen at a medium rate of cooling, both melting points can be observed. The same behavior is obtained from a fast frozen sample after a certain storage time. Such effects are explained by the occurence of stable and metastable solid phases. In our example optical evidence revealed two crystalline phases. In other compounds we identified solidified mesophases as the metastable state.

Solid and liquid phases can often be distinguished by scanning calorimetry because of the large undercooling of the former (10°C or more) and the much smaller effect in the latter (1°C or less). The thermal lags in the scanning calorimeter causes the temperature scale for cooling to be displaced with respect to the temperature calibration obtained for heating. This discrepancy can be determined as a function of scan speed by observing the temperature reversible selective reflectivity exhibited by cholesteric mesophases. Cholesteryl nonanoate of at least 99 mole percent purity was used as a sample, since we measured a temperature reversibility well below ±0.01°C and estimated a response time of less than one second for this optical effect. The compound was contained in an open capsule and covered by a mylar film to enhance the selective reflectivity*. A black dot was painted on the bottom of the capsule to increase the optical contrast. An empty capsule with a black dot of about the same size was placed in the reference holder to approximately balance the heat load of the calorimeter. A polarizing microscope with a photometer attachment was focussed on the sample through the window of the ice-water cooled jacket** enclosing the calorimeter stage. The sample was illuminated with monochromatic light (5000 Å, half width 50 Å) and viewed or monitored between crossed polarizers to reduce disturbing specular reflections. The temperature associated with the peak intensity reading of the photometer was determined on the recorder chart for heating (T_h), cooling (T_c), and for zero scan speed (T_o). The latter measurement was made with a make-shift scale attached to the temperature adjustment control of the calorimeter.

Table 1: Thermal Lag of the DSC-1

$\dot{T}_s$ [°C/min]	T_h [°K]	T_c [°K]	T_h-T_c [°K]	$T_o-1/2(T_h+T_c)$ [°K]
2.5	347.1	346.4	0.7	0
5	347.4	346	1.4	-0.1
10	348	345.4	2.6	-0.1

T_o = 346.8°K

Table 2: Temperature Reversibility of Phase Transitions

Material: Cholesteryl Nonanoate Scan Speed: 10°C/min

Transition Type	Transition Temperatures [°K] Heating	Cooling	Undercooling
Smectic-Cholesteric	347.5	344.5	3-2.6 = 0.4
Cholesteric-Liquid	363.8	361.0	2.8-2.6 = 0.2

The results in Table 1 are reproducible within ±0.2°C. As expected the thermal lag has the same value for heating and cooling and is proportional to the scan speed. The data of Table 2 were obtained from the same sample. Neither the illumination of the open sample nor the closing of this sample and its reference with the proper crimped-on aluminum top and the use of the customary radiation shields had a significant influence on the measurement of the undercooling.

*This effect is caused by surface interactions, which are not completely understood.

**This instrument modification was required to permit a 10°C per minute cooling rate down to 30°C.

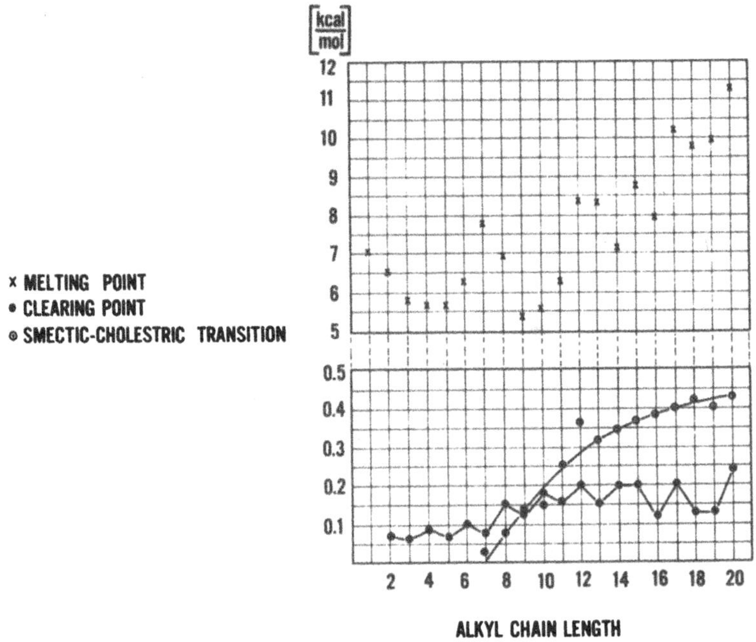

Fig. 6. Thiocholesteryl esters, heat of transition.

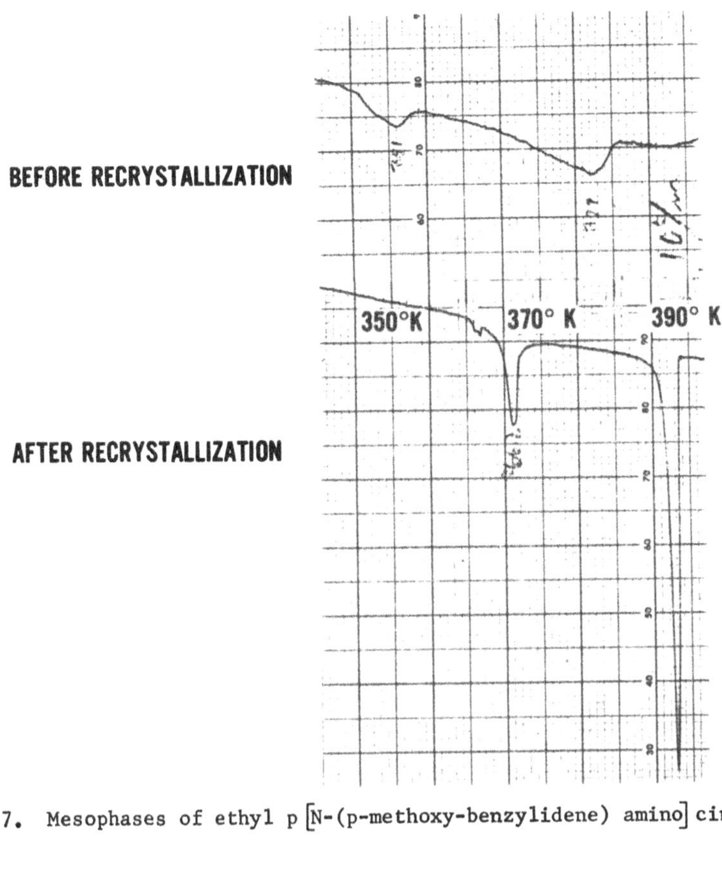

Fig. 7. Mesophases of ethyl p [N-(p-methoxy-benzylidene) amino] cinnamate.

In Figures 2 and 3 the transition temperatures (accuracy ±0.4°C) and their reversibility (uncertain to ±0.2°C) are shown for two homologous series. All values are corrected for the temperature lag as suggested by the instrument maker (13) and adjusted according to the calibration curve shown in Figure 4. For some of the monotropic phases of the cholestanyl n-alkyl carbonates the reversibility of the transition temperature cannot be checked because of intervening solidification. Such phases were identified by optical means. The enthalpy changes associated with these transitions are presented in Figures 5 and 6. The heat of fusion depends on the crystal modification and thus on the thermal history of the sample. For example, cholestanyl decyl carbonate exhibits a heat of fusion of 5.8 kcal/mol for the low (m.p. 68°C) and 10.5 kcal/mol for the high melting point modification (m.p. 71°C). The melting points and heats of fusion presented in this paper pertain to the presumably more stable high temperature crystalline states. More information about the various solid phases and about equilibrium conditions is required for a more meaningful appraisal of this data. Since the initial melting caused by impurities cannot usually be recognized (13), the measured heat of fusions may be too low by as much as 20 percent. The enthalpy changes of transitions in the melt are only known within the order of 20 percent because of the small signal and the uncertainty of the base line. In contrast to the behavior of transitions exhibited in the melt, the melting points and the corresponding transition heats for each homologous series are strongly dependent on the chain length. This indicates that small increases of molecular length affect the energy of the strict crystalline order of the solid state more than in the case of the pliable statistical order found in mesophases. The conclusion appears to be supported by the observation that enthalpy changes in the melt are approximately one order of magnitude smaller than the heat of fusion. The thiocholesteryl esters (see Fig. 6) exhibit a monotonous decrease in latent heat of the smectic-cholesteric transition for decreasing chain length. It is interesting that the smallest transition heat occurs for the shortest chain length for which smectic phases are observed. Since discussions of this type are not subject of this paper, a full interpretation of these results will be given elsewhere.

Impurities and Mesomorphic States

It is difficult to obtain mesomorphic materials containing less than 1 mole percent impurities. In some cases (e.g. p-azoxyanisole) a significant amount of additional impurities is formed due to decomposition during investigations in the melt. While the presence of larger amounts of impurities can suppress (14) or induce (15) mesophases, smaller amounts usually influence only the transition temperatures and reduce the sharpness of the transition curve. The latter effect is shown in Figure 7 for a commercial compound before and after several recrystallizations. If the minor and the major component of a compound form mixed mesomorphic states, the sharpness of the transition curve is not degraded by this impurity (see Figure 8).

Data describing the mesomorphic behavior of a compound can only be considered as complete if the type of impurities, or at least their total amount, is known. Since purified mesomorphic materials approximate the behavior of dilute solutions, the mole percentage of the impurities can be obtained by scanning calorimetry according to Van't Hoff's equation (16)

$$x = \frac{\Delta H}{RT_m^2} \Delta T$$

(ΔH molar heat of fusion, R gas constant, T_m melting point, ΔT melting point depression). The melting point depression ΔT can be determined with sufficient accuracy to permit the detection of 0.05 mole percent of impurities (16). However, the amount of sample melted at the beginning of the melting process cannot always be accurately measured (13). Therefore, the mole percentage x can only be determined for samples in which minor and major components do not form solid solutions. A coarse estimation of the impurity range is possible for a one component solid solution only (17). If more than one solid solution is formed in the sample, this calorimetric method is not applicable. Although solid solutions are rarely encountered in slowly frozen

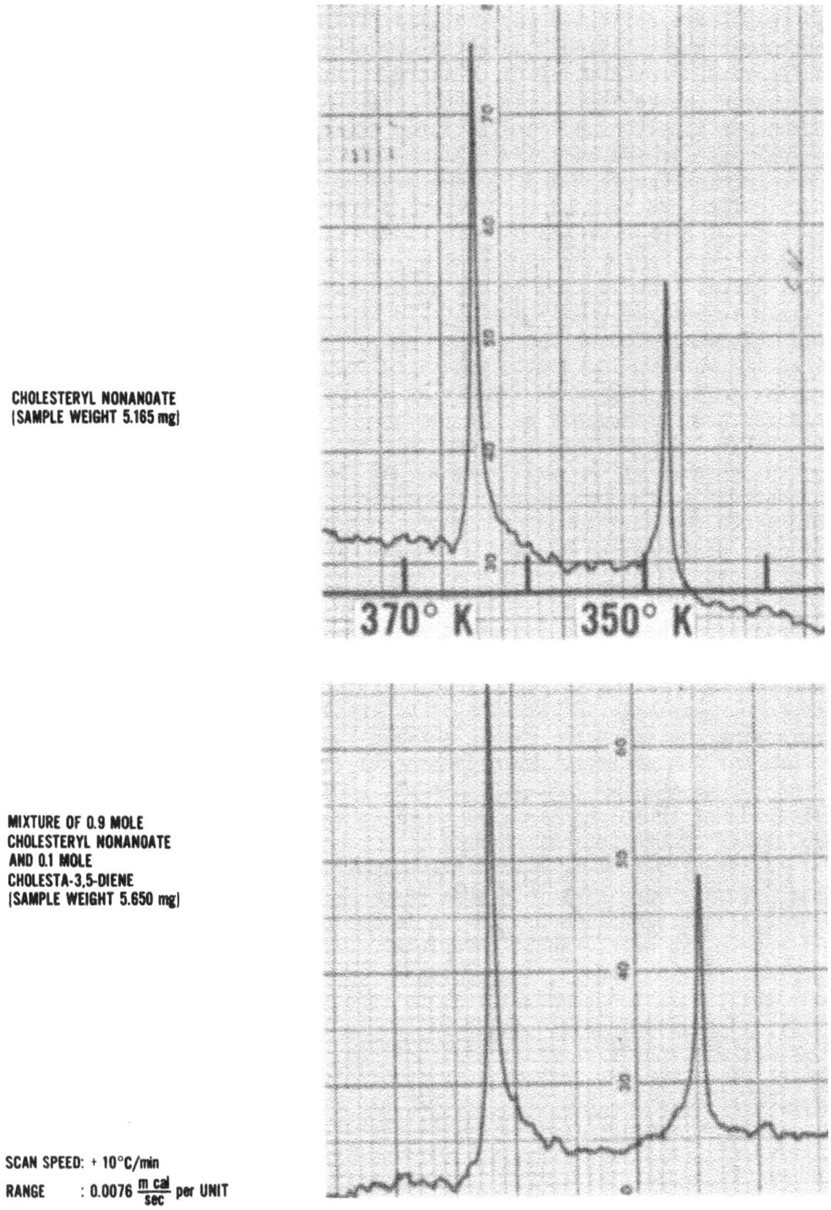

Fig. 8. Mesophases of cholesteryl nonanoate.

organic materials, this possibility should not be disregarded (19).

Table 3: Impurity Content in Mole Percent

Compound	
Cholestanyl ethyl carbonate	0.2
Cholestanyl heptyl carbonate	2.7
Cholestanyl dodecyl carbonate	1.1
Cholestanyl heptadecyl carbonate	≈ 1
Cholestanyl octadecyl carbonate	1.0

In Table 3 the impurity content is given for slowly frozen (0.625°C/minute) representative members of the cholestanyl carbonate series. Since in these computations the possible existence of solid solutions was neglected, this data represents the minimum impurity content.

According to Van't Hoff's equation, the depression of the transition temperatures is inversely proportional to the heat of transition. Therefore, the depressionsof the transition temperatures in the melt should be smaller by approximately an order of magnitude than those of the corresponding melting points. This may explain the generally smoother relationship between transition temperatures in the melt and the chain length of homologous series (18).

Conclusions

Thermal scanning calorimetry is capable of detecting transitions of mesophases and clearing points even for the smallest observed enthalpy changes. Phase transitions may be overlooked if they are separated by less than 1°C, if the heat of transition is less than 0.03 cal/g, or if a monotropic phase cannot be maintained long enough because of freezing. The temperature reversibility of transitions within the melt can be established within ±0.2°C. An improvement of thermal analysis equipment is desirable in order to detect smaller enthalpy changes, to obtain higher accuracy for the determination of small heats of transition, and for the measurement of transition temperatures.

Contrary to the behavior of the solid state, transitions in the melt are apparently not influenced by the thermal history of the sample. Therefore, transition temperatures and enthalpy changes in the melt are probably more characteristic for a given compound than melting point and heat of fusion. Amount and type of impurities can have a strong influence on the mesomorphic behavior of a compound. While, in most cases, the total impurity content can be measured or at least estimated by thermal analysis, an impurity identification is not possible by this approach. A combination of thermal analysis with other test methods is required to obtain sufficient information about impurities. The uniqueness of a description of the mesomorphic behavior can only be appraised if the essential impurities of a compound are known.

Acknowledgements

The author is grateful to Drs. W. Elser and J. Pohlmann for preparing the compounds, to Mr. A. Brown for making measurements, and to Drs. L. Cameron and J. Muller for the review and stimulating discussions.

References

1. D.G. Dervichian, Mol. Cryst. 2, 55, (1966).
2. G. Gray, Molecular Structure and Properties of Liquid Crystals, Academic Press, Inc., New York, (1962).
3. F.B. Rosevear, J. Am. Oil Chem. Soc. 31, 628 (1954).
4. A. Mabis, Private communication.
5. M.G. Friedel, Ann. Phys., (Paris) 18, 273 (1922).
6. R.D. Ennulat, Mol. Cryst. (in print).
7. P.P. Chatelain, Acta Cryst. 4, 453 (1951).
8. L. Tisza, Ann. Phys. (New York), 13, 1, (1961).
9. loc. cit. (2), pp. 5, 10, 69, 129.
10. C. Herman, Z. Kristallogr., 79, 186 (1931); c.f. A. Mabis, Acta Cryst. 15, 1152 (1962).
11. M.Z. O'Neill, Anal. Chem. 36, 1238 (1964).
12. H. Arnold, Mol. Cryst. 2, 63 (1966).
13. Thermal Analysis Newsletter, Nos. 5 and 6, Perkin Elmer Corp., Norwalk, Conn.
14. W. Elser and R. Ennulat, unpublished.

15. J. Pohlmann, Mol. Cryst. 2, 15 (1966).
16. D.D. Tunnicliff and H. Stone, Anal. Chem. 27, 73 (1955).
17. S. Mastrangelo and R. Dornte, J. Amer. Chem. Soc. 77, 6200 (1955).
18. loc. cit. (2), p. 197.
19. J.G. Aston, M.R. Cines and H.L. Fink, J. Am. Chem. Soc. 69, 1532 (1947).
W.M. Smit, Rec. Trav. Chim. Pays-Bas. 75, 1309 (1956).
H.F. Wijk and W.M. Smit, Anal. Chim. Acta, 23, 545 (1960); ibid. 24, 41 (1961).

DETERMINATION OF PURITY AND PHASE BEHAVIOR BY ADIABATIC CALORIMETRY*

Edgar F. Westrum, Jr.

Department of Chemistry
University of Michigan
Ann Arbor, Mich. 48104

Heating and cooling curves through the melting region of a substance allow calculation of the amount of impurity that is soluble in the liquid phase but insoluble in the crystalline phase as described by several investigators [*e.g.*, 1, 2, 3] and highly developed by others [*e.g.*, 4]. A more precise method of determining purity is the calorimetric procedure. In this technique, the temperature at which the crystalline and liquid phases are in equilibrium is determined as a function of the fraction of the sample liquefied. The depression of the freezing point and the enthalpy of fusion, and hence the cryoscopic constant, are determined in a single experiment. Precise results can be obtained because the sample can be held under adiabatic conditions as long as required for equilibration. The two procedures have been compared experimentally by Glasgow *et al.* [5]. The calculation of the purity of the sample from its equilibrium melting curve, which has been discussed by various investigators [6,7,8,9,10,11,12], involves straightforward application of ideal solution laws.

If an impurity forms an ideal solution with the major component in the liquid phase and is insoluble in the solid phase, the relation between the concentration of impurity and melting point depression is [3]:

$$\ln (1 - N_2) = A\Delta T(1 + B\Delta T + \ldots) \qquad (1)$$

in which N_2 is the mole fraction of impurity in the liquid phase at the temperature T_F; T_F is the equilibrium temperature corresponding to the fraction F, of the total sample in the liquid state; $\Delta T = T_{tp} - T_F$; T_{tp} is the triple-point temperature of the pure major component $A = \Delta Hm/RT^2_{tp}$ is the first cryoscopic constant; $B = (1/T_{tp} - \Delta Cm/2\Delta Hm)^{tp}$ is the second cryoscopic constant; ΔHm is the isothermal enthalpy of melting (fusion), and ΔCm is the heat capacity of the pure liquid less that of the pure solid at T_{tp}. For most samples with $N_2 << 1$, ΔT is small enough that equation (1) can be simplified to

$$N_2^* = AF\Delta T \qquad (2)$$

in which $N_2^* = N_2F$ is the mole fraction of solid-insoluble

*This work was supported in part by the Division of Research of the United States Atomic Energy Commission.

impurity in the bulk sample. Rearranging equation (2) leads to the form

$$T_F = T_{tp} - (N_2^*/A)(1/F)$$

Therefore, a plot of observed values of T_F vs. $1/F$ should be linear. Extrapolation to $1/F = 0$ and 1 gives the triple-point temperature of the pure substance and the maximum melting temperature of the actual sample, respectively. The slope of the line is N_2^*/A, from which the amount of impurity may be evaluated if A is known.

When the impurity has molecular dimensions similar to those of the major component, solid solutions may be formed, causing the curve of T vs. $1/F$ to be concave upward. Under some circumstances [13], the treatment of Mastrangelo and Dornte [14] then applies.

The limitations of the calorimetric procedure have been discussed by McCullough and Waddington [13]. Recent extensive international tests of the purity of benzene have been summarized by Saylor [15]. Indication of the precision of which the method is capable will be seen in Figure 1. This figure gives results obtained by the Bureau of Mines at Bartlesville [16] for a specially purified sample of benzene (A) and two samples made from it by adding known amounts of impurity (B and C). The value for the purity of Sample A, 99.9993 mole%, agrees within 0.0001 mole% with a value obtained on a portion of the same sample by the National Bureau of Standards [15]. The measured purities of Samples B and C obtained by workers from the Bureau of Mines and Bureau of Standards agreed with the known values to within 0.01 mole%.

Although standard low-temperature adiabatic heat capacity calorimeters in this section may be used to investigate purity during heat capacity and enthalpy of fusion determinations, they are more complex than necessary if purity determination is the primary objective. As a consequence, a number of calorimeters specifically designed for the determination of purity have been constructed. These devices include the rather complex calorimeter of Aston *et al.* [17], one permitting very precise determinations on large samples by Pilcher [18], a moderately precise, semiautomatic apparatus by Tunnicliff and Stone [19], and simpler designs permitting rapid determinations of moderate precision on samples of only several cubic centimeters, as described by Clarke *et al.* [20] and by Brooks and Pilcher [21].

A further impressive contribution of adiabatic heat capacity determinations is the delineation of phase behavior and the trend of the melting curve near congruent melting points. For example, the existence of a significant solid solution region in the NH_4F - HF system at the $NH_4F \cdot 3HF$ composition [22] is indicated by the disappearance of the eutectic transition on samples of 74.62 and 75.05 mole% as shown in Figure 2.

These data taken during equilibrium measurements are clear and unequivocal evidence for the phase behavior depicted in Figure 3. This phenomenon, involving only a small heat effect was not detected in the thermal analysis studies, but with the much more sensitive methods employed in the equilibrium low temperature calorimetry it was readily observed.

In order to obtain more accurate enthalpy of fusion data some enthalpy-type runs were also made through these regions of relatively slow equilibrium, and the enthalpy increments of these runs were compared with the integrated heat capacity over the same temperature region. The heat capacity points and the enthalpy increments of the anomalies were determined as essentially equilibrium values and were completely reproducible for a given sample. This may be verified by noting the series of repeated determinations shown in Figure 2.

The melting point of the compound $NH_4H_3F_4$ was interpolated as 296.4°K. from the calorimetric data and corresponds closely to the results of the phase studies. The entropy of fusion is 16 cal./(mole °K.). A narrow range of homogeneity is observed on both sides of 75 mole% HF, *i.e.*, for the compound $NH_4H_3F_4$. The two series of anomalies occurring below the melting point are as yet unexplained, but it is likely that they are related to a disordering of the crystal due to the presence of the ammonium ion. Further examinations of more compositions in the range 73.5 to 76.1 mole% HF are desiderata for clarification of the complex character of this system, but the wealth of detail revealed by the calorimeter is evident.

Interest in the existence of solid solutions of ammonium fluoride in ice was stimulated by a paper by Lonsdale [23] in which attention was directed to the fact that ice and ammonium fluoride are not only isomorphous but also show nearly identical Laue X-ray diagrams with strong, diffuse, star-shaped streaks attributable to thermal vibrations of the hydrogen nuclei ("proton jumps") and molecular rotation. Not only is the crystal structure of NH_4F like that of ice, but the atoms are similarly arranged; and the dimensions differ only by 3.7% (N — H···F, 2.66A.; O — H···O, 2.76A.)[24]. The hydrogen bonding in NH_4F is also comparable to that in ice [25]. These similarities in the geometry and the binding forces of ice and NH_4F led Brill and Zaromb [26] to claim evidence for the existence of a solid solution of ammonium fluoride in ice by chemical analysis of the solid phase, thermal analysis, dilatometry, etc. A thermodynamic relation applied to the liquidus curve data of Yatlov and Polyakova [27] is in accord with solidus data of Brill and Zaromb and indicates a solubility limit of approximately 5 mole% NH_4F at 246.7°K.

Heat capacities of 0.4916, 1.485, 4.603 and 19.39 mole% NH_4F aqueous solutions shown in Figure 5 were determined by direct calorimetric measurements frm 55 to 325°K. [28]. Although in heat capacity measurements outside of transition regions thermal equilibrium was usually attained within ten or

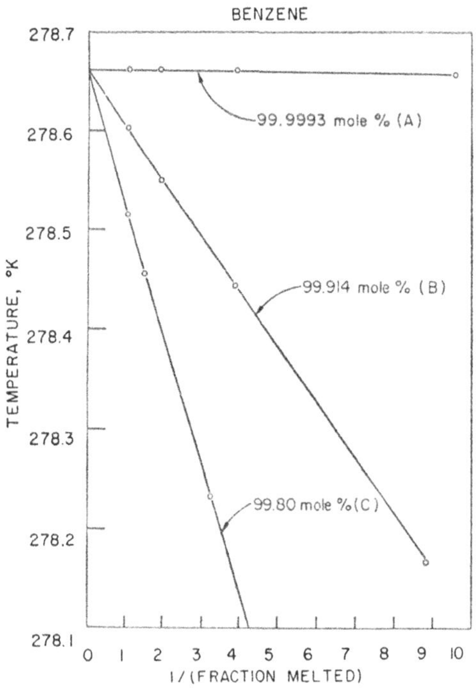

Figure 1

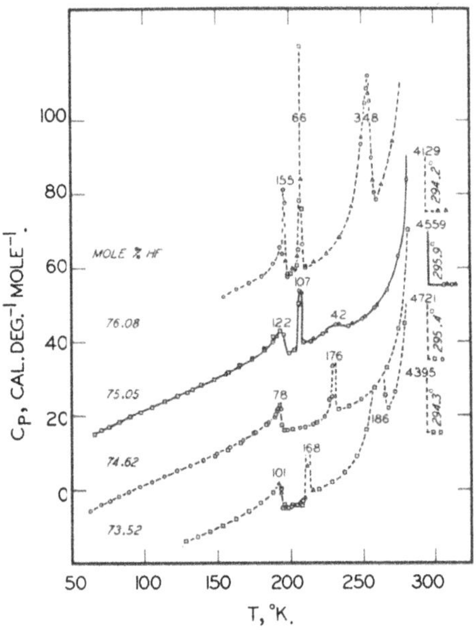

Figure 2

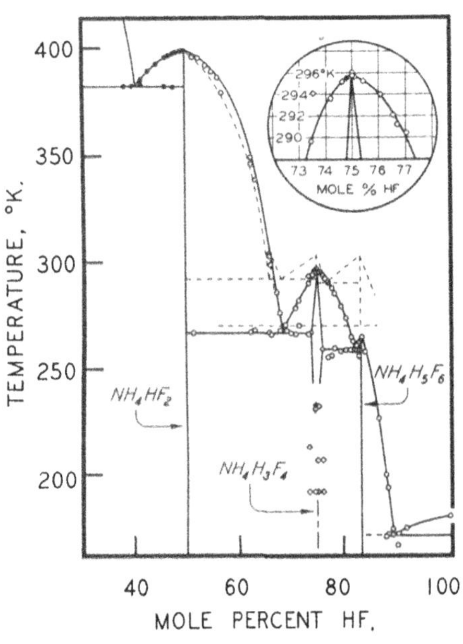

Figure 3

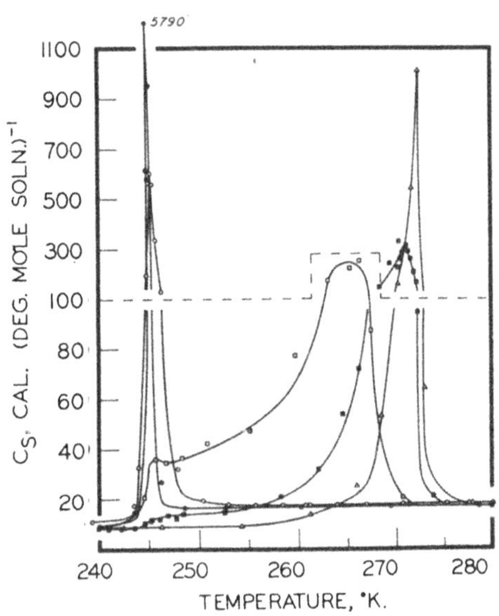

Figure 4

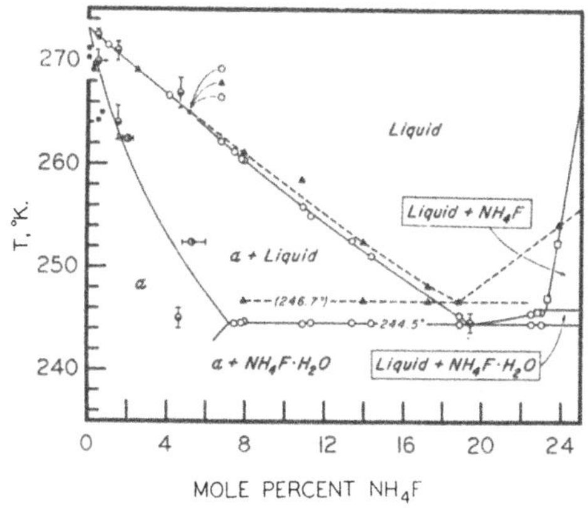

Figure 5

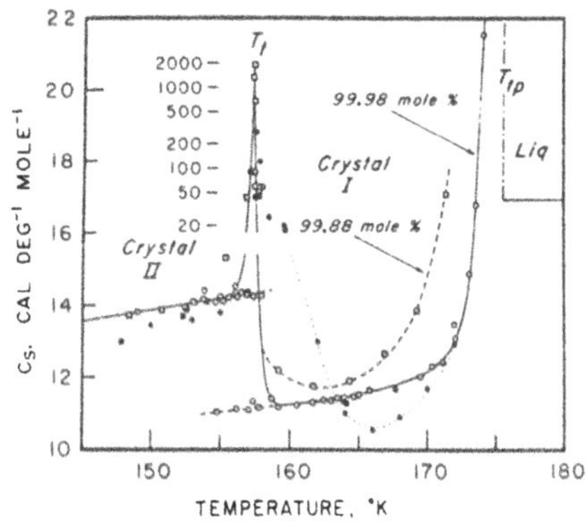

Figure 6

FIGURE LEGENDS

Figure 1.-- Melting point study of three benzene samples by adiabatic calorimetry. (A) is a highly purified sample, (B) is the same material with 0.0923 mole percent n-heptane added, (C) is similar but has 0.202 mole percent n-heptane.

Figure 2.-- Heat capacity of four samples approximating $NH_4H_3F_4$ in composition. Note that the lower curves are successively displaced by increments of 20 cal./(mole °K.) as a convenience is representing the anomalies.

Figure 3.-- Partial phase diagram for the system NH_4F -HF [22]. The open circles indicate thermal analysis data of the above research, the solid circles those of Benjamins, et al. [see 22], open squares are calorimetric data from four samples, and the solid square those from a preliminary calorimetric sample. The dotted lines shown the data of Ruff and Staube [see 22].

Figure 4.-- Heat capacities for aqueous solutions of NH_4F in the fusion transformation region. 0.4916 mole percent NH_4F, Δ, 1.485 mole percent NH_4F, ■; 4.603 mole percent NH_4F, □; 19.39 mole percent NH_4F (eutectic), •; $NH_4F{\cdot}H_2O$, ○. [22]

Figure 5.-- Phase diagram for the system $NH_4F{\cdot}H_2O$ from 0 - 24 mole percent NH_4F. The solid solution is designated by α. The data from the studies of Labowitz and Westrum [28,30] are represeented by ○ for thermal analysis, by □ for solubility measurements, by ◑ for wet residues results, and by ◒ for heat capacity determinations. The data of Zaromb and Brill [25] are given by ■, those of Yatlov and Polyakova [27] by ▲.

Figure 6.-- The heat capacity of mthanol in the fusion region [32]. Data for the higher purity sample are represented by ○ and those for a less pure sample by □ [32]. Values of other investigators are indicated by ■ and •.

twelve minutes after energy input, within the fusion and transition region, however, thermal equilibrium was usually not attained until ten or twelve hours after an energy input. The resulting partial phase diagram is shown in Figure 5.

The absence of a thermal effect at the eutectic temperature for the 0.4916 and the 1.485 mole% NH_4F solutions is conclusive evidence for the existence of the solid solution at these concentrations. It would have been possible by the technique used here to detect an effect as small as three cal./(°K. mole soln.).

Zaromb and Brill [25] derived an expression for the solidus curve by utilizing the Clapeyron equation, the liquidus curve of Yatlov and Polyakova, [27] and simplifying assumptions (such as the constancy of the enthalpy of fusion). The temperature dependence of the enthalpy of fusion of ice (ΔH_f) is considerable and may be represented by the following analytical expression over the range of interest;

$$\Delta H_f \text{ (cal./mole)} = -6997 + 63.68T - 0.1586T^2 .$$

This expression reproduces the experimentally observed values for the heat of fusion of ice within 0.2%. By incorporating this into the differential Clapeyron type equation [29] and integrating, the relationship;

$$\log (N_1/N_S) = -1530\ (1/T_0 - 1/T) + 32.02 \log (T_0/T) - 0.03467$$

was obtained. N_1 = mole fraction of H_2O in the liquid phase, N_S = mole fraction of H_2O in the solid phase, ΔH_f = the heat of fusion of ice, T_0 = the melting point of pure ice, T = the freezing point of the solution. The solidus curve calculated on the basis of this equation and on the liquidus data of the present investigation is in much better agreement with other findings of this study than is the curve calculated by Brill and Zaromb which did not take into account the temperature dependency of ΔH_f.

The solid solution line obtained from the calorimetric measurements indicating solubilities of as much a 5% NH_4F was further confirmed in the same investigation [28] (see Figure 5) by Schreinemaker's method of wet residues, by thermal analysis, and by solubility determinations.

Related studies on the resolution of the existence of the $NH_4F \cdot H_2O$ phase over a 1.5°K. range has also been done calorimetrically by Labowitz and Westrum [30] together with the thermodynamics of the peritectic reaction. The slope of the melting curve in the region of congruently melting NH_4F has been explored by Benjamins and Westrum [31]. (see data on Figure 3 insert)

An interesting example of the profound influence of purity on the thermal properties of substances may also be seen in Figure 6, which shows the striking difference in heat capacity of the crystal I phase on two methanol samples with fractional furities of 0.9998 and 0.9988 [32], compared with the results of previous investigators. Both samples readily undercooled through the transition, making it possible to measure the heat capacities of the metastable crystal I phase below its normal transformation temperature. It was exceedingly difficult to obtain the conversion of crystal II to crystal I in the purer sample; the relatively impure sample, however, readily underwent this transition.

Finally, the CCl_4 - $C(CH_3)_3$ system will be presented to show the power of the calorimetric method even at low temperatures [31]. Space, however, does not permit diagrams of the dependence of the values of thermal transformations upon composition.

REFERENCES

1. Washburn, E. W., Ind. Eng. Chem., Ind. Ed., **22**, 985 (1930).
2. White, W. P., J. Phys. Chem., **24**, 393 (1920).
3. Glasgow, A. R., A. J. Streiff, and F. D. Rossini, J. Research Natl. Bur. Standards, **35**, 355 (1945).
4. Mair, B. J., A. R. Glasgow, Jr., and F. D. Rossini, J. Research Natl. Bur. Standards, **26**, 591 (1941).
5. Glasgow, A. R., Jr., G. S. Ross, A. T. Horton, D. Enagonio, H. D. Dixon, C. P. Saylor, G. T. Furukawa, M. L. Reilly, and J. M. Henning, Anal. Chim. Acta, **17**, 54 (1957).
6. Aston, J. G., M. R. Clines, and H. L. Finke, J. Am. Chem. Soc., **69**, 1532 (1947).
7. Badley, J. H., J. Phys. Chem., **63**, 1991 (1959).
8. Pitzer, K. S., and D. W. Scott, J. Am. Chem. Soc., **63**, 2419 (1941).
9. Rossini, F. D., Chemical Thermodynamics, Wiley, New York, 1950, p. 455.
10. Scott, R. B., C. H. Meyers, R. D. Rands, Jr., F. G. Brickwedde, and N. Bekkedahl, J. Research Natl. Bur. Standards, **35**, 39 (1945).
11. Todd, S. S., G. D. Oliver, and H. M. Huffman, J. Am. Chem. Soc., **69**, 1519 (1947).
12. Weissberger, A., ed., Physical Methods of Organic Chemistry 1st ed., Vol. I, Interscience, New York, 1945, p. 350.
13. McCullough, J. P., and G. Waddington, Anal. Chim. Acta, **17**, 80 (1957).
14. Mastrangelo, S. V. R., and R. W. Dornte, J. Am. Chem. Soc., **77**, 6200 (1955).

15. Saylor, C. P., Report of the Subcommission on Physico-Chemical Data, International Union of Pure and Applied Chemistry, at International Calorimetry Conference, Ottawa, Canada, August 18, 1961.

16. U. S. Bureau of Mines, Bartlesville, Okla., unpublished.

17. Aston, J. G., H. L. Finke, J. W. Tooke and M. R. Cines, Anal. Chem., **19**, 218 (1947).

18. Pilcher, G., Anal. Chim. Acta, **17**, 144 (1957).

19. Tunnicliff, D. D., and H. Stone, Anal. Chem., **27**, 73 (1955).

20. Clarke, J. T., H. L. Johnston, and W. DeSorbo, Anal. Chem. **25**, 1156 (1953).

21. Brooks, J. H., and G. Pilcher, J. Chem. Soc., **1959**, 1535.

22. Euler, R.D., and E.F. Westrum,Jr., J. Phys.Chem., **65**, 1291 (1961).

23. Lonsdale, K., Nature, **158**, 582 (1946).

24. Pauling, L., "The Nature of the Chemical Bond", 3rd Ed., Cornell University Press, Ithaca, N.Y., 1960.

25. Zaromb, S., and R.Brill, J. Chem. Phys., **24**,895 (1956).

26. Brill, R., and S. Zaromb, Nature, 173, 316 (1954).

27. Yatlow, V. S., and E. M. Polyakova, Zhur. Obschei Khim., 15, 742 (1945).

28. Labowitz, L.C., and E. F. Westrum, Jr., J. Phys. Chem., **65**, 408 (1961).

29. Eucken, A., "Lehrbuch der chemischen Physik", Akad. Verlagsgesellschaft, Leipzig, II, 1950, p. 85.

30. Labowitz, L.C., and E. F. Westrum, Jr., J. Chem. Phys., **65**, 403 (1961).

31. Chang, Elfreda, unpublished data.

32. Carlson, H. G., unpublished data.

INSTRUMENTAL CONSIDERATIONS FOR THERMAL METHODS

Arthur F. Findeis, National Science Foundation, Washington, D. C.

The techniques of thermal measurements in chemistry have received a spectacular revival due to relatively recent developments in instrumentation for the reliable measurement of heats or rates of heat transfer associated with various types of physical and chemical changes. The commercial availability of instrumentation has in turn stimulated experimentation in other areas which provide auxilliary information to thoroughly define the chemical or physical processes taking place. The measurement of temperature or differential temperature is a direct or related important part of the process.

An evaluation of temperature from a practical measurement point of view is provided by the International practical temperature scale. This scale was adopted in 1927, revised in 1948 and again in 1960 (1). A recent change has been suggested in the assignment of the melting point of zinc as a primary standard for calibration purpose. It must be emphasized that the accuracy of temperature scales in such a practical sense must be tempered by the reality that many measurements are not made under the ideal conditions suggested for calibration purposes. Instruments, calibration points, and interpolation equations of the International Temperature Scale of 1948 have been summarized (2). A bibliography of temperature measurement for the period 1953-1960 is available (3).

It should be noted that a great deal of activity is centered around extending temperature measurements to lower and higher temperatures. The Cryogenic Data Center of the National Bureau of Standards maintains "standard" tables for thermocouple pairs frequently used at low temperatures. Such tables contain interim values until national tables can be established for cryogenic temperatures. Examples of thermocouples for low temperature work have been reported (4,5). Temperature measurements above the gold point and up to 3000° have been discussed recently (6). It is clear that the reliable measurement of temperature in terms of a practical temperature scale is indeed difficult at either extreme of the temperature scale. The standard measuring device for the International Practical Temperature Scale in the temperature range of 630.5 to 1063°C is the platinum versus platinum-10 per cent rhodium thermocouple. The use of such a thermocouple above 1500°C leaves the wires very weak mechanically. Thermocouples in which both legs are platinum-rhodium alloys are capable of reliable temperature measurements at higher temperatures. These have distinct advantages over the standard platinum versus platinum-10 per cent rhodium or platinum versus platinum-13 per cent rhodium at higher temperatures although the thermoelectric power is less (7). The use of tungsten-rhenium thermocouples for measurements to 2000°C has been reported (8). For higher temperatures, tantalum carbide versus graphite thermocouples have been suggested (9). The latter two thermocouples have not been evaluated as to long term stability.

Chromel-alumel thermocouples have been most used by chemists for differential temperature measurements in the medium temperature range. The Platinel thermocouple was developed to replace the chromel-alumel pair (10). Reference tables for this thermocouple have been published (11). Use of Platinel thermocouples has been reported by Smith and Johnson in a simultaneous DTA and TGA apparatus which utilizes multiple pans for large sample areas (12). This thermocouple is considerable more resistant to oxidation than the chromel-alumel pair and the thermoelectric stability was found to be quite good. A tungsten-graphite differential thermocouple has been used to study of the reaction between calcium oxide and carbon to form calcium carbide and carbon monoxide (13).

Implicit is all measurements with thermocouples is that the thermoelectric power, expressed in microvolts per degree, is not a constant therefore the appropriate correction must be applied in order to correct differential emf changes at one temperature to those at another in order to compare the corresponding response-time curves. The determination of the temperature at which changes in atomic or molecular arrangement take place in materials is an area of interest of great importance. This is usually done by following the temperature of a sample as it is heated or cooled and evaluating the data according to a particular method (14). Methods of interpretation of cooling or heating data were described by Burgess 60 years ago (15). These include: the time temperature method, the inverse rate method, the direct rate method, the differential method and the derived differential method. Cooling curves are often complicated by the observation of supercooling. In a very few cases the phenomena of superheating has been reported in measurements of heating curves. The latter is of some current interest in the area of both inorganic and organic polymer chemistry.

Supercooling is observed experimentally in practically every cooling measurement of a solvent which results in a solidification of a sample. Appropriate methods to avoid this effect are described (16) so that one can induce crystallization in the sample as soon as possible after the temperature has passed below the freezing point of the sample. This is virtually impossible to do with instrumentation for differential thermal measurements due to limitations imposed by size of the sample. A maximum limit for supercooling has been inferred from measurements of the size of tiny droplets of liqued solvent in a temperature environment well below the freezing point of bulk solvent. This in turn has been interpreted as a nucleation size for the solidification process in a bulk solvent.

The experimental observation of superheating of crystals has been demonstrated on high viscosity silicates (17-19), gallium (20), tin (21), p-toliudine (22), quartz (23), phosphorous pentoxide (24) and sugars (25). Recent work on linear high-polymer polyithylene crystals has been revealed the same phenomenon (26). A maximum superheating can be as high as several hundred degrees for some crystals. In all of the studies the observation was possible only by heating the interior of a single crystal or by heating faster than the melt could progress in from the edges of the crystal. Superheating occurs only if the melting velocity is slower than the heating rate of the crystal.

The name "thermal analysis" is somewhat of a misnomer since that which is actually measured is a change in resistance, as variable electromotive force, or a radiant energy per unit wavelength interval at a wavelength emitted per unit solid angle per unit area of a black body or the derivative of these signals. These changes must be related to a temperature change. A wide variety of thermocouples have been used; thermisters have been employed in a few instances (27), and resistance thermometers early to a very small extent (28) except in null balance instrumentation. The recent development of quartz thermometers (Hewlett Packard) and their incorporation in differential thermal analysis apparatus has been suggested.

At very high temperatures at which thermocouples instability and material of construction difficulties plague the methods of thermal analysis the optical pyrometer is bound to play an increasingly important role. Photoelectric pyrometers may replace the visual pyrometers as primary standard instruments. The precision is at least 15 times greater at the gold point (29-31). One example of the use of an infrared pryometer for thermal analysis has been reported (32). The sensitivity has been demonstrated by the ability to detect the gamma to delta transition in a 200 mg sample of iron a transition releasing only 0.34 cal of heat. Although no specific examples are available it is likely that the various solid state heat sensors will play an important role in thermal analysis in the near future.

Early investigation is differential thermal analysis involved a direct placement of thermocouples in the sample. The difficulties in this approach were noted in the work of Speil (33) who noted that special care was required in the fabrication of identical thermocouples with matched junctions in order to avoid troublesome base line drift. Thorough discussions of heat transfer problems appeared in the literature and a significant advance was provided by the work of Boersma (34) which the thermocouples were placed external to the sample. This was quickly followed by an evaluation of Boersma equations (35). The principal of Boersma has pervaded much of the thermal instrumentation that has been developed since that time.

The most serious errors in measurements by the differential thermal analysis method are caused by differences in heat transfer through the sample and reference and through the thermocouple wires. It has been suggested (35) that a standardization of the packing method in order the sample will have the same density and same volume will provide a reproducible heat flow pattern near the thermocouple. This has been accomplished in practice by using the reference material, which must be inert, as a diluting agent. Thermocouple effects can be minimized by calibration of each thermocouple which is used. This admittedly if often impractical for routine thermal analysis studies.

The usual experimental arrangement of sensors in a differential thermal analysis apparatus is that in which the response of a temperature sensor (thermocouple, thermister, resistance thermometer etc.) is compared with an identical sensor monitoring the temperature of a reference material. The difference in signal between the sensors is then recorded as a function of the signal from either the sample, the reference material or as a function of time or an additional sensor is placed in the system to measure the temperature of the oven. A few single sensor differential thermal analysis have been described. The differentiation has been accomplished by resistance-capacitance networks. An obvious approach for single sensor differential thermal analysis is differentiation by an operational amplifier however this has not been applied specifically to thermal analysis. Such approaches have been described for other chemical instrumentation (36).

The important consideration in differential thermal instrument is the reproducibility of measurements on replicate thermograms. It is quite clear that a knowledge of prior sample history is of the utmost importance. The difficulties inherent in measurement of thermograms on any substance other than those of the highest purity make this every apparent. Recommendations for thermal analysis of metals suggest thirteen items be known (14).

(1) Source and composition of material
(2) Complete thermal and mechanical history
(3) Size and mounting of test specimen
(4) Atmosphere
(5) Rate of temperature change
(6) Method of analysis
(7) Maximum temperature attained, and time of temperature
(8) Temperature of minimum rate of change of sample at recognized and identified inflected points
(9) Direction of change
(10) Rate of change of furnace at time of transformation, or a statement that furnace is differentially controlled
(11) Temperatures at beginning of heat effect, at maximum or at the end of the effect or both
(12) Maximum and minimum temperatures of decalescence and recalescence if either is observed
(13) Any departures from the above, or normal conditions

These recommendations cannot be disregarded for any other type of sample subject to a thermal analysis. A calorimetric method based on the overall thermal conductivity of the total sample (37) was provided by Sturm. The overall thermal conductivity is computed from data obtained when the sample is heated at a rate resulting in a constant thermal gradient. An average value of the thermal conductivity before and after a transformation yields a calibration constant for the apparatus. This approach is not applicable to differential thermal analysis requiring a multiplicity of heating rates. The apparatus is not described in that paper.

Herold and Planje (38) used a thermocouple as the sample cell to simplify heat transfer problems. A similar approach was later described by Mazieres who described a differential thermal analysis approach for very small amounts of sample (39,40). In short order Barrall, Porter and Johnson published designs of calorimetric cells in which three thermocouples were incorporated in the cell (41) and subsequently a two thermocouple cell was devised (42). These ideas have been incorporated in the du Pont calorimetric apparatus. A calorimetric approach using the external thermocouple approach has recently been the subject of U.S. patent (43). Stone and Burress described numerous thermocouple arrangements in which a separable sample pan is placed in various thermocouple wire sample holders in which the sample and reference junctions are in intimate contact with the sample holders.

Garn has reported a differential thermocouple arrangement for the thermal analysis of small samples (44). One form of the differential thermocouple comprises a platinum-10 per cent rhodium-gold-platinum-10 per cent rhodium differential junction with a temperature measuring thermocouple consisting of a platinum-platinum-10 per cent rhodium formed at one of the junctions of the differential array. It is interesting to note that a similar principle was utilized some four years later as a disc cell as marketed by du Pont which wse a chromel-constantan-chromel differential thermocouple. A polyplate system with many thin sample layers on thick metal discs surrounding a thermocouple well has been described (45). An open pan type of sample holder has been described by McAdie (46). An approach based on vacuum deposition of a thin metallic film followed by electrodeposition of a dissimilar metal which is then followed by a subsequent vacuum deposition of the differential junction is described by King et al.(47). This is not a comprehensive list but is does show that a substantial activity in instrumentation for thermal methods of analysis exists even though commerical instrumentation is well established.

The method of differential thermal analysis by its very nature is a comparative technique. Proponents of calorimetry adhere to the desirability of obtaining direct information rather than indirect thermometric data of differential thermal analysis. Detailed discussions have appeared and recently Wadso has reviewed calculation methods (50). Recent applications of a calorimetric principle have been reported by Speros and Woodhouse (51, 52) in which heats of fusion of tin, sodium nitrate, lead, aluminum and silver are reported. The application of this approach to the decomposition of solids is in progress (53). R. D. Freeman has similarly developed a precise differential calorimeter (54). Certainly the most familiar apparatus which has been used is the differential scanning calorimeter (55). This was simultaneously analyzed in detail by O'Neil (56). This is by no means an exhaustive listing of differential calorimetric approaches. The methods of differential thermal analysis have been applied to electrochemistry as well. The measurements of temperature changes associated with electrochemical processes at microelectrodes has been reported. The difference in temperature of the electrode and surrounding electrolyte were measured by two thermisters in an a.c. bridge utilizing a lock-in amplifier (57,58).

Factors Influencing Shapes of Thermograms. The basic problem in the utilization of differential thermal analysis data for the measurement of calorimetric values is the control of response time of the differential thermocouples. This is intimately associated with a large number of factors involving thermocouples, sample size, heat transfer and overall relaxation times of the differential array. Many of the arguments concerning a rigorous quantitative descriptions of thermograms are meaningless without a detailed specification of the instrumentation utilized in the measurement. Furthermore, an interpretation of area measurements depends on whether the areas are evaluated in deg^2 or deg. sec. A problem plaguing thermal analysis investigations is the inadequate specification of experimental details in published literature. In 1967 the Committee on Standardization of the International Conference on Thermal Analysis published a consensus of recommendations for reporting thermal analysis data (59). If this effort is successful then future comparison of thermograms from one instrument to another may be successful. Stabilities of molecular structure may result in multiple melting point. For example the rotational isomers of tetrabromoethane may crystallize in either the trans or gauche form depending on the method of cooling employed (60). It is difficult indeed to obtain crystals that are not mixtures of the two forms. Recent interest in liquid crystals has stimulated a substantial interest in liquid crystal phases in cholesteryl esters. To be sure this is a complicated area of interest (42).

Kinetic Considerations. The kinetics of the fusion process for a linear heating rate depend on the rate of heat transfer. If the rate of heat transfer can be experimentally determined then the slope will be equal to $2\Delta H/t_f^2$. Speros and Woodhouse have indeed showed this to be true for tin, lead and silver. Anomalies in this method were noted in the case of sodium nitrate and aluminum. Fused salts have posed a problem for some time due to the tendency to form ion-association complexes which have a number of consequences (61), one being an abnormally low volume increase on melting. The kinetics of decomposition of solids have been the subject of a vast number of publications. The specific kinetic treatments as applied to polymers have been the subject of a recent evaluation (62). Kinetic interpretations of decomposition reactions are complex because such reactions do not follow a simple reaction order or that several reactions of differing energies of activation are involved. The effects of order, activation energy, heating rate, and temperature dependence on thermograms is illustrated in the above paper. These investigators conclude that methods involving a variable heating rate or several heating rates of replicate samples may provide data which can establish the uniqueness of kinetic parameters. Ellerstein has proposed a method for obtaining kinetic parameters of the glass transition (63). This proposed method has received little attention. It is interesting to note that the approach of Kissinger (64) for differential thermal analysis data and that of Fuoss, Salyer and Wilson (65) for thermogravimetric data are identical. Garn has criticized the use of non-isothermal experiments for the measurement of data used to calculate kinetic parameters (66). He describes the ideal conditions necessary for the use of thermal analysis in kinetics investigations. It is clear that the controversy over kinetic interpretations will continue to be a matter of discussion until the resolution of discrepancies between isothermal versus programmed temperature experimental methods is resolved. The rigorous comparison of various kinetic approach has not been done yet although a large number of papers have appeared in the literature.

The interpretation of decomposition curves is complicated by the "order" of the reaction changing throughout various regions of the decomposition. The activation energy for a number heterogeneous decomposition reactions is very close to the heat of reaction (67). The use of programmed temperature rise experiments using thermogravimetric methods for the evaluation of forward and reverse energies of activation has been reported (68).

A recent interest in identification of decomposition products by infrared spectroscopy is evident. The thermal reduction of bicarbonate to formate (69) the kinetics of formulate ion pyrolysis (70) and the kinetics of oxalate ion pyrolysis (71) are examples.

The techniques of either DTA or TGA do not provide information regarding reaction intermediates thus auxilliary methods are necessary to define reaction pathways. From these experiments it is clear that free radical reactions are bound to play an important role in decomposition reaction. Similar work on the decomposition of coordination compounds has been reported (72).

Auxilliary and Related Techniques. The use of thermogravimetric analysis as a sister technique to DTA is well established. The trend to measurements on smaller and smaller samples had pushed the limits of thermogravimetric equipment. The use of quartz crystal oscillators as microbalances has not provided a thermogravimetric techniques which far exceeds the capability of DTA equipment in terms of handling tiny samples (73). The method of adiabatic cryoscopy, based on a thermal measurement on a equilibrium portion of a cooling curve of a solution has been further exploited by Bruckenstein in cryoscopic titrations (74,75). Instrumentation for thermometric titrations has developed to the point where it is now possible to study the stoichiometry and enthalpies of reactions with the accuracy of conventional solution calorimetry (76). Rapid and precise measurement of molecular weights of low molecular weight polymers (below 40,000) has been popular by the thermoelectric technique (77). Tomlinson and co-workers have received vapor pressure osmometry and calculated thermodynamic efficiencies for a number of solvent systems (78). A detailed study of such a system has recently appeared (79). It should be noted that a very similar approach has been used for measurement of water potential of plan tissue with a thermocouple psychrometer (80).

Recent applications in the field of thermal analysis have been summarized in greater detail (81) than is possible in this paper. Thermal analysis in its many forms is a powerful tool for investigations in chemistry. The interpretation of data may require the application of several techniques except for the simplest of systems. New methods will continue to appear and applications to a multiplicity of disciplines is a sure prospect for the future.

REFERENCES

1. H. F. Stimson, "International Practical Temperature Scale of 1948, Text Revision of 1960", National Bureau of Standards Monograph 37, Superintendent of Documents, U.S. Government Printing Office, Washington, D. C. 10 cents

2. W. F. Roeser and S. T. Lonberger, "Methods of Testing Thermocouples and Thermocouple Materials, National Bureau of Standards. Circular 590, Superintendent of Documents, U.S. Government Printing Office, Washington, D. C. 20 cents

3. C. Halpern and Robert J. Moffat, "Bibliography of Temperature Measurement", January 1953 to June 1960, National Bureau of Standards Monograph 27, Superintendent of Documents, U. S. Government Printing Office, Washington, D. C. 15 cents

4. R. L. Powell, M. D. Bunch and R. J. Corrucini, Cryogenics 1, 139 (1961)

5. R. L. Powell, L. P. Caywood, M. D. Bunch "Temperature, Its Measurement and Control in Science and Industry", Vol 3, Part 2 Reinhold, New York 1963

6. R. F. Walker, Rev. Hautes Temper. et Refract., 3, 301 (1966)

7. G. W. Burns and J. S. Gallagher, J. Res. Nat. Bureau of Standards, 70c, 89 (1966)

8. D. B. Thomas, J. Res. Nat. Bureau of Standards, 67c, 337 (1963)

9. L. Brewer and P. Zavitsanos, J. Phys. Chem. Solids, 2, 284 (1957)

10. D. J. Accinno and J. F. Schneider, Engelhard Industries, Inc. Tech. Bull 1, 1202, 53 (1960)

11. L. O. Olsen and P. D. Freeze J. Res. Nat. Bur. Standards, 68c, 263, (1964)

12. J. W. Smith and D. R. Johnson in "Proceeding of the Second Thermal Toronto Symposium on Thermal Analysis" H. G. McAdie, Editor Toronto Section, Chemical Institute of Canada 1967

13. W. J. Thoburn and L. M. Pidgeon, Can. Met. Quart. 4, 205 (1965)

14. "Recommended Practice for Thermal Analysis of Metals and Alloys", ASTM Designation E 14-63

15. G. K. Burgess, Scientific Paper No 99, Nat. Bureau of Standards (1908)

16. Standard Method of Test for Freezing Points of High Purity Hydrocarbons ASTM Designation D 1015-55.

17. (1) A. L. Day and E. T. Allen, Carnegie Inst. Wash. Publ. No 31 (1905)

18. (2) C. Doelter, Z. Elektrochem. 12, 413 (1906)

19. (3) J. W. Greig and T. F. W. Barth, Am. J. Science,35A, 93 (1938)

20. (4) M. Volmer and O. Schmidt, Z. Phsik, Chem. 35B, 467 (1937)

21. (5) S. E. Khaikin and N. P. Bene, Compt. Rend. Acad. Sci. USSR 23, 31 (1939)

22. (6) G. W. Sears, J. Phys. Chem. Solids, 2, 37 (1957)

23. (7) N. W. Aimslie, J. D. Mackenzie, and D. Turnbull, J. Phys. Chem., 65, 1718 (1961)

24. (8) R. L. Cormia, J. D. Mackenzie, and D. Turnbull, J. Appl. Phys. 34, 2239 (1963)

25. (9) G. Tammann, Z. Physik. Chem. 68, 257 (1910)

26. E. Helmuth and B. Wunderlich, J. App. Phys. 36, 3039 (1965)

27. J. M. Pakulak and G. W. Leonard, Anal. Chem. 31, 1037 (1959)

28. N. A. Nedumov, Russ. J. Phys. Chem. (English Translation) 34, 84, (1960)

29. D. R. Lovejoy, Can. J. Phys. 36, 1397 (1958)

30. R. D. Lee, "Temperature, Its Measurement and Control in Science and Industry, Vol 3, Part I Reinhold, New York p 587

31. I. I. Kirenkov. All Union Sci. Res. Inst. for Metrology, Leningrad, Transactions, No. 71 131 (1963)

32. J. L. Wittenberg and G. R. Grove. Reactor Fuels and Materials Development Plutonium Research: 1965 Annual Report MLM-1328 Jan 20, 1967

33. S. Speil, L. H. Berkelhamer, J. A. Pask and B. Davies, U. S. Bur of Mines Tech. Paper, No. 664 81 pp (1945)

34. S. L. Boersma, J. Am. Ceram. Soc. 38, 281 (1955)

35. G. de Josselim de Jong, J. Am. Ceram. Soc. 40, 42 (1957)

36. C. N. Reilly, J. Chem. Educ. 39, A853 (1962)

37. E. Sturm, J. Phys. Chem. 65, 1935 (1961)

38. P. G. Herold and T. J. Planje, J. Am. Ceram. Soc. 31, 20 (1948)

39. C. Mazieres, Comp. Rend. 248, 2990 (1959)

40. C. Mazieres, Anal. Chem. 36, 602 (1964)

41. E. M. Barrall, R. S. Porter and J. F. Johnson Anal. Chem. 36, 2172 (1964)

42. E. M. Barrall, R. S. Porter and J. F. Johnson J. Phys. Chem. 70, 385 (1966)

43. R. L. Stone and G. T. Burress, U.S. Patent 3,298,220 Jan 17, 1967

44. P. D. Garn "Thermal Analysis of Small Samples, Proc. Intern Microchem. Techniques State College, Pennsylvania 1961 (N. D. Cheronis, ed.) pp 1105-1109 Wiley (Interscience) New York, 1962

45. J. Paulik, F. Paulik and L. Erdey, Anal. Chim. Acta. 34, 419 (1964)

46. H. G. McAdie, Anal. Chem. 35, 1840 (1963)

47. W. H. King, A. F. Findeis, C. T. Camilli, Paper presented this A.C.S. Meeting

48. J. Coops, R. S. Jessup and K. Van Nes in "Experimental Thermochemistry" F. D. Rossin, Editor, Interscience, New York 1956

49. J. M. Sturtevant in "Physical Methods in Organic Chemistry, Part I, Third Ed., A. Weissberger, Editor, Interscience, New York 1959

50. J. Wadso, Science Tools, LKB Inst. J. 13, 22 (1966)

51. D. M. Speros and R. L. Woodhouse, Nature, 197, 1261 (1963)

52. D. M. Speros and R. L. Woodhouse, J. Phys. Chem. 67, 2164 (1963)

53. D. M. Speros, Personal Communication

54. R. D. Freeman, Personal Communication

55. E. S. Watson, M. J. O'Neil, J. Stein and N. Brenner Anal. Chem. 36, 1233 (1964)

56. M. J. O'Neil, Anal. Chem. 36 1238 (1964)

57. S. L. Cooke, Personal Communication

58. B. B. Graves, Ph.D. Dissertation, University of Louisville, Louisville, Kentucky (1966)

59. H. G. McAdie, Anal. Chem. 39, 543, (1967)

60. R. E. Kagarise, J. Chem. Phys., 24, 300 (1956)

61. W. J. Davis, S. E. Rogers, A. R. Ubbelohde, Proc. Roy. Soc. (London) 235 A, 469 (1956)

62. J. H. Flynn and L. A. Wall, J. Res. Nat. Bur. Standards, 70A 487 (1966)

63. S. M. Ellerstein, J. Phys. Chem. 67, 2471 (1965)

64. H. E. Kissinger, J. Res. Nat. Bur. Standards 57, 217 (1956)

65. R. M. Fuoss, I. O. Salyer and H. S. Wilson, J. Polymer Sci. 2, 3147 (1964)

66. P. D. Garn, "Thermoanalytical Methods of Investigation, Academic Press, New York 1964

67. Fishbeck and Snaidt, Z. Electrochem. 38, 199 (1932)

68. T. R. Ingrahm and P. Marier, Can. J. Chem. Engr. Aug 1, 1961 (1964)

69. I. C. Hisatsune, Science, 145, 1455 (1964)

70. K. O. Hartman and I. C. Hisatsune, J. Phys. Chem. 70, 1281 (1966)

71. K. O. Hartman and I. C. Hisatsune, J. Phys. Chem. 71, 392 (1967)

72. R. M. Wing and G. M. Harris, J. Phys. Chem. 69, 4328 (1965)

73. W. F. Fisher and W. H. King Anal. Chem. 39, 1265 (1967)

74. S. Bruckenstein and N. E. Vanderborgh, Anal. Chem. 38, 687 (1966)

75. S. Bruckenstein and M. Kopanica. Abstracts of Papers, Automation in Analytical Chemistry XXI International Congress of Pure and Applied Chemistry, Prague Sept 4-10, 1967

76. J. J. Christenson, R. M. Izatt and L. P. Hansen, Rev. Sci. Instr. 36, 779 (1965)

77. M. J. Cantow, R. S. Porter and J. F. Johnson, J. Polymer Sci. A2, 2547 (1964)

78. C. Tomlinson, C. Chylevski and W. Simon, Tetrahedron, 19, 99 (1963)

79. A. Adicoff and W. J. Murbach NOTS TP 4247 U.S. Naval Ordinance Test Station, China Lake, California 1967

80. P. J. Kramer, Botany Department, Duke University, Personal Communication

81. H. G. McAdie, Editor, "Proceedings of the Second Toronto Sympsoium on Thermal Analysis"

DIFFERENTIAL ENTHALPIC ANALYSIS AS A CALORIMETRIC METHOD: EVALUATION BY A STATISTICAL DESIGN

by

Robert F. Schwenker, Jr.*, Personal Products Company
Division of Johnson & Johnson, Milltown, New Jersey
and
John C. Whitwell, Textile Research Institute and
Department of Chemical Engineering, Princeton University
Princeton, New Jersey

Quantitative data on the heat or enthalpy of crystallization, fusion, crosslinking, decomposition and other polymer transformations are invaluable for the elucidation of reactions and reaction mechanisms as well as for quantitative characterization. However, there is a great paucity of calorimetric data on polymer reactions in the literature because previous investigators were forced to rely largely upon the relatively cumbersome and tedious technique of adiabatic calorimetry for such information. As a consequence, polymer scientists generally eschewed the calorimetric approach, relying upon other types of data in studies of polymer transitions and reactions. Therefore, the successful application of differential thermal methods as rapid and reliable calorimetric techniques represents an important breakthrough in polymer analysis. Differential enthalpic analysis[1], DEA, or differential scanning calorimetry as it is also known, provides the most direct information and a commercial instrument is available[2]. Since the DEA curve reflects heat flow in millicalories per second as a function of the programmed temperature, it follows that the area of an exothermic or endothermic peak is directly proportional to the enthalpy of transformation.

Despite the present advanced development of statistical methods, the great majority of papers still present numerical data without qualification as to precision and accuracy. Not infrequently, when a new instrumental method appears, measurements are made on an arbitrary number of samples with data points obtained by the time honored arithmetic mean approach. The resulting numbers in too many cases take on a disconcerting absolutism and are subsequently cited as definitive values. The major objective of this work was the statistical evaluation of the precision and accuracy with which enthalpy measurements could be made by DEA.

*This work was carried out at Textile Research Institute while the author was a member of the TRI staff.

EXPERIMENTAL

The instrument used was the Perkin-Elmer Differential Scanning Calorimeter, Model DSC-1. The primary materials for calibration were pure metals, selected to provide melting transitions encompassing a wide operational temperature range from 156° to 420°C and described in Table I.

TABLE I
Metal Standards

Metal	M.P., °C	ΔH_f, cal/g.*
Indium (Comm. Reagent)	156.4	6.79
Tin (NBS Std.)	231.88	14.24
Lead (NBS Std.)	327.42	5.55
Zinc (NBS Std.)	419.50	24.39

*Literature values[3].

The variables of (1) heating rate, °C/min., (2) sample size, mg., (3) instrument sensitivity level and (4) metal, were incorporated into a factorial design shown schematically in Figure 1. In order to keep the total number of experimental runs within reasonable limits and to minimize design complexity, three metals, indium, tin, and lead, were used. Zinc was subsequently run separately in a reduced design. In the main statistical design, each of the variables of heating rate, sample weight, and sensitivity or range was used at two levels as follows.

Heating Rates, °C/min.:	5 and 10
Sample Weights, mg.:	5 and 10
Sensitivity Levels:	8 and 16

Thus, in all there were four fixed populations to provide a 3×2^3 design. Within each cell the random variables representing sample and curve replication appear twice each, requiring a total of 96 experimental runs. Peak areas were measured by means of a planimeter and data points were the average values of two measurements. The data were subjected to analysis of variance to determine significant interactions, components of variance for the random variables, and precision.

For each metal, individual samples were run in each of 8 possible combinations of weight, heating rate, and sensitivity level for a total of 16 separate samples and 32 runs. Thermograms for analysis were obtained on samples prepared by melting metal shavings and the melt cooled in the sample pan. This provided known thermal history in addition to eliminating curve "noise" or other perturbations due to original sample form.

		LEAD		INDIUM		TIN	
		5 °C/min	10 °C/min	5 °C/min	10 °C/min	5 °C/min	10 °C/min
SENSITIVITY = 8	5 mg	a					
		b					
		c					
		d					
	10 mg	a					
		b					
		c					
		d					
SENSITIVITY = 16	5 mg	a					
		b					
		c					
		d					
	10 mg						a
							b
							c
							d

Fig. 1 THE FACTORIAL DESIGN:
Plan of Experimental Runs for Metal Standards

a – A sample of indicated weight, 1st area measurement
b – Same sample as at a, 2nd area measurement
c – A 2nd sample of indicated weight, 1st area measurement
d – Same sample as at c, 2nd area measurement

Thus, the experimental procedure was to run each sample twice using the curve from the second run for measurement. Initially, three runs were made and areas of the last two curves measured. However, it was found that there was no significant difference in peak areas so that the two-run procedure was deemed adequate. The metal samples were weighed to 0.1 mg. on a standard analytical balance.

RESULTS AND DISCUSSION

The results of preliminary work to explore the influence of sample weight and heating rate on the heat of fusion, ΔH_f, are shown in Table II.

TABLE II

Effect of Sample Weight and Heating Rate on ΔH_f of Lead

Sample Weight (mg.)	ΔH_f*	Heating Rate** (°C/min.)	ΔH_f*
5.4	5.74	2.5	5.55
10.2	5.54	5	5.52
22.1	5.60	10	5.58
32.6	5.45	20	5.50

*Means of five measurements
**Sample weight - 10.2 mg.

The data suggested small or no significant effects in essential agreement with results reported earlier[2] on the ΔH_f of Indium.

Instrument calibration constants were calculated from experimental data using the following equation.

$$C = \frac{A \times R \times 60}{\frac{1}{2} W_c \times L_c \times W} \cdot \frac{1}{\Delta H_{lit}} = \frac{\Delta H_{exp}}{\Delta H_{lit}} \qquad (1)$$

where, C = calibration constant
A = peak area, sq. in.
R = instrument sensitivity, millicals/sec.
W_c = chart width, inches
L_c = chart speed, in./min.
W = sample weight, mg.

Calibration constants thus obtained may then be used to calculate the heat of transformation, ΔH, of an unknown sample as follows.

$$\Delta H_X = \frac{A \times R \times 60}{\frac{1}{2} W_C \times L_C \times W \times \overline{C}} \qquad (2)$$

where, ΔH_X = enthalpy of transformation, cals./g
$\overline{C}$ = average calibration constant

The calibration constants obtained for the test metals are given in Table III.

TABLE III
Calibration Constants for Metal Standards

Metal	Constant, C*	M.P., °C
Indium	1.0436 ± 0.0189	156.4
Tin	1.0191 ± 0.0189	231.9
Lead	0.9930 ± 0.0189	327.4
Zinc	1.0427 ± 0.0482	419.5

*Estimated 95% Confidence Level (C.L.)

The values were all relatively close to unity, but it is possible that a temperature effect was operative in view of the higher values found at the upper and lower temperatures, whereas values very close to unity were found in the middle range of temperature. The lower precision evident in the case of zinc is due to the fact that only 8 separate samples were used instead of 16 samples.

The variance of a calibration constant includes (1) variance due to the particular instrument, (2) the variance of the separately weighed samples, (3) the variance in area measurements, and (4) all other random, uncontrollable and unidentifiable effects. The irreducible, minimum variance would be obtained by recycling the same sample. A variance so measured would not, however, pertain to a realistic experimental situation where all contributions to variance are involved. Since any instrumental variations are involved in these calculations, each different instrument will have its own constants for different materials so that a calibration constant obtained on one instrument should not be used to determine Δ H values using a different instrument.

The predicted influence of the number of samples taken and the number of peak area measurements on calibration constant precision, is shown by the data in Table IV. The data indicate that the best procedure is to replicate samples rather than planimeter measurements. Thus, the standard errors show that no significant improvement in precision results from making four area measurements instead of two, whereas precision is markedly improved by increasing the number of samples. These standard errors were calculated based on the variances obtained in the experimental design, as shown in the Appendix.

TABLE IV
Predicted Precision of Calibration Constants

No. of Measurements Samples	Area	Std. Error	Estimated C.L., % (95% C.L.)
1	1	0.0286	-----
	2	0.0277	± 5.5
2	1	0.0202	-----
	2	0.0196	± 3.9
4	1	0.0168	-----
	2	0.0139	± 2.8
	4	0.0137	-----
8	1	0.0101	-----
	2	0.0098	± 2.0
	4	0.0096	-----

Using equation (2), the results obtained on the heats of fusion, ΔH_f, of the test metals are shown in Table V.

TABLE V
Heats of Fusion of Metal Standards

Metal	$\Delta H_{exp.}$, cal/g.	Est. Precision,% (95% C.L.)	$\Delta H_{lit.}$, cal/g*
Indium	7.09 ± 0.13	± 1.89	6.79 ± 0.17
Lead	5.51 ± 0.10	± 1.89	5.55 ± 0.14
Tin	14.51 ± 0.28	± 1.89	14.24 ± 0.25
Zinc	25.44 ± 1.23	± 4.82	24.39 ± 0.46

*Literature values[3].

Using a dynamic adiabatic calorimeter the heat of fusion of lead was recently reported as 5.6 cals/g. with a precision of ±2% and an accuracy of ±2%.[4] In the case of the metals which figured in the main design, indium, lead, and tin, precision as good as that reported by conventional calorimetry was obtained. With respect to accuracy, the results agreed with corresponding literature values to within 1% to 4.5%. The precision in the case of the experimental value for zinc was reduced because fewer samples were involved.

The results obtained on the metals can be used to predict a combination of calibration and unknown sample runs required to obtain a given level of precision. The effect of the number of samples used to determine the calibration constant and the number of samples for determining the ΔH of an unknown transformation on precision, expressed as the 95% confidence limits, is shown in Table VI.

TABLE VI
Predicted Confidence Limits on ΔH

No. of Samples Calibration	Unknown	Est. 95% C.L. in percent
	2	± 6.52
1	4	± 6.22
	8	± 5.90
	2	± 5.56
2	4	± 4.80
	8	± 4.36
	2	± 4.86
4	4	± 3.96
	8	± 3.42
	2	± 4.40
8	4	± 3.46
	8	± 2.80
30	30	± 1.00

The results of these calculations indicate that for the higher levels of precision several samples are required for calibration. This conclusion implies correctly that the variance of the calibration constant, which reflects the effect of sample weighing, is a large contributor to error.

In the case of fibers and other polymeric materials, the levels of precision predicted by Table VI represent only approximations since such materials are perforce less homogeneous than pure metals. In many instances the transformations, such as cross linking and decomposition for example, are more complex than simple fusion. Thus, the precision attainable will depend on sample purity and the complexity of the transformation. It would be expected that for polymers the best precision would obtain for fusion.

Some preliminary results on a low molecular weight organic compound and two polymers are given in Table VII.

TABLE VII
Heats of Transformation of Organic Materials

Material	Transformation	ΔH, cal/g	Est. 95% C.L.,% Predicted	Found
Dextrose	Fusion	42.58 ± 1.72	± 3.5	± 4.3
Nylon 66	Fusion	14.38 ± 1.06	± 3.5	± 7.3
Polyacrylonitrile	Cross linking and/or cyclization	116.9 ± 8.2	± 3.5	± 7.1

These results, from four samples of each compound, show that in the case of pure dextrose, the predicted precision based upon the analysis in Table VI was much better than that for the polymer transformations.

ACKNOWLEDGEMENT

The authors are grateful to Textile Research Institute for permission to publish these results. They also wish to thank Robert DiPietro for assistance in the laboratory phase of this work.

REFERENCES

1. Eyraud, C., Comptes Rendus 238, pt. 2, 507 (1954).
2. Watson, E. S., O'Neill, M. J., Justin, J. and Brenner, N., Anal. Chem. 36, 1233 (1964).
3. Kubaschewski, O. and Evans, E. L., Metallurgical Thermochemistry, 4th Edition, Pergammon Press, N. Y. 1958.
4. Wendlandt, W. W., Delta Therm Application Notes, No. 58, May 13, 1966, Technical Equipment Corp., Denver, Colo.

APPENDIX

The constants calculated for all 96 points are reported in Table VIII. The analysis of variance, for these data, is given in Table IX. Conclusions drawn from this last table were as follows.

1. The constants differ for the metals as would be expected since they depend upon literature values of unknown accuracy.

2. There is a possibility of a small heating rate effect which might be attributed to pen response time in the recorder. This possibility has been ignored in calculating predicted precisions in Tables IV and VI. Consequently, the values reported in these tables may be applicable only to a single heating rate.

3. The average values of constants reported in Table III include both heating rates. Although the effect is apparently discernible in all 96 runs, it appears to be insignificant for each of the three metals.

4. No mean squares for two-factor, three-factor, or four-factor interactions are of a size to cause concern.

5. The variance due to replicate sample preparation is appreciably greater than that due to replicated area measurements. This effect would be anticipated due to the problems of weighing small samples on a standard analytical balance. This situation can be improved if a balance of greater sensitivity is used.

TABLE VIII
Individual Calibration Constants

		Lead		Indium		Tin	
		H_1	H_2	H_1	H_2	H_1	H_2
S_1	W_1	0.9506	0.9787	1.0692	1.0080	1.0120	1.0379
		0.9694	0.9975	1.0230	1.0150	1.0050	1.0084
		0.9694	1.0070	1.0538	1.0769	1.0306	1.0379
		0.9600	1.0352	1.0307	1.0692	1.0196	1.0084
	W_2	1.0117	0.9553	1.0460	1.0150	0.9965	1.0012
		1.0117	0.9741	1.0460	1.0610	1.0250	1.0453
		0.9600	1.0352	1.0538	1.0308	1.0417	1.0399
		0.9694	1.0023	1.0230	1.0308	1.0288	1.0197
S_2	W_1	0.9787	1.0540	1.0300	1.0460	0.9903	1.0420
		0.9599	1.0540	1.0150	1.0310	1.0049	1.0350
		0.9189	1.0351	1.0461	1.1230	0.9610	1.0417
		0.9369	1.0163	1.0615	1.0461	1.0490	1.0563
	W_2	0.9320	0.9974	1.0540	1.0610	1.0050	0.9940
		0.9410	0.9881	1.0540	1.0380	0.9973	1.0014
		1.0539	1.0257	1.0154	1.0155	1.0087	1.0420
		1.0539	1.0445	1.0615	1.0463	1.0050	1.0200

TABLE IX
Analysis of Variance for the Calculated Constants

Source of Variation	Mean Square	Degrees of Freedom	F
Metals (M)	0.020479	2	14.4
Weights (W)	0.000007	1	
Sensitivity (S)	0.000379	1	
Heating Rate (H)	0.008518	1	6.0
MW	0.000486	2	
MS	0.000756	2	
MH	0.002802	2	<2
WS	0.000170	1	
WH	0.005489	1	3.9
SH	0.003000	1	2.1
MWS	0.000261	2	
MWH	0.000966	2	
MSH	0.000173	2	
WSH	0.001249	1	
MWSH	0.000014	2	
Replicated Sample Prep.	0.001423	24	3.9
Replicated Area Meas.	0.000362	48	

$F(0.05, 2, 24) = 3.40$
$F(0.05, 1, 24) = 4.26$
$F(0.05, 24, 48) = 1.75$

THIN FILM THERMOCOUPLES FOR DIFFERENTIAL THERMAL ANALYSIS

W. H. King, Jr.
Esso Research and Engineering Company
Linden, New Jersey

A. F. Findeis
University of Alabama
University, Alabama

C. T. Camilli
Esso Research and Engineering Company
Linden, New Jersey

SUMMARY

Thin film thermocouples supported on quartz were developed for differential thermal analyses (DTA) over the range -125°C to 500°C. The low mass of the thermocouples, the small sample size, together with other design features practically eliminates problems from changes in heat transfer coefficient between sample and sensor. This results in very flat baselines and high precision measurement. It is this high degree of repeatability that distinguishes this DTA system from others. The system allows calorimetric values to be obtained from peak areas. Calibrations based only on indium yielded calorimetric data which agree well with the established literature.

INTRODUCTION

The precise measurement of thermal transients by the technique of Differential Thermal Analysis (DTA) has recently attracted a number of investigators. In this paper, a new approach to DTA is described which results in the highly reliable measurement of thermal processes taking place in a wide variety of samples. This approach involves the use of vacuum evaporated thermocouples on a quartz substrate. These thermocouples provide a sensing element possessing a number of advantages over previously described differential measuring devices.

Thermocouple fabrication by the deposition of thin metal films has been described many times. Harris et al (1,2,3,4) have made single and multijunction thermocouples by sputtering bismuth-antimony and bismuth-tellurium on thin cellulose ester substrates (films). Rapid response thermocouples for use a radiation detectors in the infrared region of the spectrum have been described (5). Hornig and O'Keef (6) have thoroughly discussed the design of fast thermocouples for use with interrupted radiation in infrared spectroscopy. The work reported herein is the first known application of such rapid response thermocouples to DTA.

Differential thermocouples, thermistors, and resistance thermometers as applied to DTA have been described. Different problems are associated with each sensing method and the problems vary with the temperature range of the measurement. Ordinary differential thermocouples can readily be made, but precise matching of the junctions is difficult. Precise DTA measurements depend largely on matched thermocouples. Intimate coupling of the sample to the sensor is of equal importance. Thermistors may be matched precisely only over limited temperature ranges, and their upper temperature limits are severely restricted. Resistance thermometer techniques for DTA measurements imply that matched resistance thermometers be fabricated and this is indeed difficult. Furthermore, the differential sensor should ideally contribute no heat to the system. Resistance thermometers fabricated for small sample holders and incorporated in the usual differential bridge type of circuit fail in this regard due to their necessarily small dimensions. The very nature of the resistance change with temperature results in a nonlinear response. Some of these disadvantages have been overcome in the null-balance design of a differential scanning calorimeter by (11) Watson, et al. The ideal sensing element should have high thermal impedance paths for all possible modes of heat flow except for that to the sensing elements, and the sensing elements for sample and reference should match perfectly. This ideal would require that the sensor, sample, and sample holder have not mass; no leads should be connected to the sensing element, and heat flow to the sample should be by a high thermal impedance path.

This report presents a method of fabrication of a sensing element having nearly ideal characteristics for use in the technique of DTA. Lightweight, precisely matched thermocouples were produced by the evaporation of thin films of dissimilar metals which overlap to form the junctions. In order to maintain high purity gold films the vacuum evaporation equipment was used for gold only. Thus, possible contamination from other metals in the system was reduced. Nickel was subsequently electroplated onto the gold to form the thermocouple junction. In this way precisely matched differential thermocouples may be fabricated. Gold and nickel have a favorable thermoelectric power which is intermediate between the standard Pt-Pt (19% Rh) and the chromel-alumel pairs.

EXPERIMENTAL

Quartz slices, AT cut, 0.66 x 0.55 x 0.027 inches, similar to those used in quartz crystal oscillators for frequency control, were used as a substrate for the thermocouples. The quartz slices were first lapped with #500 carborundum abrasive. They were then cleaned in an ultrasonic detergent bath followed by a wash with water and finally with reagent grade methanol and then air dried. The quartz slices were then masked with aluminum foil and placed in a vacuum deposition apparatus to receive a gold film. The gold deposit on the quartz slice was then electroplated with nickel from a stirred bath containing 280 g. of $NiSO_4 \cdot 7H_2O$, 30 g. of H_3BO_3, and 20 g. of KF per liter of solution. A pure nickel anode was used in the plating bath. The thickness of the nickel plate was carefully controlled with a coulometer. The gold deposit was usually around 100 μg./cm.^{2} and the nickel plate was about 200 μg./cm.^{2}. Both nickel-gold-nickel and gold-nickel-gold differential thermocouples were produced by this procedure. The sequence of steps in the fabrication of the thermocouples is shown in Figure 1. The thermocouples were electrically connected by the appropriate nickel or gold wires to avoid an additional thermal junction. A small drop of silver print paint was placed on each connection to ensure good electrical contact. Copper wires were most often used in place of gold because the thermal EMF between Cu and Au is small (0.15 μv/°C).

Both glass microscope slides and quartz plates were very satisfactory as substrates. Both are excellent insulators and good adherence of the metal films was obtained by roughening the surface. Quartz is free of transitions up to 575°C. Mica was not satisfactory because adhesion was poor. It was thought that the thickness of the quartz plates would affect results by providing better insulation under the thermocouple. This was tested by adding extra insulation (up to 1/4") under the quartz plate. The added insulation had no observable effect on peak shape or amplitude in n-C_{32} transition.

ALUMINUM SAMPLE PANS

The sample pans are considered an important part of the system. We used the Perkin-Elmer DSC pans, covers, and pan crimper. These pans are 1/4" diameter aluminum and together with a cover weigh 21.10 mg. with a standard deviation of 0.14 mg. Some samples in the form of films can be deposited on the thermocouples and the pan eliminated. Tests were made with and without pans. Repeatability is best when the samples are enclosed in the pans. Sample weights from 0.5 to 25 mg. were tested and it was found that 5 to 10 mg.

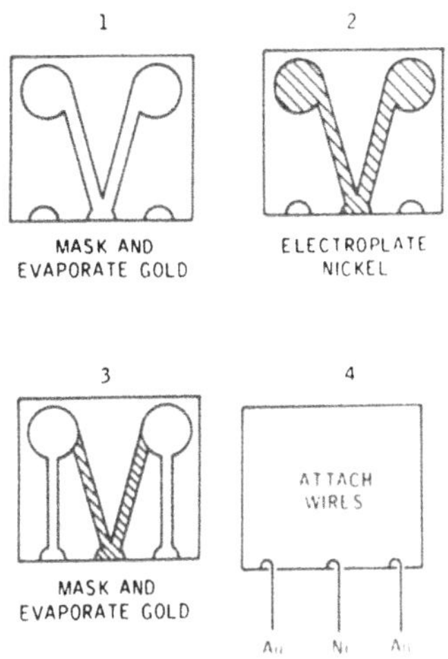

Fig. 1. Steps in making thermocouples.

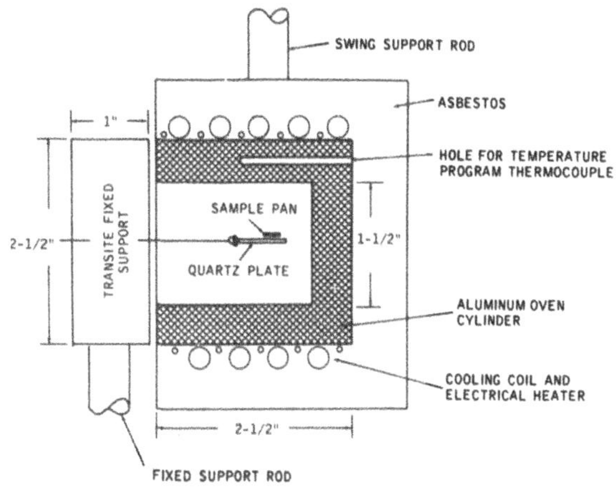

Fig. 2. Temperature programmed oven. The sensing elements and sample pans are attached to a fixed support. The oven assembly swings into position pendulum fashion.

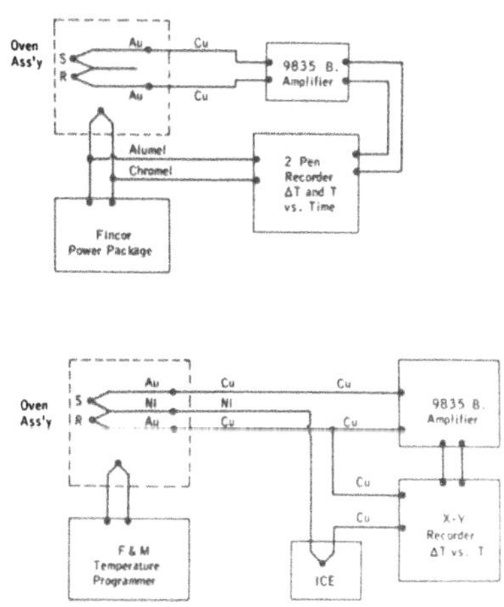

Fig. 3. Electric hook-up of DTA apparatus for strip chart and X-Y recorder presentations.

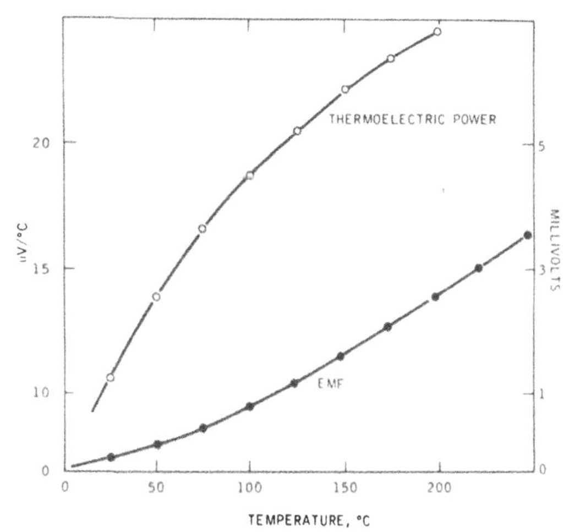

Fig. 4. Thermocouple calibration. Because the thermoelectric power changes with temperature, calorimetric measurements must be appropriately corrected for this effect.

is a good compromise between resolution, signal strength, and repeatability. Making the pans from a better conductor than aluminum was considered; however materials with higher conductivities also have higher heat capacities thus offsetting any advantage. Further, a subsequent test showed the aluminum pans to have excellent heat distribution properties.

OVEN DESIGN

The design and shape of the oven is not critical. The first oven design was made from a tin can 2-1/2" diameter by 1-3/4" height. An electrical heater was wound on the outside of the can and then covered with 1/2" of asbestos insulation. A rectangular slot in the can permitted entry of the junctions. A more sophisticated design made from a 2-1/2" diameter x 2-1/2" long aluminum cylinder is shown in Figure 2. No difference in performance could be detected when the center hole was 3/4" diameter or 1-1/2" diameter. A third design having a 1/2" x 3/4" x 1-1/2" rectangular shaped hole in an aluminum block was also used. The oven experience indicates that the thermal impedance between the oven walls and the samples is very high and doesn't change much with the hole size. The high thermal impedance coupling of the oven to the sample is an important design consideration of this DTA system. The advantage of not using a heat sink to homogenize the temperature increases sensitivity. This was pointed out by Mazieres (7,8).

The oven designs allow programmed temperature operation from -120°C to over 500°C. Cooling coils were fabricated from 1/4" diameter tubing. The oven was cooled by applying a partial vacuum to one end of the stainless steel coil and drawing cool nitrogen vapor and subsequently liquid nitrogen through the coil. Controlling the flow of nitrogen vapor either up or down, linear temperature programming was possible.

THE OVERALL SYSTEM

Figure 3 shows the electrical circuits for two of the systems used in this work. The first system used for calorimetric data employs a two pen strip chart recorder. A chart speed of 1.5 inches per minute was used. One pen records the differential temperature signal (ΔT) obtained from the L+N 9835-B amplifier. The other pen records the temperature signal (T) from a small chromel-alumel thermocouple placed near the differential junction. A Fincor Power Package programs the oven temperature from a cam cut to provide linear temperature rates. The system, except for oven and thermocouples, was obtained as a package from the Leeds & Northrup Company, Philadelphia, Pennsylvania.

The second system shown in Figure 3 was used to obtain thermograms displaying differential temperature as a function of reference or sample temperature. This plot is convenient for research work because the same size graph is obtained regardless of heating rates. In this set-up, an X-Y recorder was employed, and the oven temperature was controlled by an F & M Model 240-M Linear Temperature Programmer.

PERFORMANCE

The thermocouples tested in the absence of sample holders or samples showed a drift of 0.5 μv. from room temperature to 400°C. With temperature raise rates of 200°C per minute, baselines were straight. Under normal run conditions of 10°C/ minute a baseline drift was typically 2×10^{-9} volts/°C. In the few cases where drift and noise appeared, the trouble was usually due to poor electrical contacts. Drift is much less with this system than has been obtainable with the conventional type of differential thermocouples. The improvement is probably due to the exact match of composition of the junctions resulting from the method of fabrication. The noise level of the recorder trace could be maintained at 0.05 μv. with a time constant of about one second.

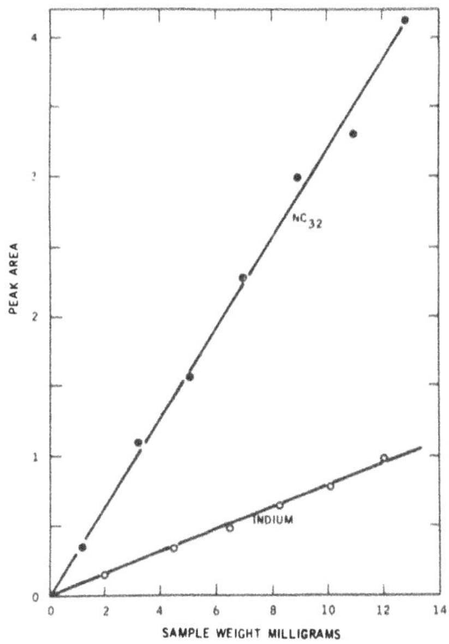

Fig. 5. The peak areas are linear functions of sample weight. This is a good indication for a quantitative system.

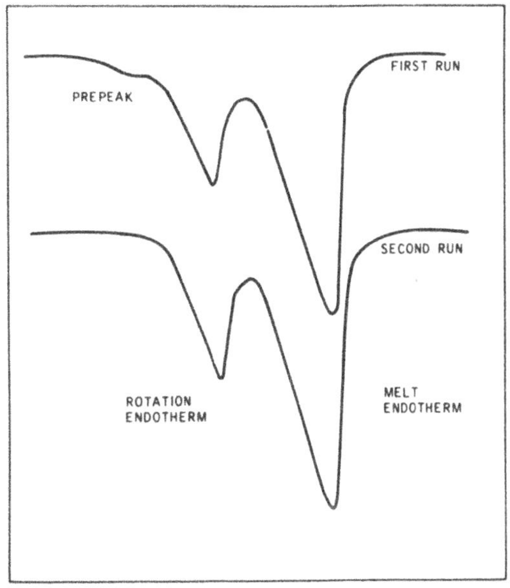

Fig. 6. Thermograms of n-C_{32}. The first run shows a small peak before the rotational transition peak. This peak is removed from the second run by slow cooling of the sample from the first run.

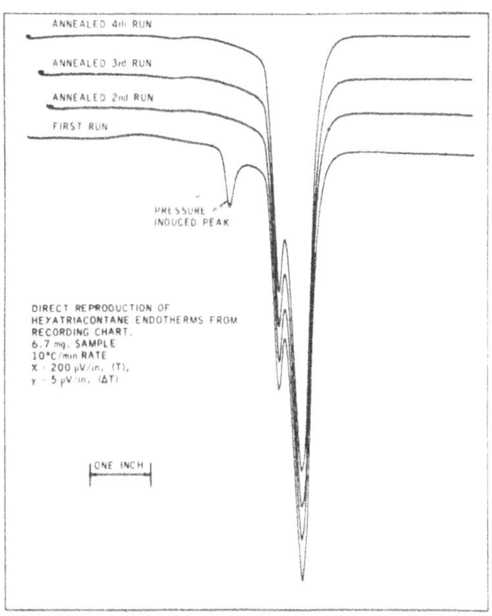

Fig. 7. Repeat thermograms of a sample of n-C_{32}. Slow cooling after the first run removes the "pressure peak" and the curves show a very high degree of repeatability. This figure is a direct copy from the recorder chart so the noise level and flatness of the baseline are revealed.

The voltage response and the thermoelectric power of the thermocouple are shown in Figure 4. Temperature measurements were made with a chromel-alumel thermocouple placed adjacent to the gold-nickel thermocouple in the DTA oven. The oven was programmed at a rate of 2°C/min. and the data recorded with an X-Y recorder. The thermoelectric power was calculated from the slope of the gold-nickel thermocouple vs. temperature curve. The thermoelectric power of the gold-nickel junction increases with temperature and it must be known for accurate calorimetric measurements with this system.

The areas under DTA thermograms were shown to be a linear function of sample size for a wide range from 1/2 mg. to 25 mg. A small positive deviation from linearity for exothermic reactions and a negative deviation for endothermic reactions would be predicted from the shape of the thermoelectric power curve but area measurements with the planimeter are not precise enough to observe such deviations. A response curve for the melt transitions in indium and n-dotriacontane are shown in Figure 5. The deviations from the curve were subsequently found to be the result of errors in the measurement of the mass of the sample and in area measurements with the planimeter. Repeat measurements on the fusion of a precisely weighed sample of indium resulted in area measurements reproducible to $\pm$ 1.2% for one standard deviation.

CALORIMETRIC MEASUREMENTS

Numerous investigators have noted the utility of peak area measurements for calorimetric values from DTA thermograms (9). This approach has been strongly criticized by the classicists as thermometric rather than calorimetric. The authors choose the calorimetric term. With the sensing element described in this paper precise calorimetric values may be obtained. The calorimetric values reported here were obtained using indium as the only standard. Area measurement were made with a K&E compensating polar planimeter. The thermoelectric power of the gold-nickel thermocouples is an increasing function of temperature, thus a correction must be applied based on the ratio of the thermoelectric powers at the transition temperatures of the sample and the standard. In the case of indium, the thermoelectric power at the fusion temperature is 22.6 μv./deg. while that at the fusion temperature of n-dotriacontane is only 16.2 μv./deg. Thus, for an equivalent response, the peak area for indium would be 22.6/16.2 times as large as the n-dotriacontane area. Incorrect values of n-dotriacontane would result without such a correction. This correction is not unique to the gold-nickel system. Chromel-alumel thermocouples, in the range of 100-400°C have a relatively constant, thermoelectric power. Thus the correction is often ignored since it is small. At lower temperatures, i.e., -100°C, the thermoelectric power of the chromel-alumel pair is lower by about 30%; hence such a correction is necessary.

The problem of area measurement deserves further comment. Planimeter measurements are precise for well defined areas and reproducibilities of 0.5% are easily achieved. Precision deteriorates when decisions about the establishment of baselines or the resolution of overlapping peaks are involved. The measurements can be made more reproducible from observer to observer only if one defines precisely both where to draw the baseline, and the method of resolution of peak areas for overlapping transitions. A heat capacity change is usually observed in melting transitions; hence the baseline is not cotinuous. The baseline both before and after the transition must be extrapolated. The first deviation from the extrapolation on both the lower and higher temperature ends of the transition defines two points between which the baseline should be drawn. The points may also be established by a preliminary run at much greater sensitivity. In the case of overlapping transitions, the valley between the transitions was divided on the basis of the ratio of the peak amplitudes of the transitions. The region of overlap was divided assuming the trailing edge of the first transition was identical in normalized shape with the trailing edge of the second transition.

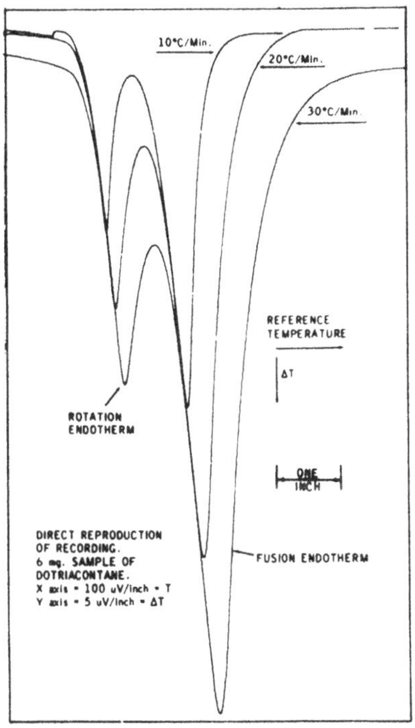

Fig. 8. n-C_{32} thermograms showing the effect of heating rate on size, shape, and peak position.

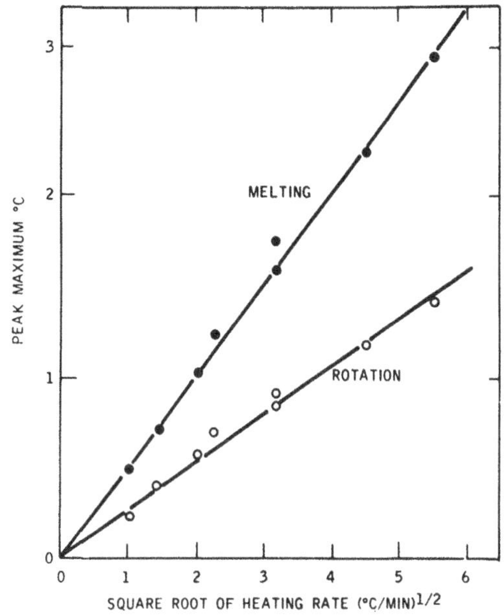

Fig. 10. The amplitudes of n-C_{32} peaks are straight-line functions of the square root of the heating rate.

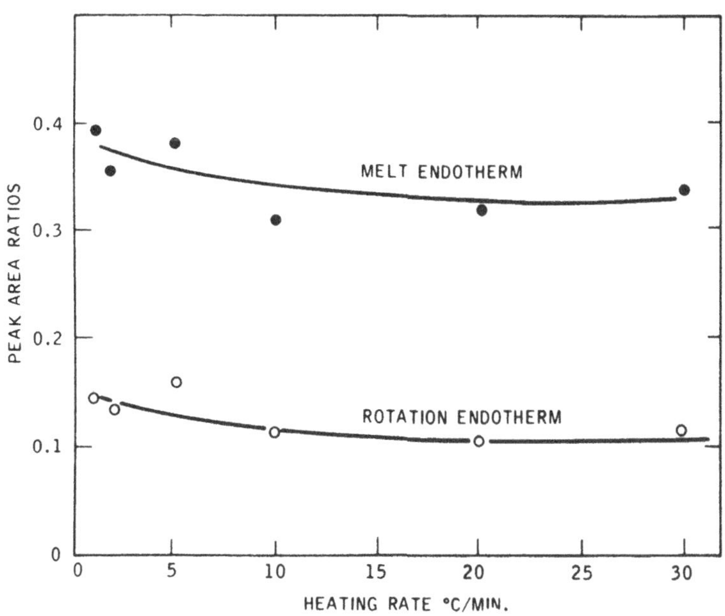

Fig. 9. The ratio of peak area to heating rate is almost a constant.

The value obtained for the heat of fusion of n-dotriacontane from the line average of Figure 5 was 38.9 cal/gm. More precise measurements on carefully weighed 7-9 mg. samples of n-dotriacontane yielded a value of 38.4 ± .46 cal.g. This value is consistent with that found by others (10). The heat of fusion of silver nitrate was determined to be 16.29 cal./g., a value in excellent agreement with an accepted value of 16.216 (11).

Thermograms of n-dotriacontane obtained on crystalline material show the existence of a pre-rotational endotherm. This was described by Broadhurst as the monoclinic-orthorhombic transition (12). The presence of this transition is strongly dependent on the thermal and mechanical history of the sample as well as purity. This endotherm appears in freshly recrystallized dotriacontane and can be revealed again by thermal or mechanical treatment of the sample. A sample, after rapid quenching, upon reheating shows this prepeak endotherm. Squeezing the sample pan with the crimper will also cause it. This transition was not observed on a slow freezing cycle, nor is it observed on a heating cycle following the slow cooling. The prepeak transition can complicate calorimetric measurements of the rotational transition since the two are not always well defined. Thus, the ratio of melting to rotational transitions in samples prepared from crystalline n-dotriacontane was 2.46 ± 0.17 when programming up in temperature, and 2.87 ± .10 after the sample had been melted and slowly cooled. The latter value is in good agreement with the average ratio of 2.78 reported previously (10). A thermogram of n-dotriacontane showing the "pre-rotational" transition and the same sample re-run immediately after slowly cooling is shown in Figure 6. The calorimetric value of the rotational transition (orthorhombic-hexagonal) is thus 13.4 cal./g., a value in excellent agreement with Differential Scanning Calorimeter (10) data. A value for the pre-rotational transition, based on the average ratios of the melt and rotational transitions, is approximately 2.2 cal./g.

The existence of a pre-rotational peak in other normal paraffins has also been observed. For example, it is readily observed in n-hexatricontane on the first run with a sample in a crimped container. The peak subsequently disappears in succeeding thermograms of the same sample after 10°C/min. cooling. This n-C_{36} pre-rotational peak may also be made to appear again by quenching or recrimping the same sample. This is shown in the repetitive scans in Figure 7.

HEATING RATE VS. RESPONSE

An increase in heating rate results in an apparent sharpening of the transition if one records the differential temperature vs. time. The increased rate actually makes the resolution of peaks poorer, which becomes apparent when one records the differential temperature vs. temperature. In Figure 8 the broadening of peaks with increased heating rate is observed together with movement of the peaks to an apparent higher temperatures. The area of the peaks must increase directly with the heating rate since the transition must absorb the same quantity of heat regardless of the speed. To absorb this energy in a shorter time, the temperature difference between sample and reference must be greater, thus the peaks are higher at higher rates. In Figure 9 the peak areas were divided by the heating rates. Agreement to the expected horizontal line is reasonably good.

The amplitude of the peaks is a linear function of the square root of the heating rate. Excellent agreement with straight correlation lines is shown in Figure 10. This correlation is better than that of the area measurements because the peak amplitude can be measured more precisely than peak area. Agreement of heating rate data with theory provides an extra measure of confidence in the calorimetric values determined with this new DTA system.

CONCLUSIONS

This new DTA probe has been successfully applied to waxes, polymers, inorganic salts, metals, and a wide variety of other materials. No correlation is necessary for response due to heating rate and calorimetric data superior to any previously described DTA system may be obtained by direct area measurement followed by a correction for thermoelectric power. It is believed that this system constitutes a DTA approach superior to any previously reported technique.

LITERATURE CITED

1. Harris, L., Phys. Rev., 45, 636 (1934).
2. Harris, L. and Johnson, E. A., Rev. Sci. Inst., 4, 454 (1933).
3. Harris, L. and Johnson, E. A., Ibid, 5, 153 (1934).
4. Harris, L. and Scholp, A. C., J. Opt. Soc. Am., 30, 519 (1946).
5. Roess, L. C. and Dacus, E. N., Rev. Sci. Inst., 16, 164 (1945).
6. Hornig, D. F. and O'Keefe, B. J., Rev. Sci. Inst., 18, 474 (1947).
7. Mazieres, C., Compt. Rend., 248, 2990 (1959).
8. Mazieres, C., Anal. Chem., 36, 602 (1964).
9. Barrall, E. M., Porter, R. S., Johnson, J. F., Anal. Chem., 36, 2172 (1964).
10. Watson, L. S., O'Neil, M. J., Justin, J., Brenner, N., Anal Chem., 36, 1233 (1964).
11. Kelley, K. K., U.S. Bureau of Mines Bulletin 601, pg. 105, U.S. Government Printing Office, Washington, D.C., 1962.
12. Broadhurst, M. G., J. of Research, Nat. Bu. Stds., 66A, No. 3, May-June, 1962.

PURITY DETERMINATIONS USING A DIFFERENTIAL SCANNING CALORIMETER

by

G. L. Driscoll, I. N. Duling, and F. Magnotta

Research & Development
Sun Oil Company
Marcus Hook, Pennsylvania

INTRODUCTION

A Perkin-Elmer Differential Scanning Calorimeter Model 1B (DSC) has been described as useful for many varied thermal measurements,[1] including measurement of the purity of organic compounds.[2] Calculation of purities from the experimental data by the suggested method was not completely satisfactory. A program was undertaken, therefore, to better define the applicability of the method and its accuracy. A computer program has also been developed to aid the calculations.

DESCRIPTION OF METHOD

The DSC system consists of a sample and a reference pan, each equipped with a platinum resistance thermometer, a temperature programmed heater, and an auxiliary heater. The temperature programmer heats the pans at a rate selected in the range of .625°/minute to 80°/minute. A sensing circuit detects any difference in temperature between the pans and supplies current to the auxiliary heaters to maintain a zero differential. This auxiliary energy input is measured, amplified, and displayed on a recorder. The scan, therefore, is time (linear with temperature) against millicalories per second supplied to whichever pan is requiring more energy.

Starting with the van't Hoff equation, which relates the melting point depression to the amount of impurity, equation I can be derived.

$$T_s = T_o - \frac{RT_o^2 X_2}{\Delta H_f} \cdot \frac{1}{F} \qquad \text{(I)}$$

Where:

ΔH_f = heat of fusion of the solvent (major component)
R = gas constant, cal/deg.-mole
T_o = melting point of infinitely pure solvent
T_s = instantaneous temperature of the sample
X_2 = mole fraction of the impurity
F = mole fraction melted at T_s

Thus, plotting T_s versus $1/F$ should give a straight line with an intercept of T_o and a slope of $-\frac{RT_o^2 X_2}{\Delta H_f}$.

Since ΔH_f can be determined from the total peak area and F from the ratio of the partial area at T_s to the total area, sufficient information is obtained from the melting peak to calculate the mole fraction impurity (X_2).

A typical plot of T_s versus $1/F$ is shown on Figure 1. It is readily apparent that the raw data do not define a straight line. This is due to the fact that some peak area is missed before the instrument departs measurably from the baseline due to the noise levels and sensitivity involved. Therefore, trial and error additions of small increments to both

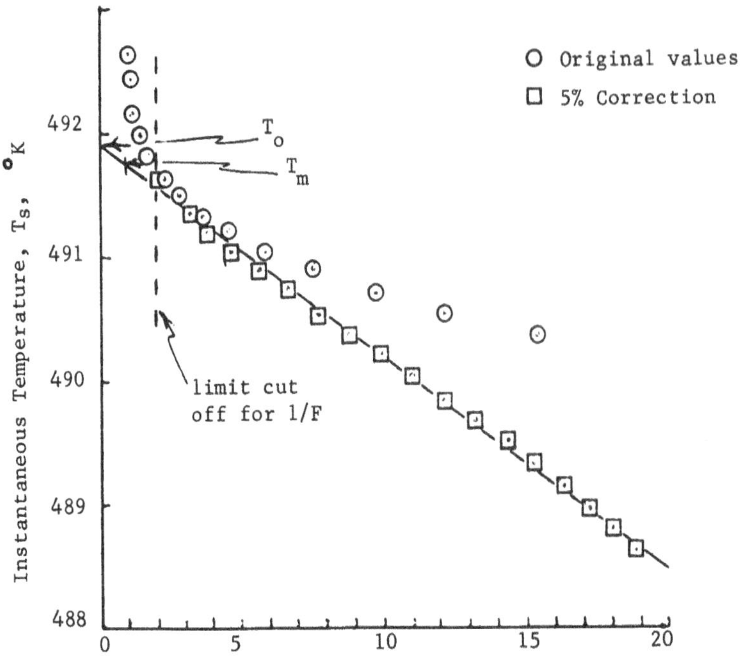

FIGURE 1 PLOT OF SAMPLE TEMPERATURE VS. RECIPROCAL FRACTION MELTED FOR 5,7-DIMETHYL-1,3-ADAMANTANE DIOL

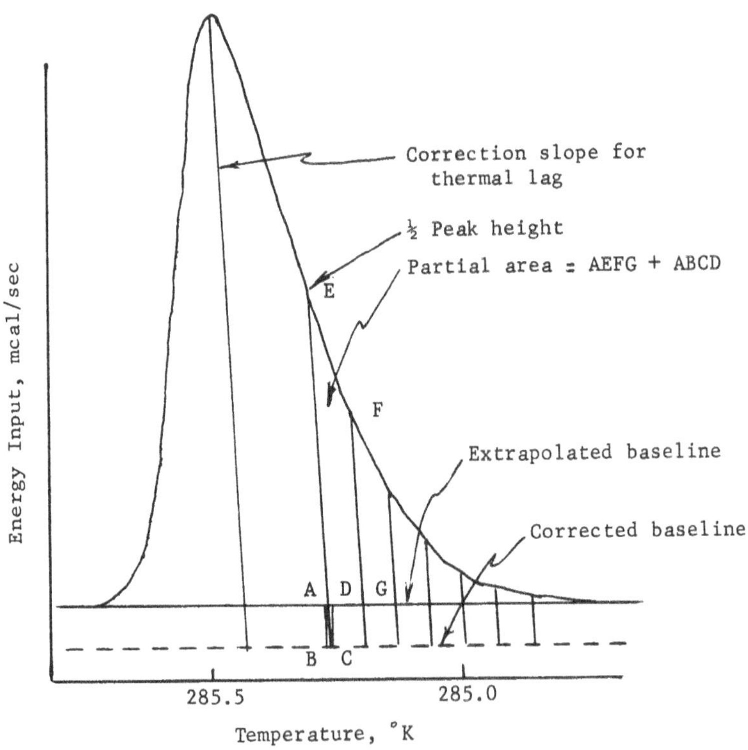

FIGURE 2 TYPICAL SCAN FOR PURITY DETERMINATION ON CYCLOHEXANE

the partial areas and total area is carried out until a straight line is obtained.

Two small corrections are necessary, however, in order to properly analyze the scan. The correct sample temperature (T_s) is the curve temperature extrapolated (on a slope characteristic for thermal lag in the system) to a true baseline which is about 5.5 millicalories/second below the extrapolated baseline for the instrument used. In addition, a small area measurement (ABCD) must be added to the partial area under consideration. This is illustrated in Figure 2.

One of the most serious limitations on the method is the assumption that no solid solutions are formed. A proper treatment for solid solution systems has been given where equation I is replaced by equation II. [3]

$$T_s = T_o - \frac{RT_o X_2^2}{\Delta H_f} \cdot \frac{1}{\frac{K}{1-K} + F} \qquad \text{(II)}$$

The distribution ratio of the impurity between the liquid and solid phases is $K = k/k'$, and is zero in the absence of solid solution formation. There is no criterion, however, which allows solid solutions to be detected in thermal experiments such as these. In fact, some literature [4] criticizes use of equation II to correct curves from adiabatic calorimeters, suggesting other causes for the curvature.

A number of assumptions are contained in the derivation of equation II. The van't Hoff equation is derived from the Clausius-Clapeyron equation by assuming that the melting point is the triple point and that the impurity gives ideal solution (specifically that Raoult's law holds).[5] The heat of fusion is assumed to be independent of the amount of impurity present, and it is further assumed that equilibrium is maintained during freezing and melting. The latter condition, however, is never met due to the design of the equipment.

EXPERIMENTAL

Experimental procedures for operation of the DSC have been adequately described. [1] For purity determinations sample size was kept as small as practical (1-3 mg. for samples with $\Delta H_f > 3{,}000$ cal/mole, 3-5 mg. for $\Delta H_f < 3{,}000$ cal/mole). Scan rates were .625 or 1.25°/min. Since samples often supercooled extensively, they were partially melted and cooled slowly in order to obtain a more nearly equilibrium crystallization. When running below room temperature, dry inert gas was purged through the chamber a few minutes to prevent ice formation, and the liquid nitrogen container filled before a heating cycle was begun. Sealed sample pans were used in all determinations to avoid volatilization and contamination problems.

A computer program was developed to simplify the purity calculations. The required input is the sample mass and molecular weight, instrument constants, a reference temperature at a point where the scan is still on the baseline, and ordinate and abscissa measurements on the curve. Abscissa (time, temperature) intervals should be short enough so that linear interpolation between points will introduce no significant error. One measurement should be at the peak of the melting curve, but the intervals need not be of uniform size. A maximum of ninety-nine pairs of readings can be accommodated in our program. A transparent plastic grid marked in 0.1 inch intervals is convenient for making these measurements from the recorder chart.

RESULTS AND DISCUSSION

Computer Program for Purity

Although in theory all of the melting curve or any portion of it could be used for purity calculations, in practice it was necessary to apply severe limits. A set of linearized 1/F values are plotted on Figure 1. The general "S" curve is typical of all samples when sufficient points are available. It is apparent that if the lower 1/F limit had been 4 rather than 2, somewhat less correction would have been needed. This would result in a lower slope and a considerably lower calculated impurity. On the other hand, if the limits had been such as to cut off those points above 1/F = 15, the line would be undercorrected giving a higher calculated impurity. Thus, the calculated impurity is a function of the particular set of 1/F limits chosen. That this is the case is clearly shown in Table I. These results were all calculated from the same set of input data obtained on a sample of NBS octane (impurity = .06 $\pm$ .04%). The extreme cases show variation by a factor of 170. This sample was found to be particularly sensitive to the 1/F limits used, but was in no sense unique.

TABLE I

1/F Limits	Calculated Impurity, mole %
10 - 50	.016
5.6 - 50	.23
2.8 - 40	.52
2.5 - 40	.59
2.0 - 40	.78
2.0 - 10	2.69

The method found to be most reliable uses a sliding limit based on the fraction melted at the point where the rate of heat input reaches 1/2 its maximum value (1/2 peak height). This effectively changes the lower limit from about 2.5 for impure samples to about 8 for very pure samples. The upper limit used is 1/F = 50.

The computer program, therefore, divides the curve into 99 equal temperature intervals and integrates to determine ΔH_f. The temperature correction and baseline area correction are determined for each interval and the partial area is calculated. The program then applies successive one-half percent area corrections on each partial area and the total area and calculates the 1/F value. A least squares regression analysis is run on each corrected line until a minimum standard deviation of the points about the calculated line is reached. The "best value" is then used in the subsequent calculations according to equation I. The output from the computer is the ΔH_f, T_o, T_m, mole percent impurity, the 1/F limits used, the percent correction applied, and the cryoscopic constant. A corrected mole percent impurity assuming solid solution behavior is also calculated using equation II. Provision is made to put in the heat of fusion for the pure material if it is known since this gives much better results at impurity levels above .5 mole %. Using computer calculation, a standard deviation of .03 mole % impurity was found for twenty determinations involving eight different compounds. This, however, includes data on n-octane and n-decane which are known to give difficulty in this type of analysis.[6,7] Excluding this data, gives a calculated precision of $\pm$ .02% which is consistent with results obtained on the "doped" sample series described below. The precision of hand calculations is $>\pm$ .04 mole % impurity.

The vitally important question of accuracy is difficult to handle. Problems of homogeneity are extremely important to a method which uses a very small sample. Two different calibration techniques were used. First, four samples which were of certified impurity level were run. The results are listed in Table II, and on inspection would indicate that the method is rather unsatisfactory. But as indicated above, octane and decane are known to give difficulties in these determinations. For all samples, however, it was found necessary to assume solid solution behavior in order to get acceptable results and all calculations are reported on this basis.

TABLE II

Sample	Impurity, mole %	Average*	Certified Impurity
	.02		
API Cyclohexane	.01	.015	.003 ±.002
NBS Decane	.41		
	.32	.36	.04 ±.02
NBS Octane	.28		
	.34		
	.22	.28	.06 ±.04
	.26		
API 1,2-Dimethyl-3-ethyl-benzene	.15	.15	.26 ±.02

* Of computer calculated impurities.

A second method frequently used (8) is to measure the purity of a sample, add known amounts of impurity, and measure the new purity level. The measured original purity should be the same as the extrapolated purity at zero added impurity. This method was employed with the systems benzil-benzophenone, adipic acid - benzoic acid, and with benzene using carbon tetrachloride, toluene, and naphthalene as impurities. Except for the naphthalene series where eutectic formation interfered with the analysis, the agreement was good, Table III. It should be noted that these results were calculated using experimental heats of fusion determined from the pure material before "doping".

TABLE III

Sample	Added "Impurity"	Impurity, mole % Measured	Extrapolated
Benzene	Carbon tetrachloride	.125 (avg.)	.14
Benzene	Toluene	.125 (avg.)	.15
Benzene	Naphthalene	.14	.24
Benzil	Benzophenone	.07	.08
Adipic Acid	Benzoic Acid	.19	.21

Figure 3 shows a plot of measured impurity versus true impurity for a variety of samples. The "true" impurity is the measured impurity of the base compound plus the amount of impurity added. The line is a 1:1 direct correspondence. This shows that the method of calculation gives good correlation from 99.9% purity to a purity level of at least 97.0%. The scatter is probably due in large part to mixing and sampling difficulties.

Special Considerations

It has been stated that considerable improvement was obtained where the data were calculated as though the plot curvature was due to solid solution behavior and that heat of fusion corrections were necessary. Treatment of all samples as cases of solid solution can be rationalized even if not predicted. The equipment used causes samples to recrystallize

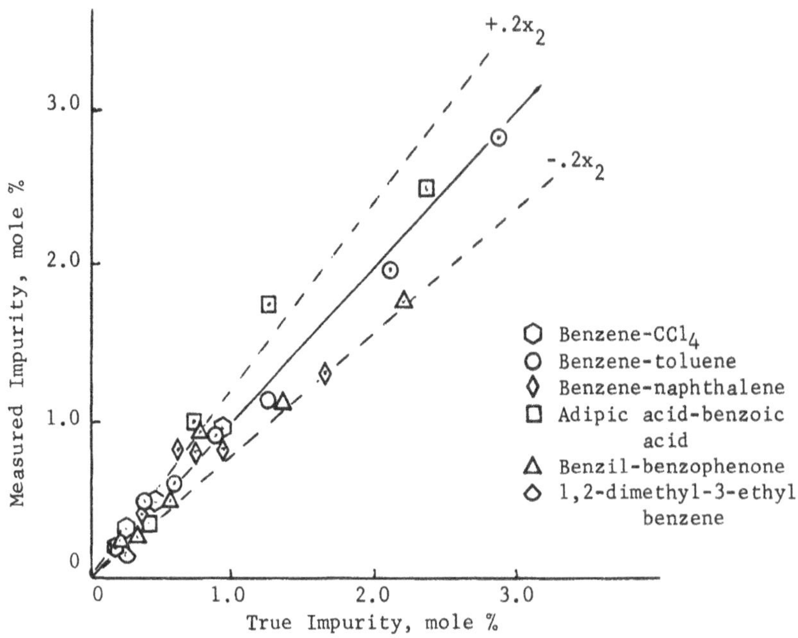

FIGURE 3 CORRELATION BETWEEN MEASURED IMPURITY AND TRUE IMPURITY

under extremely non-equilibrium conditions which can cause the impurity to freeze in the solid phase instead of concentrating in the melt. The curvature of experimental data has also been ascribed to non-equilibrium solid solution formation by other authors.[9,10,11]

Systems which form true solid solutions, however, cannot be handled by this method of analysis. Table IV gives the results of a series of doped samples prepared from benzene-thiophene and naphthalene-β naphthol, systems known to form solid solutions. In these runs the measured impurity is insensitive to the amount of impurity actually added.

TABLE IV

Sample	Impurity	Added Impurity Mole, %	True Impurity Mole, %	Measured Impurity Mole, %	ΔH_f Cal/Mole
Benzene	Thiophene	0	-	.11	2352
		.07	.18	.21	2209
		.20	.31	.13	2364
		.44	.55	.13	2370
		1.27	1.38	.30	2438
		3.04	3.15	.43	2291
		5.27	5.38	.97	2468
Naphthalene	β-Naphthol	0	-	.11	4554
		.26	.37	.38	4467
		.60	.71	.21	4518
		.71	.82	.18	4475
		1.37	1.48	.15	4504
		3.60	3.71	.20	4543

The necessity for correcting the heat of fusion is illustrated in Table V. It was assumed in the derivations that the heat of fusion of the pure substance did not change on addition of purity; this is not the case, however.

TABLE V

Sample	Added Impurity	Measured Purity, %	Heat of Fusion (cal/mole)	Difference, %
Benzene		99.8	2352	0
Benzene	CCl_4	99.05	2237	-4.9
Benzene	Toluene	99.10	2131	-9.4
Benzene	Toluene	97.14	1788	-24.0
Benzil		99.93	5764	0
Benzil	Benzophenone	98.38	5273	-8.5
Adipic Acid		99.81	9221	0
Adipic Acid	Benzoic Acid	99.64	9032	-2.0
Adipic Acid	Benzoic Acid	97.52	7974	-13.5

With few exceptions, all the data fell into such a pattern, and clearly are outside the $\pm 5\%$ heat of fusion accuracy limits established for other materials.[1] Since the calculated impurity is directly proportional to the heat of fusion, a correction is easily made if a correct heat of fusion is available, preferably from measurement of a pure (99.5+%) sample of the same material. Without correction, the calculated impurities can be low by as much as 20-30%. It should be noted that this dependence of the measured heat of fusion on the sample purity does not occur if true solid solutions are present as shown in Table IV, and may account for some of the results reported in the literature[1].

Using solid solution calculations and correct heats of fusion, the best assessment of accuracy based on present data is that a result can be stated as impurity = $X_2 \pm .2X_2$ for a single result or $X_2 \pm .1X_2$ for average results with a minimum error of $\pm .02$ mole percent due to precision. Most probable errors will result in low impurity values. Among the various samples run only decane and octane are not described by these considerations.

References

1. E. S. Watson, M.J.O'Neill, J.Justin, and N.Brenner, Anal.Chem., 36, 1233 (1964).
2. Perkin-Elmer Corp., Norwalk, Conn., Thermal Analysis Newsletter, Nos. 1-6.
3. S.V.R.Mastrangelo and R.W.Dornte, J.Am.Chem.Soc., 77, 6200 (1955).
4. G.Pilcher, Anal.Chem.Acta., 17, 144 (1957).
5. E.F.Westrum, S.Chang, J.Phys.Chem., 64, 1547 (1960).
6. W.M.Mazee, Anal.Chem.Acta., 17, 97 (1957).
7. G.W.Smith, Int.Sci. and Tech., No. 61, 72, (1967).
8. E.F.G.Herington, Anal.Chem.Acta., 17, 15 (1957).
9. G.S.Ross and H.D.Dixon, J.Res.Natl.Bur.Stds., 67A, 247 (1963).
10. N.I.Lyashkevich, Zh.Fiz.Khim., 37, (10), 2365 (1963).
11. J.P.McCullough and G.Waddington, Anal.Chem.Acta, 17, 80 (1957).

THE USE OF DYNAMIC DIFFERENTIAL CALORIMETRY FOR ASCERTAINING THE THERMAL STABILITY OF POLYMERS

Stuart M. Ellerstein
Thiokol Chemical Corporation
780 N. Clinton Avenue
Trenton, N.J. 08607

INTRODUCTION

The use of thermal analytical methods for the study of the thermal stability (and/or the thermal oxidative stability) of high polymeric materials has enjoyed increasing popularity. Traditionally, thermogravimetric analysis (TGA) or differential thermal analysis (DTA) have been used either separately or in complementary studies[1-3].

A third method[4,5], dynamic differential calorimetry (DDC)[6], while having found acceptance for the measurement of polymer transitions at lower temperatures, e.g., glass transition, has appeared to meet with only limited appeal in quantitative thermal degradation studies[7]. One obvious reason for this is that the one commercially available instrument for this type of measurement, i.e., the Perkin-Elmer differential scanning calorimeter, has a high temperature ceiling of only 500°C, which precludes its use with many materials of current interest. On the other hand, there are a great many polymers which have degradation intervals well below the present limit and, therefore, could be used. Another reason is that if one does not take special precautions with the method of preparation of a given sample, then one will generally encounter one or more of the following:

1. Excessive noise.
2. Excessive base line drift.
3. Temperature lags and gradients across the sample.
4. Lack of sufficient atmosphere control.
5. Volatile product entrapment.
6. Possible catalysis.

It is therefore the purpose of the present paper to describe a method which has been developed to optimize the use of DDC.

EXPERIMENTAL

I. Materials

The instrument used was a Perkin-Elmer differential scanning calorimeter Model DSC-1B which was used in conjunction with a Texas Instrument Servoriter II potentiometric recorder. Instrumental calibration was performed periodically in the manner described in the instruction manual.

Chart areas were obtained using a polar planimeter.

The polymeric materials used herein were as follows:

Polystyrene, (PS) a narrow molecular weight distribution sample of $M_n \sim 10^4$ which was supplied by Waters Associates as a gel permeation chromatography standard.
Poly(methyl methacrylate), (PMMA) a high molecular weight $M_n \sim 10^6$ benzoyl peroxide initiated polymer prepared by the author.
Poly(ethylene sulfide) (PES) a medium high molecular weight polymer $M > 10^5$ prepared in these laboratories.

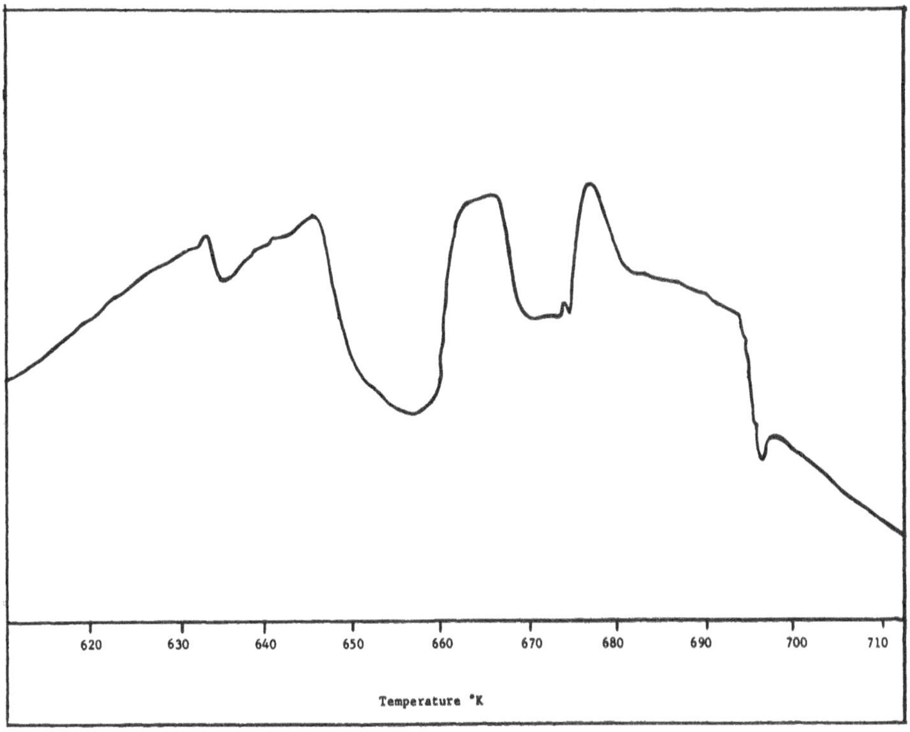

Fig. 1 DSC Thermogram. 1.47 mg of polystyrene in an open pan vs. an open pan reference Range 32 - heating rate 10°K/min - Chart speed 1"/min. atmosphere - notrogen. (Endo upward)

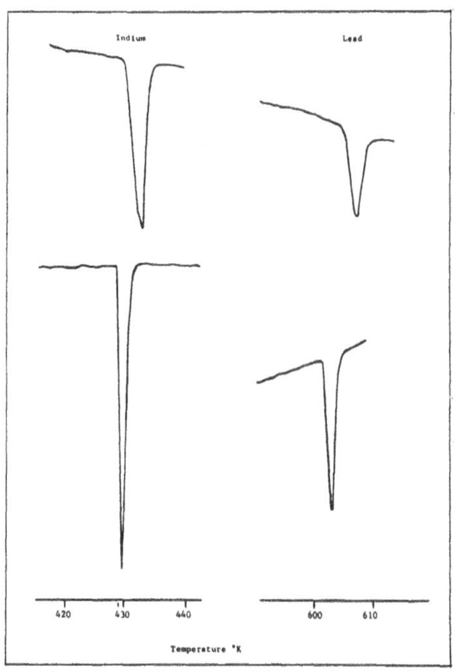

Fig. 2 DSC Thermograms of Indium and Lead fusion transitions - (a) Indium (8.78 mg), Lead (5.59 mg.) in a pan between two screens - (b) Indium (7.55 mg.), Lead (5.96 mg) a pan with an aluminum cover. (Endo downward).

II. Sample Preparation

The use of an aluminum (or other metal foil) cup together with a crimp sealed aluminum cover disc which tends to encapsulate the sample has been found to be very undesirable for most degradation studies. This is especially true where most of the polymer eventually volatizes leaving little or no residue in the sample pan. The violent bubbling caused by the escaping volatiles together with the close thermal coupling of the sample and sample enclosure tended to produce a very noisy thermogram. Also, the comparatively tight compartmentation did not allow the atmosphere around the sample to be adequately purged by the gas (usually nitrogen) flowing through the system, and also tended to retard the escape of the volatiles. An open cup system was tried as shown in Figure 1 for a polystyrene sample, but this too showed an irregular thermogram. It does, however, allow the sample to be swept by the gas flowing through the system. In this type of system there is also the disadvantage of possible thermal gradients across the sample which because of the poor thermal conductivity of the polymeric materials involved might be significant even when relatively small samples are used. The use of 90 mesh aluminum oxide powder was also tried as a superstrate, but this tended to retard volatilization, as well as, apparently causing some oxidation by the occlusion of air. It did produce a reasonably quiet thermogram, and a readily recognizible base line return.

The method finally settled on was that of placing the sample between two fine mesh inert metal screens, e.g., 100 mesh stainless steel.

In practice the screens are stamped out as circular discs (an ordinary loose-leaf paper punch is suitable), and one is placed on the bottom of the aluminum pan. A small amount of the polymer sample (usually 1-2 mg. was used) is then weighed in and the second screen disc is placed on top and crimped in place. The reference was usually an aluminum pan with two screens crimped in place.

As is seen in Figure 2 the use of such an arrangement produces only a slight lag 1-3°K in the onset of a fast transition, such as the fusion of a metal standard, which may be corrected for. It also causes a slight decoupling of the sample and the sample enclosure which should tend to damp out the effect of rapid transients. This should have little effect on most polymer degradation processes. This configuration also allows intimate contact between the environmental atmosphere and the sample, while reducing the thermal gradients across the sample. It is noted that a slight amount of noise is desirable since it makes the location of the onset of degradation easier to find. Some representative thermograms are shown in Figures 3-5.

Calculations

The calculations are based upon the method of Borchardt and Daniels[8] adapted to a reaction occurring in a bulk condensed phase.

If one starts with the assumed rate law

$$\frac{dH}{dT} = \frac{k}{q} H_r^{\,n} \qquad (1)$$

where dH/dT is the instantaneous change of enthalpy with absolute temperature, q is the constant heating rate dT/dt in °K/min, k is the rate constant which is assumed to be of the Arrhenius type, n is an apparent order of reaction, and H_r is that part of the total enthalpy of reaction remaining at temperature T. Now both dH/dT and H_r are related to measurable quantities on the thermogram chart by the following relations:

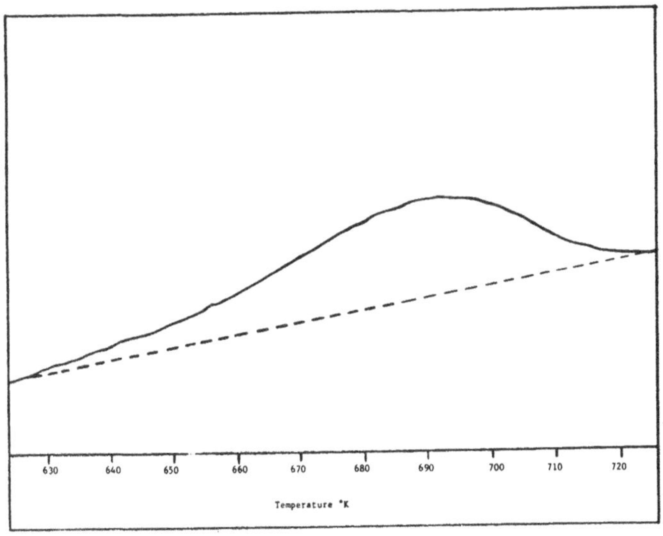

Fig. 3 DSC Thermogram of 1.47 mg. polystyrene in a pan between two screens vs. a reference pan containing two screens. Range 32 - heating rate 10°K/min. (Chart speed 1"/min. atmosphere - nitrogen. (Endo upward).

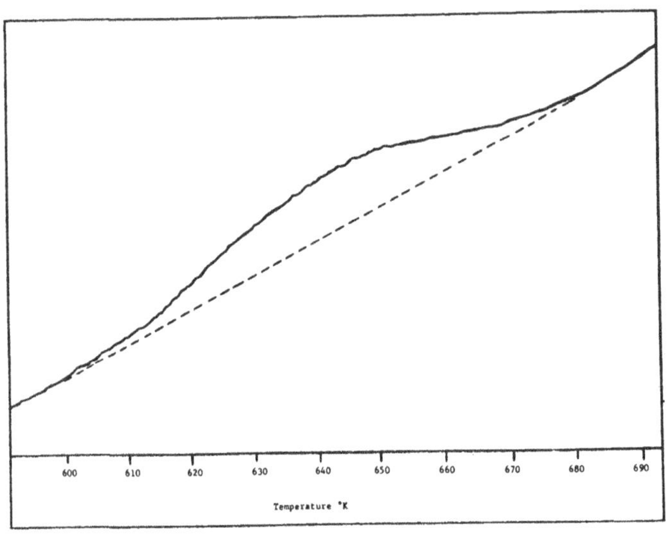

Fig. 4 DSC Thermogram of 1.70 mg Poly(methylmethacrylate) in a pan between two screens vs. a reference pan containing two screen range 8 - heating rate 10°K/min - chart speed 1"/min. atmosphere - nitrogen. (Endo upward).

$$\frac{dH}{dT} = \ell \left(\frac{R' \times 60}{L \times q} \right) \tag{2a}$$

$$H_r = A_r \left(\frac{R' \times 60}{L \times S} \right) \tag{2b}$$

where ℓ is the ordinate displacement between the base line and curve and A_r is the corresponding area remaining at temperature T, R' is the calibrated instrumental full scale range setting in millicalories/sec, L is the length of a full scale displacement, and S is the linear chart per minute (note ℓ, L, A_r, and S should all have compatible units).

Using equations (1), (2a) and (2b) one may now use the method of Freeman and Carroll[9] which gives the equation

$$\frac{\Delta \log \ell}{\Delta \log A_r} = \frac{-E^*}{2.303R} \left[\frac{\Delta (1/T)}{\Delta \log A_r} \right] + n \tag{3}$$

where E* is the Arrhenius energy of activation, and R is the gas constant. If one plots the left hand side of equation (3) vs. the bracketed expression, then the slope will be given by -E*/2.303R and the intercept will be equal to n. An IBM Quiktran program has been written to calculate the terms of equation (3). These are then fed into an IBM Quiktran common library program (FITLIN) which gives a "best" line fit of the calculated points.

An alternate method is based upon taking the second derivative of H with respect to T, and setting this equal to zero, cf.,[10,11]. Using equations (2a) and (2b) one obtains the relationship

$$E^*/n = \frac{RT^2 \ell S}{A_r q} \tag{4}$$

where all quantities are evaluated at the peak temperature (ℓ is a maximum). This, however, does not give E* and n explicitly, and is of value chiefly where n is known or assumed to be unity.

RESULTS AND DISCUSSION

The numerical results for the three polymers shown in the accompanying thermograms are given in Table I. All of the samples (powders) had been purged with nitrogen for thirty to forty minutes at 313°K before being heated up at 10°K/min. A fresh reference consisting of a crimped aluminum pan containing two screen discs was used for each run.

Generally calculations were based upon five or six temperatures, which were spaced at five degree intervals.

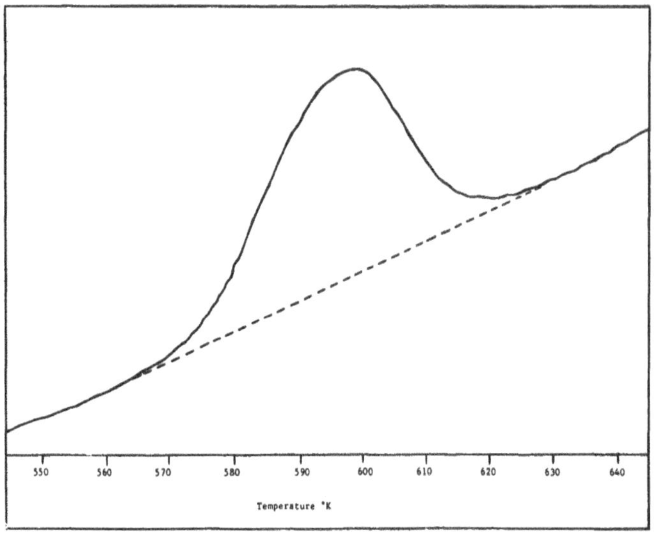

Fig. 5 DSC Thermogram of 1.54 mg Poly(ethylene sulfide) in a pan between two screens vs. a reference pan containing two screens. Range 16 - heating rate 10°K/min. - Chart speed 1"/min. (Endo upward).

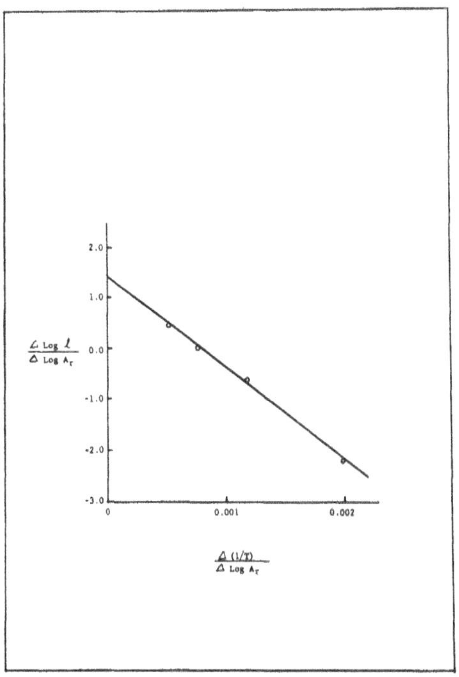

Fig. 6 Freeman-Carroll Plot based upon the thermogram shown in Figure 5.

Table I

Polymer	Sample Weight in mg.	Apparent Temperature Range of the Reaction Zone in °K	Peak Temp. °K	Method of Calculation Freeman-Carroll r	n	E* kcal/mol	Peak E*/n kcal/mol
PS	1.47	625-725	688	-0.999	0.95	48.0	46.8
PMMA	1.70	600-680	642	-0.990	1.02	44.0	40.5
PES	1.54	565-630	597	-0.999	1.42	83.6	56.3

where r is the correlation coefficient. It is seen that the method of Freeman and Carroll appears to give an excellent self-consistent straight line fit for a given experimental run, viz., Figure 6. One difficulty encountered in this type of diffuse transition is the location of the base line in the reaction interval.

In this study the base lines were simply drawn straight across the apparent reaction zone. The optimization of the instrumental slope control setting used in conjunction with a slower heating rate, and a higher attenuation should reduce the base line curvature sufficiently to permit interpolation error to be minimized.

While the energies of activation of PS and PMMA are well within the range of diverse values found in the literature[3] (obtained by other methods). The value for PES was somewhat higher than was anticipated. On the other hand, it should be remembered that the method of Freeman and Carroll has been subject to criticism by Doyle[12], and depends upon whether or not the rate law is well represented by a power function as given in equation (1). The fact that for PS and PMMA the apparent order of reaction n is near unity is not unreasonable at these temperatures, and probably represents random scissions. In contrast, the PES has a value of n which tends to indicate a more complicated reaction scheme. More work will have to be done to elucidate this point, as well as, complete analyses of the products formed before any mechanisms can be postulated with any certainty. The precision of both E* and n appear to give an error estimated to be less than 10%.

It is emphasized that the primary purpose of this study was to examine the feasibility of using DDC as an experimental method for obtaining kinetic data, and that the method of calculation used and interpretation of the parameters are left open to the individual investigators.

CONCLUSION

Dynamic differential calorimetry appears to be capable of being used to obtain the kinetic parameters of a polymer undergoing thermal degradation. The sample preparative technique described appears to have many advantages, while keeping the disadvantage to a minimum. The method of Freeman and Carroll appears to give reasonable results. More work is in progress, and the variation of heating rate, sample size, atmosphere (including vacuum), and molecular weight dependence are parameters that will be investigated as soon as possible.

ACKNOWLEDGEMENT

Thanks are given to Dr. Riad H. Gobran, Director of Research and Development of the Thiokol Chemical Corporation, Chemical Division. Also to Mrs. Arline Grove for doing the experimental part of this study.

BIBLIOGRAPHY

1. Murphy, C. B., Anal. Chem., 34, 298R (1962).

2. Wendlandt, W. W., "Thermal Methods of Analysis", Interscience, New York, 1964.

3. Madorsky, S. L., "Thermal Degradation of Organic Polymers", Interscience, New York, 1964.

4. Watson, E. J., O'Neill, M. J., Justin, J., and Brenner N., Anal. Chem., 36, 1233 (1964).

5. O'Neill, M. J., Anal. Chem., 36, 1238 (1964).

6. Ellerstein, S.M., (Appl. Polymer Symposia, No. 2), 111 (1966).

7. Mack, C. H., and Hobart, S. R., (Appl. Polymer Symposia, No. 2), 133 (1966).

8. Borchardt, H. J., and Daniels, F., J. Am. Chem. Soc. 79, 41 (1957).

9. Freeman, E. S., and Carroll, B., J. Phys. Chem., 62, 394 (1958).

10. Friedman, H. L., J. Polymer Sci. C, 6, 183 (1964).

11. Fuoss, R. M., Salyer, I. O., and Wilson, H. S., J. Polymer Sci. A, 2, 3147 (1964).

12. Doyle, C. D., "Techniques and Methods of Polymer Evaluation", Ed., Slade, P. E., Jr., and Jenkins, L. T., Dekker, New York, 1966, pp. 175 - 177.

APPENDIX

Another method has been derived for the explicit calculation of the kinetic parameters. Taking the first derivative of equation (1) with respect to temperature yields the relationship

$$\frac{d^2H}{dT^2} = n \frac{k}{q} H_r^{n-1} \frac{dH_r}{dT} + \frac{k}{q} H_r^n \frac{E^*}{RT^2} \qquad (5)$$

If one multiplies and divides the first term on the r.h.s. by H_r, remembers that $dH_r/dT = -dH/dT$, and substitutes equation (1) appropriately, then one may obtain the operational equation

$$\left[RT^2 \frac{d\ell/dT}{\ell}\right] = -n\left[RT^2 \frac{\ell\ S}{A_r\ q}\right] + E^* \qquad (6)$$

where $d\ell/dT$ is the derivative of the ordinate displacement with respect to temperature. This last quantity may easily be obtained graphically by drawing a tangent to the thermogram, and calculating its slope with respect to the base line at each temperature used.

It may be seen that a plot of the bracketed expression on the l.h.s. vs. the bracketed expression on the r.h.s. of equation (6) should give a straight line of slope (-n) and intercept E*.

THERMAL ANALYSIS OF FRICTION MATERIALS

by

Michael G. Jacko
Bendix Research Laboratories
Southfield, Michigan 48075

DESCRIPTION

A polymer-bound friction material such as an automotive brake lining is a complex system (Figure 1) consisting of (a) an organic resin binder, such as a phenolic or a modified phenolic; (b) asbestos fibre, which is a reinforcing agent; (c) friction property modifiers. A typical photomicrograph (Figure 2) shows the physical appearance of the ingredients in the bulk of the lining. Also shown is the layer affected by heat generated at the lining-drum interface.

USES OF THERMAL ANALYSIS

Differential Thermal Analysis (DTA)

In previous papers,[1,2,3] the application of DTA to the quality control of organic brake linings has been described, along with its utility for determining fingerprint of formulation, degree of resin cure, solvent retention, missing components, tramp components and abnormal quantities of components. In developing the DTA test, Bendix Research Laboratories have checked the procedure for the following: variables affecting results, repeatability, reproducibility, discrimination between formulations and discrimination between process changes.

Thermogravimetric Analysis (TGA)

A comparison of TGA results in air and inert atmospheres reveals the following information: thermal stability of ingredients, oxidative stability of ingredients, degree of resin cure, thermal stability of inorganic components in the composite, and inorganic residue content.

MATERIALS

- R-1: an oil-modified phenolic that is a solid at room temperature and melts above 60°C.
- R-2: an epoxy-modified phenolic that is a solid at room temperature and melts above 65°C.
- F-1: a cashew-type friction modifier that is used "as received." It is classified as a medium-hard modifier.
- F-2: a slightly harder cashew-type friction modifier.
- F-3: a mixture of 20- to 320-mesh zinc powder that is used as a higher-temperature friction modifier.
- A-1: a short-fibre asbestos.
- M-1: a ground rubber stock that is carbon-filled.
- M-2: barytes mineral used in a very fine particle size.
- M-3: copper dust used as a friction modifier.

Sample Size

A sample of 15.0 ± 0.1 mg normally fills about 1/4 to 1/3 of a 4-mm O.D. sample tube after tapping for uniform packing density. The thermocouple is placed in the sample with the ceramic shields moved up so that they do not touch the sample. Silicon carbide (15.0 ± 0.1 mg) is used as reference (and can be used repeatedly).

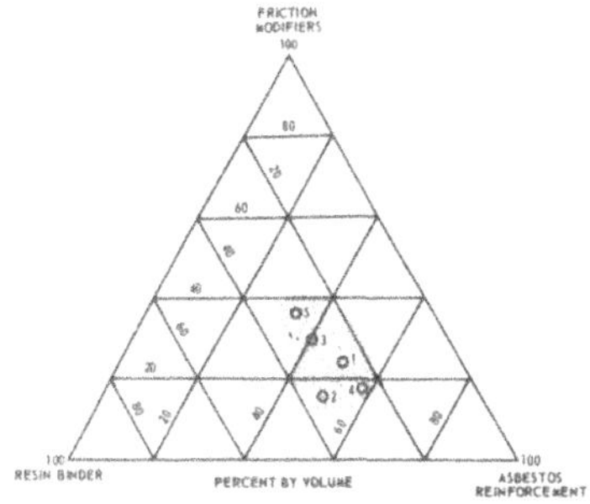

Figure 1 - Ternary Composition Diagram for Polymer-Bound Friction Materials

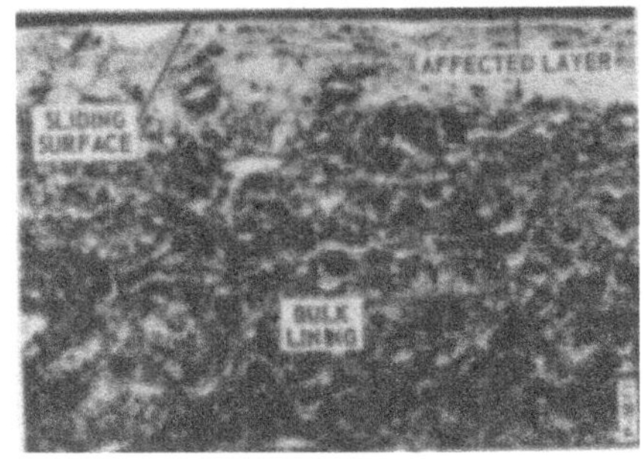

Figure 2 - Surface and Affected Layer of an Automotive Brake Lining After 8550 Vehicle Miles

Figure 3 - Sample Grinding

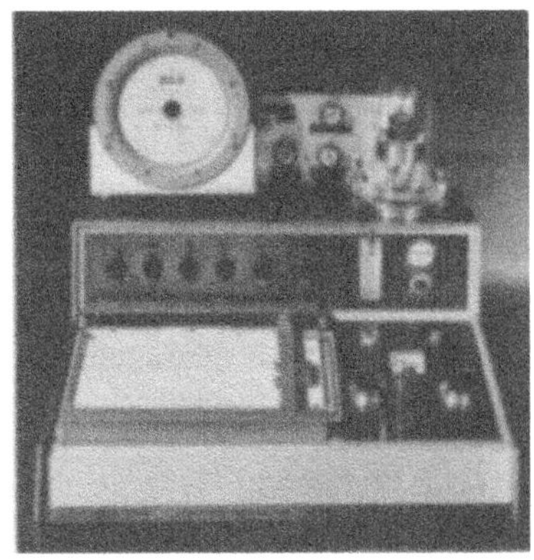

Figure 4 - Modified DuPont 900

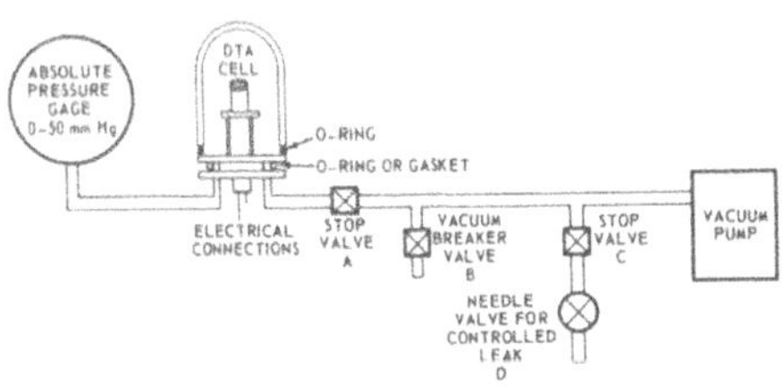

Figure 5 - Schematic of Improved DTA Vacuum System

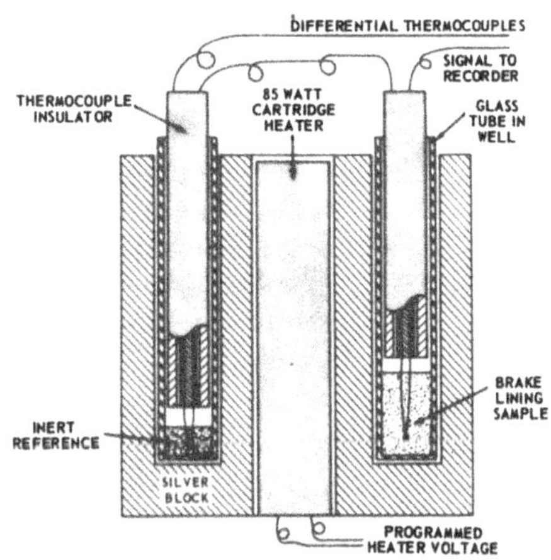

Figure 6 - Modified DTA Block

In some situations, a sample such as phenolic resin gives overly large ΔT signals. The sample size is reduced for two reasons: 1) to keep the ΔT signal on the recorder paper, and 2) to closely approximate the amount of resin in a normal friction material. In cases where reduced sample size is used, the remaining weight is made up to 15.0 mg with silicon carbide.

EQUIPMENT

A. Modification of the DuPont 900 DTA Apparatus

The original vacuum gage on the DuPont DTA was unsuitable for obtaining the degree of pressure control required; consequently, a Wallace and Tiernan 0-50 mm vacuum gage was added (Figure 4). The schematic of the improved DTA vacuum system (Figure 5) shows the DTA cell between the absolute pressure gage and the controlled leak (a needle valve). Valve A is used to isolate the cell and the gage from the remainder of the system to check vacuum tightness; valve B is simply a vacuum breaker; valve C is used to minimize manipulation of the controlled leak. The vacuum tightness of the cell base plate was improved by sealing up the existing system and using the two spare holes with metal plumbing. The system was considered vacuum tight when it could be pumped below 1.0 mm and would show a pressure increase of less than 0.1 mm in 30 seconds.

B. Modification of the DTA Cell

The DTA heating range selected is 25 to 600°C, recorded at 50°/inch. The desired heating rate was 50°C/min; however, the 55-watt heater furnished with the instrument did not produce this rate between 400 and 600°C. The DTA cell was therefore modified by drilling out the standard cell (Figure 6) to accommodate an 85-watt cartridge heater. Using this modified DTA cell, a uniform programming rate from 100 to 600°C can be obtained at 50°C/min.

In some instances, the DuPont intermediate cell to 850°C was used at 20°C/min.

RESULTS AND DISCUSSION

A. Typical DTA and TGA Thermograms

Typical DTA thermograms are shown in Figure 7. Test discrimination is indicated by these thermograms for five different linings. Figure 8 is still another formulation along with the TGA (50% suppression) run in both air and nitrogen (40cc/min). In this presentation, an attempt will be made to show how the thermal analysis of each individual ingredient tends to give certain features to the thermograms of the composite formed by the ingredients.

B. Thermal Analysis of Asbestos

Asbestos (chrysotile) is a preferred reinforcing agent in friction materials because it has the following properties: strength, high temperature stability, suitable frictional characteristics and low cost.

Figure 9 shows the DTA and TGA curves for a common short-fibre grade of asbestos. The DTA shows that a 15.0-mg sample at ΔT=1.0°C/inch and a 7.5-mg/7.5-mg SiC sample at ΔT=0.5°C/inch have similar thermograms.

The transition at 350 to 400°C in both DTA and TGA thermograms represents the dehydration of brucite (magnesium hydroxide), which is an impurity in asbestos [4]. The transition at 650 to 750°C in both thermograms represents the major dehydroxylation reaction in chrysotile asbestos. At 850°C a sharp exotherm, which is not associated with any weight change, represents the conversion of the dehydroxylated chrysotile to forsterite or olivine [$2MgO \cdot SiO_2$ or $2(Mg_{1.00-x}Fe_x)O \cdot SiO_2$ respectively].

A series of samples of shortfibre asbestos were heated to 200, 400, 600, 800, 1000, and 1200°C in the TGA apparatus and the residues submitted for analysis by x-ray diffraction. The diffraction patterns are given in Figure 10: samples heated to 200, 400, and 600°C were confirmed as chrysotile, sampkes heated to 800 and 1000°C were confirmed as forsterite (olivine), sample heated to 1200°C shows 4 peaks (arrows), which are probably enstatite peaks.

C. Thermal Analysis of Friction Property Modifiers

i) Cashew Friction Dust

Cashew friction dust is usually the product of a condensation reaction between an aldehyde and cashew nut shell liquid[5] normally called "cardanol." The material is then ground to less than 20 to 40 mesh. It has good friction and wear properties along with thermal stability.

Figure 11 shows the DTA thermograms of two friction dusts based on cardanol. Also seen is the TGA in air and nitrogen for F-1.

ii) Rubber Stock

Rubber stock (M-1) from a variety of sources has a DTA thermogram that exhibits two major endotherms (Figure 12): 275 to 325°C and 450 to 500°C. However, it is usually added in small quantities (less than 5%) and will contribute very little to the over-all thermogram of the composite.

iii) Barytes Mineral

Barytes (barium sulphate) (M-2) shows no major exotherms or endotherms in the 100 to 600°C region (Figure 12).

iv) Copper Dust

Copper (M-3) has no major DTA exotherm or endotherm in the 100 to 600°C region (Figure 12).

v) Zinc Dust

Zinc (M-4) is known to be added to some commercial friction materials and is characterized by its strong DTA melting endotherm (Figure 12) at 429°C (419°C corrected).

D. Thermal Analysis of Phenolic Resins

Synthetic resins such as the phenolics and oil or epoxy-modified phenolics are commonly used as friction material binders. Figure 13 gives the DTA curves for an uncured and cured 205°C (400°F), 2000 psi, 2 hours oil-modified phenolic resin. The early peaks in the thermogram for the uncured resin are associated with the following phenomena:[6] endotherm below 100°C, broad melting point range of the pre-polymer, endotherm between 140 to 170°C, heat of solution of curing agent in pre-polymer, exotherm between 170 to 200°C, heat of reaction for rapid initial curing, exotherm between 210 to 300°C, heat of reaction for slower final curing; endotherm above 400°C, thermal degradation of the cured polymer. The cured resins do not show the same large peaks between room temperature and 300°C.

The TGA curves for this resin in air and nitrogen (40cc/min) show a 2% weight loss below 200°C, probably due to adsorbed moisture along with phenol and curing-agent evolution. Thermal decomposition sets in at about 400°C.

Figure 14 shows similar DTA and TGA thermograms for another modified phenolic that is not quite as stable as the one discussed above.

E. Optimizing Stability of Phenolic Resins

The mechanisms involved in the hardening and curing of phenolics

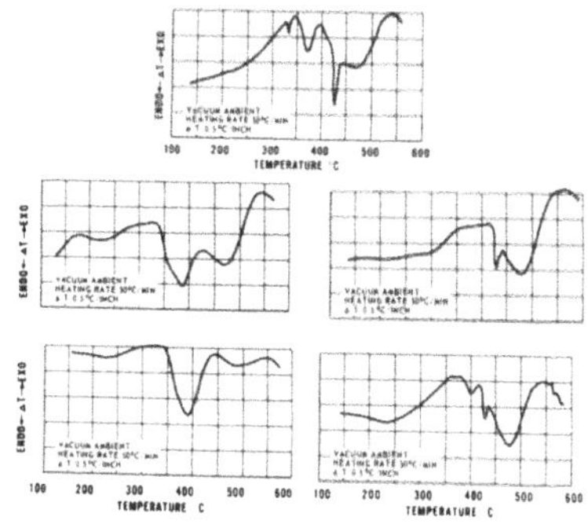

Figure 7 - DTA Curves Showing Discrimination Among Formulations

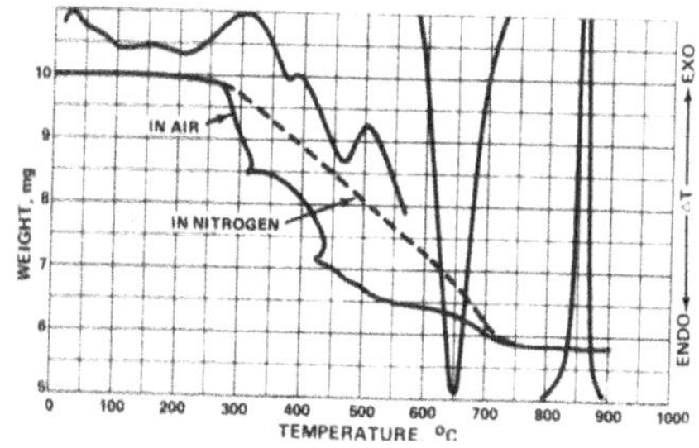

Figure 8 - DTA and TGA Thermograms of a Typical Friction Material

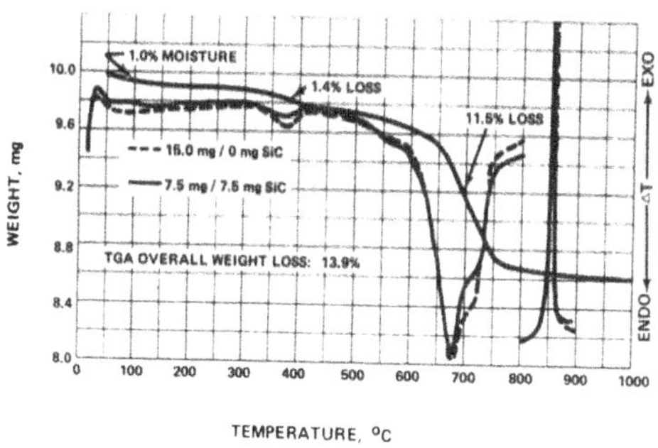

Figure 9 - DTA and TGA Thermograms of Asbestos

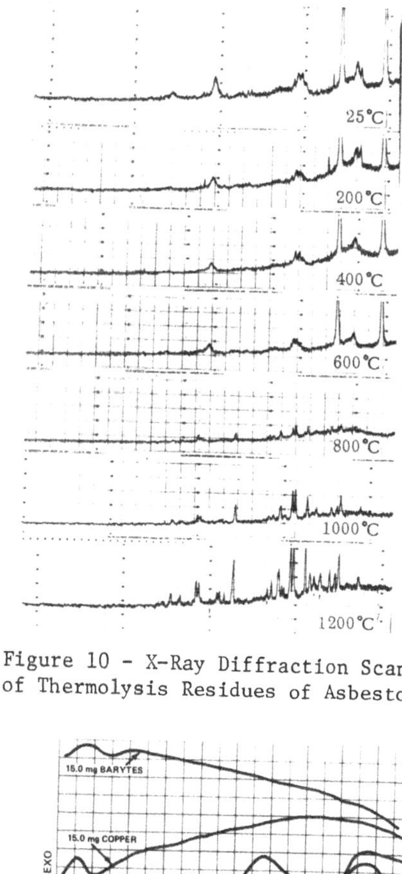

Figure 10 - X-Ray Diffraction Scans of Thermolysis Residues of Asbestos

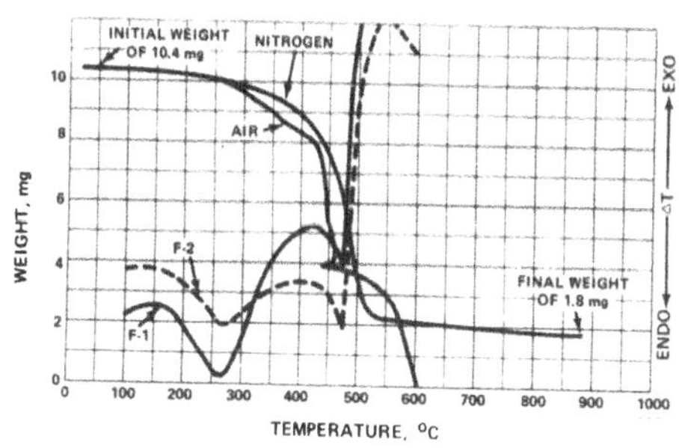

Figure 11 - DTA and TGA Thermograms of Cashew Friction Modifiers

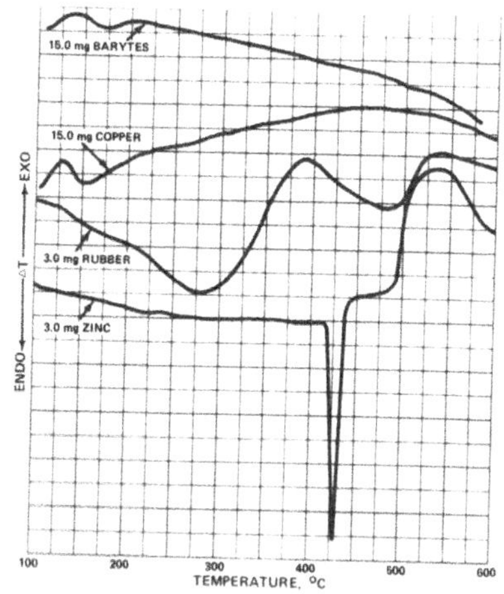

Figure 12 - DTA Thermograms of Various Friction Modifiers

have been studied by thermal analysis previously in numerous reports. Studies on the optimization of resin stability, on the other hand, have been neglected.

Phenolic resin (R-1) was studied extensively by preparing samples cured for 2.0 hours at 4000 psi at the following temperatures: 150°F (66°C), 200°F (93°C), 250°F (121°C), 300°F (149°C), 400°F (204°C), 450°F (232°C), and 500°F (260°C). The DTA thermograms (Figure 15) show the change in the thermal analysis curve of a resin from the uncured to the cured state. The curing exotherms between 150 and 250°C disappear, leaving only the decomposition endotherm near 450°C. At this point there are two competing effects: an exothermic reaction with the formation of gaseous products and the endothermic volatilization of these products as indicated by TGA. The magnitude of this interaction can change the shape of the endothermic portion. Figure 16 shows the DTA thermograms for R-1, R-2, and F-1. R-1 is slightly more stable than R-2 while both are more stable than F-1. Thus the DTA thermograms of R-1 and R-2 have less differential signal than F-1 because the rate of decomposition of cashew friction particles is much faster than the decomposition rates of the synthetic phenolics. The TGA traces also show less carbonaceous material formation for the cashew friction particles than for the synthetic phenolics.

The same series of R-1 resins was subjected to thermal stability tests in the TGA (nitrogen, 40cc/min); results are shown in Figure 17. Samples cured below 450°F show a weight loss of 4% at 200°C, whereas those cured at higher temperatures show a 1% loss at 200°C. Samples cured below 450°F show a 10% weight loss between 370 to 400°C, while those cured at higher temperatures show the corresponding loss at 415 to 425°C; samples cured above 450°F have a residue of 46% at 800°C while samples cured below this temperature show residues of 45 to 44%. Figure 17 also indicates oxidative stability in the TGA test (air, 40cc/min). The results are comparable to those shown above up to about 460°C. Above 460°C, uncured resin samples appear to have lower weight losses than cured samples, possibly as a result of oxidation.

F. Synergistic Effects

To obtain the maximum of information from the thermal analysis of a composite, it is necessary to know the thermograms of each ingredient and the synergistic effects produced on compounding.

A test friction material having the following formulation was blended till the mixture was "homogeneous," then molded and cured for 4 hours at 400°F (204°C) and 4000 psi: phenolic resin, 12 wt %; asbestos fibre, 67wt %; cashew particles, 5 wt %; zinc, 7 wt %; barytes, 9 wt %. The sample was prepared for DTA by the conventional grinding technique. The DTA thermogram for a 15.0 mg sample is shown in Figure 18, curve a.

This same resin was independently cured under the above conditions and ground to the proper mesh size for the next part of the experiment. The proper relative amounts of the ingredients were weighed into a vial to give a total of 30.0 mg. The resulting thermogram is that labeled curve b. Curve c is the thermogram of a 15.0 mg sample of the blended ingredients (using only the "fines").

Curve c is very similar to curve a. The shape of the DTA thermogram is dependent on the materials present and the method of their individual preparation. There appears to be little or no interaction on compounding. The slight lag of the asbestos endotherm in curve c is probably due

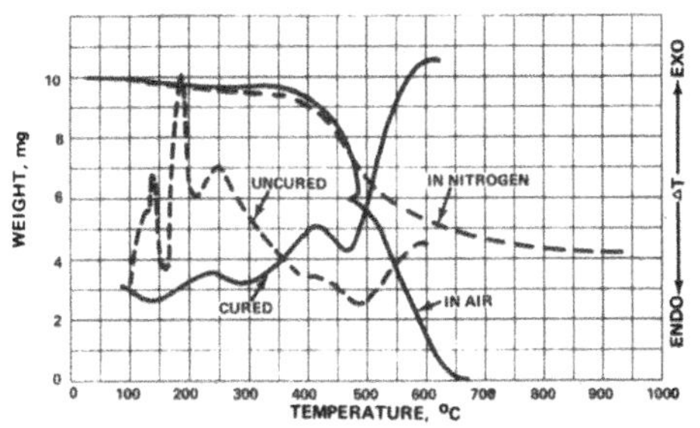

Figure 13 - DTA and TGA Thermograms of a Modified Phenolic Resin (R-1)

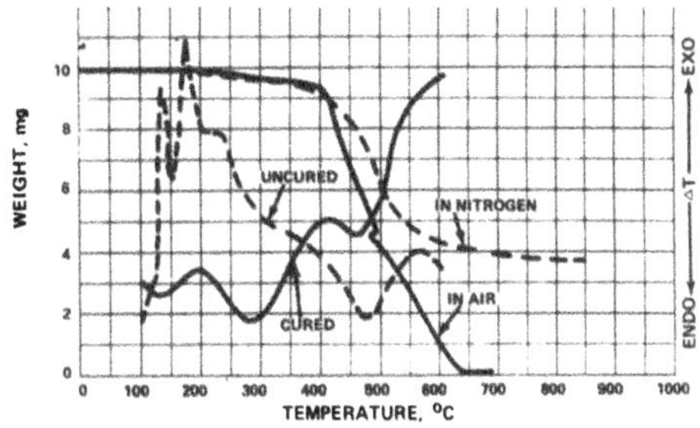

Figure 14 - DTA and TGA Thermograms of a Modified Phenolic Resin (R-2)

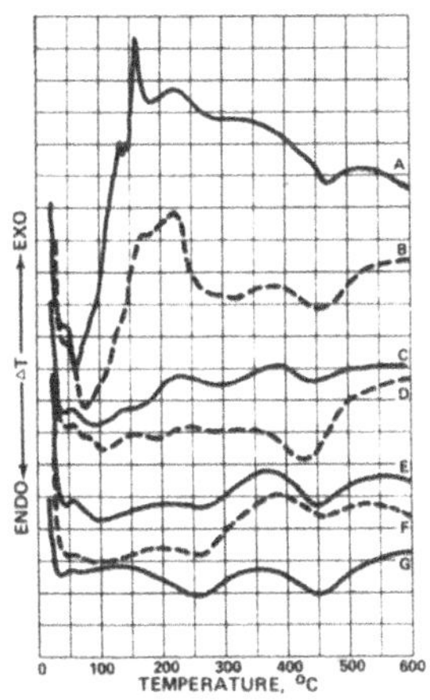

Figure 15 - DTA Thermograms Showing Progressive State of Resin Cure

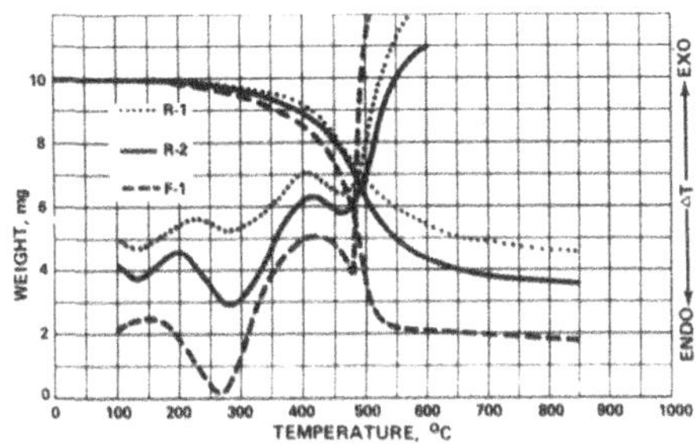

Figure 16 - Differences in Stability of Phenolic Resins

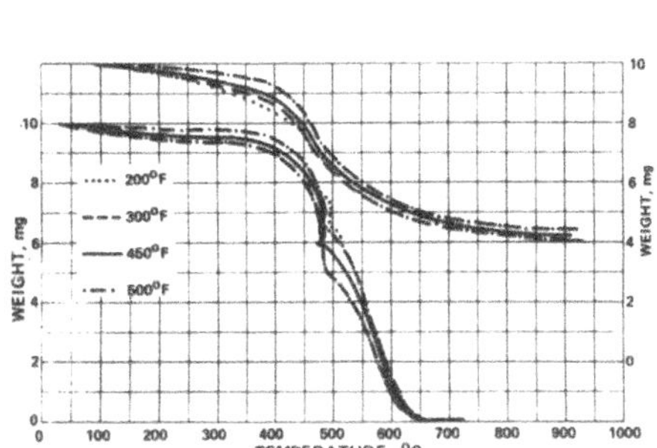

Figure 17 - TGA Thermograms of a Phenolic Resin With Increased Cure Rates

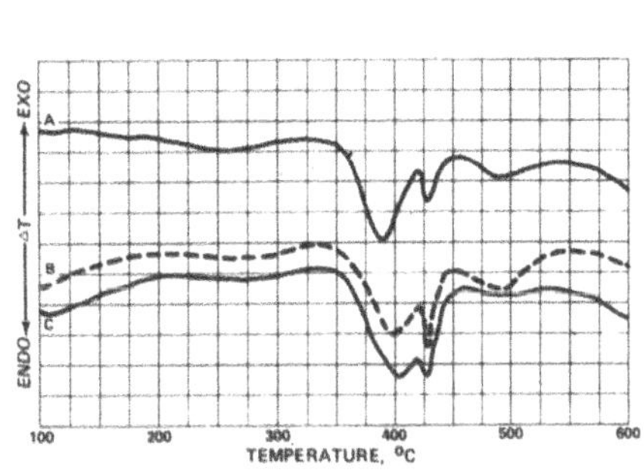

Figure 18 - DTA Thermograms Showing No Synergistic Effects of Components

to a lower degree of openness than in the finely ground sample.[8]

CONCLUSIONS

A DTA method originally specified for insuring consistency of quality of automotive brake linings has been expanded by combination with TGA. The ingredients and the resultant composites of friction materials lend themselves to characterization by thermal methods. Organic components are identified by both DTA and TGA; inorganic components such as asbestos are also characterized by both techniques.

The DTA thermogram of a composite can be related to the DTA thermogram of each individual component. Little or no synergistic effect occurs. Although not emphatically shown in this report, the same conclusion is applicable to the TGA thermograms.

A DTA thermogram changes markedly with the state of undercure for a phenolic resin; less than about 50% of total undercure can be detected. Considerable difficulty is experienced in determining the final curing stages.

The thermal stability of phenolic resins increases with state of cure. An exact comparison of thermal stabilities by TGA is difficult because of such factors as particle-size variations, oxidative versus thermal degradation effects, the atmosphere in the hot zone, and the resins used. However, disregarding these factors, increased thermal stability can be expected for modified phenolic (R-1) samples cured above 450°F (232°C). Above this temperature, stability increases very little with temperature.

REFERENCES

1. A.R. Spencer, W.M. Spurgeon, and J.L. Winge, "Four Tests for Consistency of Automotive Brake Linings," Paper No. 660142, presented to the Society of Automotive Engineers, Detroit, Michigan, June, 1966.
2. A.R. Spencer and W.M. Spurgeon, "A Review of Progress with Some Recently Developed Techniques for Friction Materials Evaluation," Paper No.670081, presented to the Society of Automotive Engineers, Detroit, Michigan, January, 1967.
3. M.G. Jacko, A.R. Spencer, and W.M. Spurgeon, "Optimizing DTA Peak Definition for Quality Control of Automotive Brake Linings," presented to the 1967 Anachem Conference, Detroit, Michigan, October, 1967.
4. A.A. Hodgson, "Fibrous Silicates," Lecture Series No. 4, Royal Institute of Chemistry, London, England, 1965.
5. P.K. Sanyal, K.K. Rohatgi, and A.C. Sinha, "Commercial Cashew Nutshell Liquid and Its Standardisation," J. Proc. Inst. Chemists (India), Vol. XXVII, 208 (1955).
6. R. Burns and E.W. Orrell, "A Thermal Analytical Study of Phenol Formaldehyde Resins," J. Materials Sci. 2, 72 (1967).
7. R.H. White and W.T. Rust," Cure Rates of Phenolic Resins by Differential Thermal Analysis," J. Appl. Polymer Sci. 9, 777 (1964).
8. E. Martinez, "The Effect of Particle Size on the Thermal Properties of Serpentine Minerals," Am. Minerologist 46, 901 (1961).

HEATS OF REACTION FOR CARBONATE MINERAL DECOMPOSITION

By

Keith L. Reddick
Sun Oil Company Production Research Laboratory
Richardson, Texas

INTRODUCTION

Earlier thermal analysis investigations (1, 3, 10) of carbonate system were conducted under static conditions where decomposition progressed in a variable atmosphere of reaction products. These studies have made fundamental contributions to the field of carbonate mineralogy.

During the present study differential thermal analysis (DTA) in dynamic gas atmospheres (11), with supporting thermogravimetric (TGA), X-ray diffraction, and chemical analysis, has been utilized for the determination of kinetic data related to thermal decomposition of selected carbonate minerals. This study has clarified some of the thermal properties of iron and manganese substituted dolomites.

Selected samples of calcite ($CaCO_3$), magnesite ($MgCO_3$), rhodochrosite ($MnCO_3$), siderite ($FeCO_3$), dolomite $CaMg(CO_3)_2$, and ankerite ($Ca(Mg, Mn, Fe)(CO_3)_2$) have been examined in dynamic atmospheres of air, helium, and carbon dioxide. These data have been compared with thermal analyses in static atmospheres. All thermal analyses were conducted at atmospheric pressure.

Endothermic heats of reaction (ΔH_r) for carbonate decomposition reactions have been determined by a simplified calibration technique. Correlation of sample weight with differential thermocouple response has permitted determination of reaction areas for 100.0 milligram samples of each material. Heats of reaction for the decomposition reactions of other carbonates have been determined by using the magnesite decomposition area as a standard of comparison.

ANALYTICAL PROCEDURES

All of the carbonate samples were ground to a fine powder in a tungsten carbide mill. After five minutes of grinding, the powdered carbonates were sieved. The 150 to 250-mesh portion was retained for analysis. This portion was oven-dried for three days at $104^{\circ}C$. Dried samples were stored either in a desiccator, using Drierite as the desiccant, or in sealed plastic vials.

Each sample was dissolved in 1:1 hydrochloric acid for chemical analysis. Metal cation concentrations were determined in these solutions by atomic absorption analysis, flame spectrophotometry, and wet chemical methods. Carbon dioxide was determined by TGA, alkilometer analysis, and carbon/hydrogen analyzer data. Acid insoluble residues were determined gravimetrically.

Differential thermal and thermogravimetric analyses were made with an integrated system manufactured by the Robert L. Stone Company. DTA traces were made using a well-type encased sample holder (SH-12BE2), which was designed to permit dynamic gas flow. The holder contains two wells; one for the sample, and one for the reference material. Dynamic gas flows through both wells simultaneously. The holder was constructed of stainless steel 446. Hydrochloric acid washed alumina (Al_2O_3, 150 to 250-mesh) was used as the inert reference material.

Samples were placed in the sample well by a sandwich technique. Layers of the reference material were placed above and below the sample, in a manner which would locate the thermocouple bead in the center of the sample. Differential thermocouples were constructed of Platinel II alloys, while the reference thermocouple was constructed of platinum-platinum/13% rhodium. The sample holder was heated by F I-D furances which were controlled by a KA-2H recorder controller.

SH-12BE2 sample holders permit gases to enter at the bottom, flow through the sample, and exit from the top. Gas flowed through the sample for 15 minutes prior to heating. During this period the exit flow rate was adjusted to 0.05 standard cubic feet per hour (0.05 SCFH = 25 cc/minute).

The present study was divided into three phases. During the first phase the decomposition behaviors of calcite, magnesite, rhodochrosite, siderite, dolomite, and ankerite samples were observed in dynamic atmospheres of air, helium, and carbon dioxide. These data were compared with decomposition behavior in static atmospheres. In the second phase variations in the decomposition behavior of selected dolomites and ankerites were observed for a dynamic carbon dioxide atmosphere. Variations in differential thermocouple response with sample weight were examined during the third phase of this study. Response variation was determined by correlation of area under a reaction trace with sample weight.

Operating conditions for the thermal analysis equipment are presented in Table 1.

Table 1
THERMAL ANALYSIS OPERATING CONDITIONS

Initial Exit Gas Flow Rate	0.05 SCFH	Atmospheric Pressure
Microvolt Setting	DTA 1200	TGA 1000
Heating Rate	$10^{o}C$/minute	
Sample Weight Mgms.	DTA 60-240	TGA 100.0
Preheat	DTA 60V, 30 seconds	TGA None
Initial Operating Voltage	DTA 20	TGA 16

EXPERIMENTAL RESULTS

The mechanisms of thermal decomposition for carbonate minerals depend upon their composition, associated impurities (9), and the atmosphere in which they are heated. Representative chemical analyses of these minerals are presented in Table 2.

Table 2
CHEMICAL COMPOSITION OF CARBONATE MINERALS

Weight %	Iceland Spar Calcite (1)	Austria Magnesite (2)	Montana Rhodochrosite (2)	Connecticut Siderite (2)	Haley, Ontario Dolomite (3)
CaO	54.00	-	-	-	30.23
MgO	-	45.77	-	-	21.73
FeO	-	0.40	1.58	56.13	0.02
MnO	-	-	56.02	3.05	0.00
CO_2	43.17	50.12	36.32	36.27	47.45
TOTAL	97.17	96.29	93.92	95.45	99.43
Impurities:					
HCl Soluble	1.48	1.15	1.77	2.80	0.00
HCl Insoluble	1.35	2.56	4.31	1.75	0.57

Weight %	Austria Dolomite (4) 10057	Czechoslovakia Dolomite (4) 102953	Massachusetts Ankerite (4) R2432	Colorado Ankerite (4) 15143	Czechoslovakia Ankerite (2)
CaO	29.71	27.71	27.23	27.62	27.00
MgO	20.97	17.10	19.52	14.67	9.88
FeO	0.66	4.98	2.65	4.86	15.59
MnO	0.03	0.02	1.16	4.31	1.37
CO_2	46.64	43.50	45.03	43.36	42.38
TOTAL	98.01	93.31	95.59	94.82	96.22
Impurities:					
HCl Soluble	1.71	5.31	3.10	4.15	2.95
HCl Insoluble	0.28	1.38	1.31	1.03	0.83

Sources: (1) E. H. Sargent and Company, (2) Wards Natural Science Establishment, Inc., (3) Dominion Magnesium, Ltd., Haley, Ontario, (4) the Smithsonian Institution.

A summary of temperature ranges over which selected carbonate minerals decomposed in different gas atmospheres is presented in Table 3.

Table 3
TEMPERATURE RANGES IN °C FOR CARBONATE DECOMPOSITION REACTIONS IN DIFFERENT ATMOSPHERES

	Air	Helium	CO_2	Static Air
Calcite	600-900	630-900	900-1000	720-960
Magnesite	400-800	400-800	500- 800	400-800
Rhodochrosite	400-600	400-600	440- 660	400-600
Siderite	400-600	400-600	470- 620	500-600
Dolomite	530-880	600-900	550- 950	560-960
Ankerite	550-900	650-850	650- 950	750-925

Variations in the thermal decomposition behavior of eight dolomites and three ankerites were studied in a dynamic carbon dioxide atmosphere. This gas was selected because the decomposition behavior of complex carbonates is distinct, and more clearly defined, in a CO_2 atmosphere. In addition, CO_2 tends to inhibit oxidation reactions.

The mechanisms of dolomite and ankerite decompositions have been described in detail (4, 5, 6, 7, 8) by previous authors. In CO_2 atmospheres dolomite decomposes with two major endothermic reactions, while ankerites exhibit three major endothermic reactions.

An additional small, but distinct, endothermic reaction was observed for all of the dolomite decompositions in a CO_2 atmosphere. This reaction, as shown in Figure 1, occurs between 550° and 650°C, and precedes the first major endothermic reaction.

Ankerites are generally considered as iron or iron-manganese substituted dolomites. A correlation of chemical and thermal analysis data indicates that ankerite is formed as a result of isomorphous substitution of both iron and manganese for magnesium in the dolomite. The Czechoslovakian dolomite (Table 2) is considered a ferroan dolomite. It decomposes as a dolomite in carbon dioxide atmospheres, while the Massachusetts ankerite with half the FeO content of the Czechoslovakian dolomite decomposes as an ankerite. The difference in these two substituted dolomites appears to result from the additional manganese substitution in the lattice. The extent of manganese substitution necessary to form ankerites has not been investigated. All dolomites in this study contained less than 0.03% MnO, while ankerites contained more than 1% MnO.

The first and third ankerite decomposition reactions correspond to the two major decomposition reactions observed in dolomites. The intermediate reaction is a ferrite (4, 5, 9) reaction wherein calcium ferrite, 2 CaO · Fe_2O_3, is formed from the reaction of Fe_2O_3 (from the first decomposition) and residual $CaCO_3$. The magnitude of this reaction, and the temperature range of its occurrence, appears to be related to the combined iron and manganese substitution in the sample. Dolomite and ankerite decomposition reactions occur over distinctly different temperature ranges in a dynamic carbon dioxide atmosphere. Table 4 summarizes these temperature ranges.

Table 4
TEMPERATURE RANGES FOR DOLOMITE AND ANKERITE DECOMPOSITION REACTIONS IN A DYNAMIC CO_2 ATMOSPHERE

Reaction	Dolomite	Ankerite
1	550° to 650°C	680° to 780°C
2	650° to 810°C	780° to 880°C
3	900° to 970°C	880° to 950°C

Heats of reaction (ΔH_r) values were estimated from reaction areas of carbonate decompositions in carbon dioxide atmospheres. Tables 5, 6, and 7 present a summary of the steps required to determine experimental ΔH_r values by the proposed procedure.

Areas under reaction curves have shown a nearly linear increase with increasing sample weight up to sample charges of 175 milligrams. The variation in reaction area for different sample weights of magnesite is shown in Figure 2. Reduction of reaction area for each sample weight to a common base of reaction area (in square inches) per 100.0 milligrams provides a basis of comparison. Table 5 presents a summary of the averaged areas for each decomposition, and a range of sample weights used to derive the data.

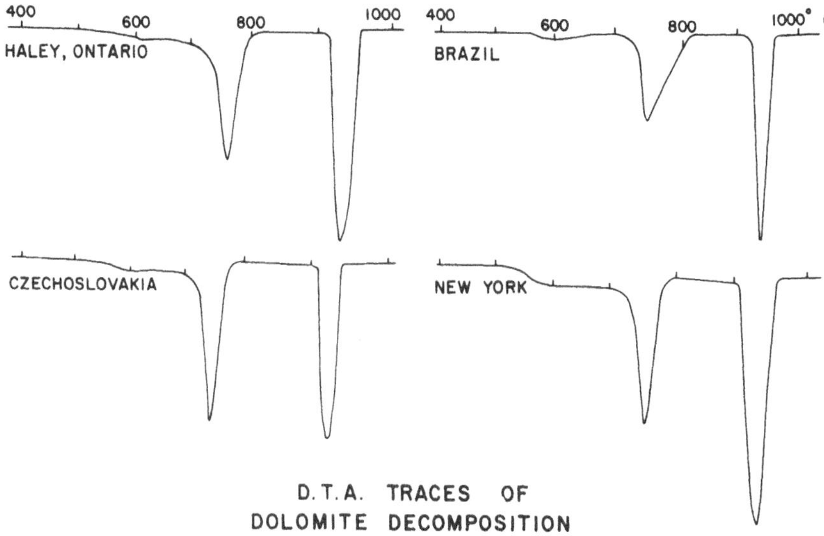

D.T.A. TRACES OF DOLOMITE DECOMPOSITION IN A CO_2 ATMOSPHERE

Figure 1

RESPONCE CALIBRATION FOR MAGNESITE DECOMPOSITION IN A CO_2 ATMOSPHERE

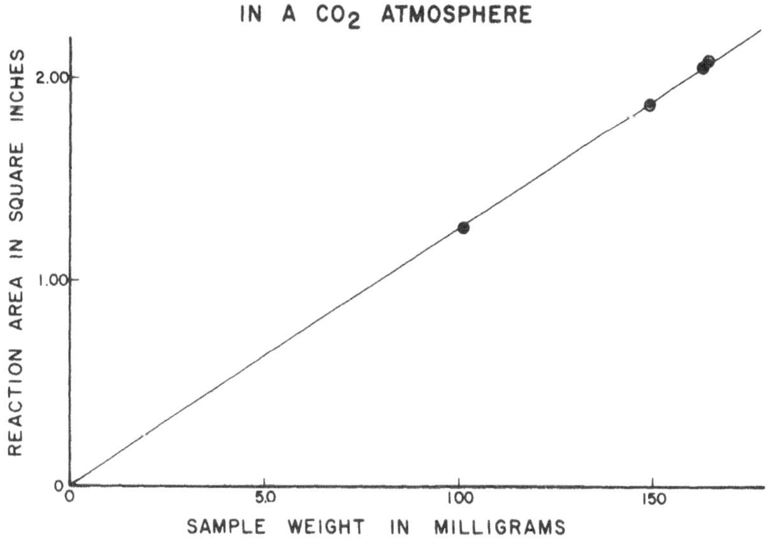

Figure 2

Table 5
VARIATION IN REACTION AREAS FOR CARBONATE DECOMPOSITIONS

Carbonate	Sample Weight Range in Mgms.	Averaged Area Reaction 1	Reaction 2	Reaction 3
Calcite	110-136	1.335 ± .005		
Magnesite	102-165	1.257 ± .005		
Rhodochrosite	128-159	1.270 ± .003		
Siderite	127-149	0.902 ± .005		
Temperature Range °C		550-650 **	650-810	900-970
Dolomites*	77-225	0.074 ± .066	0.518 ± .006	0.657 ± .009
Temperature Range °C		680-780	780-880	880-950
Ankerites				
Colorado	95	0.734 ± .010	0.184 ± .005	0.781 ± .005
Massachusetts	to	0.652 ± .002	0.182 ± .002	0.431 ± .001
Czechoslovakia	165	0.505 ± .010	0.361 ± .010	0.315 ± .005

* Average of eight samples
** Variation between dolomites

Experimental values for ΔH_r were estimated from data in Table 5 by two methods. In the first method the reaction area for the calcite decomposition was set equal to -42.50 K cal/mol. (One square inch equals -31.84 K cal/mol.) For the second method the reaction area for the magnesite decomposition was set equal to -28.10 K cal/mol. (One square inch equals -22.35 K cal/mol.) A summary of results of these two methods is presented in Table 6.

Table 6
EXPERIMENTAL ΔH_r (K CAL/MOL.) VALUES FOR THE DECOMPOSITION OF SIMPLE CARBONATES IN CARBON DIOXIDE ATMOSPHERES

Sample	ΔH_r Free Energy Calculations	ΔH_r Calcite Area	ΔH_r Magnesite Area
Calcite	-42.50	-42.51	-29.84
Magnesite	-28.10	-40.02	-28.10
Rhodochrosite	-27.85	-40.44	-28.38
Siderite	-20.02	-28.72	-20.15

Use of the calcite area method produced consistently high ΔH_r values for magnesite, siderite, and rhodochrosite when compared with ΔH_r values determined from free energy of formation values. The rapidity with which residual CaO from the calcite decomposition reverts back to $CaCO_3$ suggests that an equilibrium between the CO_2 atmosphere and the decomposing calcite may exist in the sample holder. If an equilibrium does exist, then decomposition would be incomplete. Incomplete reactions would appear in DTA traces as a reaction peak of smaller area. Good agreement between the experimental and free energy ΔH_r values occurred for siderite and rhodochrosite when the magnesite area method was applied. Because of this agreement, the magnesite area method was used to determine experimental ΔH_r values of dolomite and ankerite reactions. A summary of these results is presented in Table 7.

Table 7

EXPERIMENTAL ΔH_r VALUES FOR DECOMPOSITION OF DOLOMITE AND ANKERITE IN CARBON DIOXIDE ATMOSPHERES

Carbonate	Experimental ΔH_r (K Cal/Mol.)			
Temperature Range °C	550-650 **	650-810	810-970	Total
Dolomites*	-1.65 ± 1.32	-11.58 ± 0.13	-14.69 ± 0.18	-27.92 ± 1.40
Temperature Range °C	680-780	780-880	880-950	Total
Colorado Ankerite	-16.41	-4.23	-17.46	-38.10
Massachusetts Ankerite	-14.58	-4.07	- 9.65	-28.29
Czechoslovakia Ankerite	-11.29	-8.07	- 7.04	-26.40

* Average of eight samples

** Variation between dolomites

CONCLUSION

Differential thermal analysis techniques have been used to determine reaction kinetic data on the decomposition of natural carbonate minerals in a dynamic carbon dioxide atmosphere. Comparative reaction data have been obtained by correlation of sample weight and area within a reaction peak over a sample range of 0 to 175 milligrams.

Endothermic heats of reaction (ΔH_r in K cal/mol.) for decomposition of calcite, rhodochrosite, siderite, dolomites, and ankerites can be determined when the reaction area for decomposition of magnesite is used as a reference standard. Experimental ΔH_r values for rhodochrosite and siderite were in good agreement with ΔH_r values determined from free energy of formation data for these two minerals.

ACKNOWLEDGMENT

The author wishes to thank Dr. S. B. McCaleb for his constructive criticism and guidance in the preparation of this paper, and to Sun Oil Company for permitting the publication of the report. Additional recognition should go to the Dominion Magnesium, Ltd., Haley, Ontario, Canada, and the Smithsonian Institution for their generosity in contributing samples for the present study (Table 2).

LITERATURE CITED

1. Beck, C. W., American Mineralogist 35, 985, 1950.
2. Bradley, W. F., Burst, J. F., and Graf, D. L., American Mineralogist 37, 207, 1953.
3. Cuthbert, F. L., and Rowland, R. A., American Mineralogist 32, 111, 1947.
4. Dasgupta, D. R., Mineralogical Magazine 35, 634, 1965.
5. Dasgupta, D. R., Mineralogical Magazine 36, 138, 1967.
6. Haul, R. A. W., and Heystek, H., American Mineralogist 37, 166, 1952.
7. Haul, R. A. W., and Wilsdorf, H. G. F., Nature 167, 945, 1951.
8. Kulp, J. L., Kent, P., and Kerr, P. F., American Mineralogist 36, 643, 1951.
9. Martin, R. J., American Mineralogist 43, 649, 1958.
10. Smykatz-Kloss, W., Beit. zur Mineral. und Petrog. 9, 481, 1964.
11. Stone, R. L., J. Amer. Ceramic Soc. 33, 46, 1954.

STUDY OF THE THERMAL BEHAVIOR OF SOLID PROPELLANTS BY DIFFERENTIAL SCANNING CALORIMETRY

by

George D. Sammons
Member of Technical Staff
Solid Rocket Division, Rocketdyne
A Division of North American Rockwell
McGregor, Texas

The complex combustion of composite (heterogeneous) solid propellants has not as yet been described in a practical combustion model. The main interest in establishing such a model is the need for *a priori* knowledge of the effect of changes in physical properties and chemical composition of solid propellants on the ballistic parameters of a rocket motor. If one contends that solid phase reactions are important and possibly even controlling in this complex combustion process, a study of the thermal behavior of solid propellants and their individual ingredients is essential to understanding their combustion process.

The differential scanning calorimeter (DSC) was selected as a tool for study of thermal behavior of solid propellants in view of its proven usefulness in the studies of thermal decomposition of insulators that have been conducted in our laboratories since mid-1962. Extensive descriptions of the DSC and its quantitative nature have already appeared in the literature [1,2]. The quantitative nature of this instrument eliminates much of the difficulty in interpreting data obtained by conventional differential thermal analysis (DTA) of propellants. The effects of sample geometry, packing, and self-heating on the shape of thermograms normally seen by DTA are essentially eliminated. The shape of DSC thermograms is far more reproducible and more significant of properties of the material being tested. Additionally, the type of sampling required for the DSC lends itself uniquely to the study of thermal processes in heterogeneous materials, such as composite propellants.

Ultimate purpose of this investigation was to provide new insight into the control of burning rate of solid propellants. Propellants of primary interest are the composites containing ammonium perchlorate oxidizer and a hydrocarbon binder such as one of the linear carboxy terminated polybutadiene polymers available commercially.

Experimental Procedure

The DSC has proven to be quite insensitive to sample geometry and essentially free from thermal lags experienced with DTA equipment. Composite propellants are heterogeneous in nature and decompose by a heterogeneous mechanism. It has been our experience that propellant decomposition is sensitive to sample geometry. Since we were interested in relating the DSC thermograms to actual rocket motor combustion, special sample geometries were selected. The most useful appears to be a disc of propellant about 3--5 millimeters in diameter and 280--840 microns in thickness. These samples were prepared by slicing propellant on a microtome and punching out the discs with cork borers. All samples run on the DSC were encapsulated in aluminum. In runs where large amounts of gas were produced, holes were punched in the capsule lid. Ammonium perchlorate studies were generally conducted on material of 200-micron average particle size; however, checks were made by drastically varying particle size and essentially the same thermograms were obtained. Temperature standardization and power calibrations were made by the methods described in the instrument maker's manual; although we found these seldom needed adjustment except when changing sample heads.

A number of methods were available for reduction of thermograms to obtain kinetic parameters[3,4,5,6]. The DSC is particularly well suited to most of these methods. The method of Rogers and Morris is the simplest and requires only that the logarithm of the height of the thermogram curve be plotted

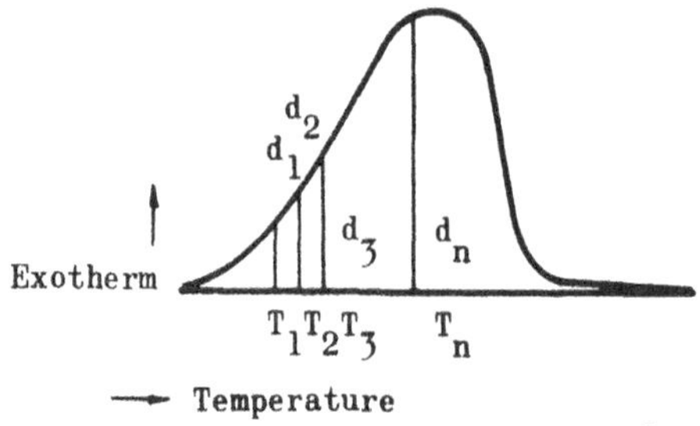

From Arrhenius plot of ln d - vs - $\frac{1}{T}$

$$- E - R \frac{\ln d_1 - \ln d_n}{\frac{1}{T_1} - \frac{1}{T_n}}$$

Figure 1. Method of Rogers and Morris for Kinetic Parameters

against the reciprocal of the temperature at which these heights occur to give an Arrhenius plot. This procedure is illustrated in Figure 1.

This method assumes zero reaction order and should be used with care. It does apply quite well to pure explosives and in some cases propellants, particularly for initial decomposition. A modification is shown below, based on the theory of Borchardt and Daniels(4).

$$\ln d/(1 - \alpha)^x = \ln CA - E/RT$$

α = area of exotherm to d/ total area of exotherm

C = calibration factor relating d to rate of heat produced times (weight of sample)^x

X = order of reaction

A = frequency factor

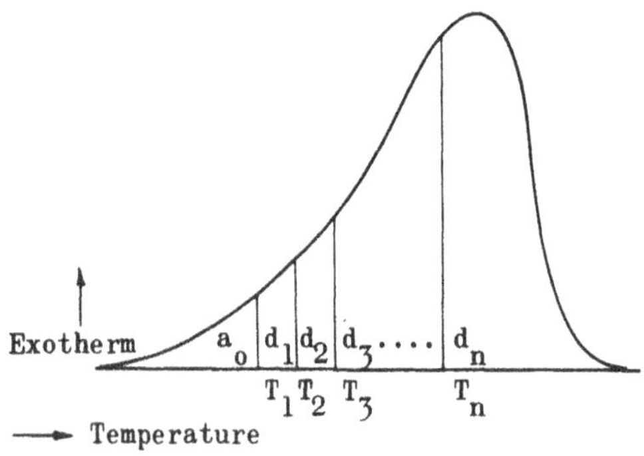

$\Delta T_1 = T_2 - T_1 \qquad \Delta T_2 = T_3 - T_2 \qquad$ etc.

$\Delta T_1 \left(\frac{d_1 + d_2}{2}\right) \cong$ area between d_1 and d_2

for $\Delta T_1 = \Delta T_2 = \Delta T_{n-1} = 0.1$

$$a_n = a_o + 1/20\ (d_1 + 2d_2 + 2d_3 + \ldots d_n)$$

$$\alpha_n \doteq \frac{a_n}{A}$$

Figure 2. Modified Method of Borchardt and Daniels for Kinetic Parameters

The modification of the Rogers and Morris method still uses curve heights for the Arrhenius plot; but the heights are corrected for reactant consumption. This is done by a ratio of a stepwise numerical integration of the thermogram to the total integral as shown in Figure 2. A reaction order must be assumed for this method.

Other methods of much more complexity(6,7) can be used to determine reaction order and even variation in reaction order during the course of reaction. The choice of methods for use in heterogeneous combustion studies depends on the intended use of the kinetic parameters. Some will give kinetic parameters of individual processes that are a component of the heterogeneous combustion and these can be used to obtain a clearer understanding of the heterogeneous process. The methods that sum all the individual processes give kinetic parameters that can be considered the effective macroscopic parameters of the process. These values would contain the effect of physical controls such as diffusion. Macroscopic kinetic parameters are of value in describing propellant thermal stability and its decomposition during steady-state burning.

Experimental Results

Propellant Binder

The propellant binder selected for this study is a linear polybutadiene polymer. Since hydrocarbon decomposition is normally endothermic, this polymer would be expected to decompose endothermically. However, thermograms run at a number of heating rates (Table I, Figure 3) show the first reaction to be exothermic beginning at 570 K.

TABLE I

ACTIVATION ENERGIES BY METHOD OF ROGERS AND MORRIS

Carboxy-Terminated Polybutadiene Binder	
φ[a]	E[b]
5	30
10	32
20	36
20	34
20	32
40	44
80	51

(a)Heating rate in deg C/minute
(b)Activation energy in kcal/mole

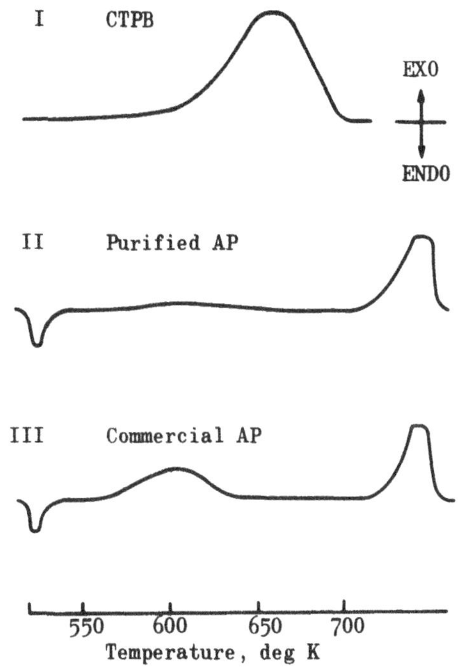

Figure 3. Binder and AP Thermograms (Heating Rate 20 C/min)

Weighing samples before and after the occurrence of this exotherm showed about 30% of the sample had been consumed. Mass spectrum of off-gas of a sample heated at 575 K showed the following composition:

	Mole %
CO_2	41.53
N_2	55.75
Butenes	0.93
Hexenes	0.18
H_2O	0.53

The gas composition suggests decarboxylation and breakup of the cure linkages. It is likely that the exothermic nature of this decomposition is due to vinyl polymerization.

Rogers and Morris method gave an activation energy of 33 ± 3 kcal/mole. This value was obtained by averaging data obtained from a number of heating rates, as shown in Table I.

Ammonium Perchlorate

Although decomposition of ammonium perchlorate (AP) has received considerable attention in the past decade, it is still not well understood. A typical DSC thermogram of a commercially available AP is shown in Figure 3. Pertinent features are: A crystal phase change at 516 K from orthorhombic to cubic; a low temperature decomposition at about 600 K, which usually terminates at 30% decomposition; and a high temperature decomposition at 725 K, which results in complete consumption of the sample.

A controversial result obtained in DSC work is the amount of consumption of sample in the low temperature (600 K) decomposition based on heat evolution. The ratio of the low temperature to the high temperature exotherm shows the consumption of AP during the low temperature exotherm to vary from 15 to

50%. Thermograms obtained by thermogravimetric analysis do not agree with this. Although the extent of the low temperature reaction varies from sample to sample, it has always been found to be equal to or less than 30%, which is an unexpected but understandable result. The amount of consumption is reproducible among samples from one lot of AP. Small variations are found in different lots from the same manufacturer and large variations are found between lots from different manufacturers. High-purity samples of AP show little or no low temperature decomposition (Figure 3); this is most likely due to an induction time, since isothermal studies show a 30% decomposition in even the purest samples[8]. As shown in Figure 4, AP decomposing in a relatively stable aromatic hydrocarbon is also initially endothermic. This result tends to corroborate Jacobs' theory[10] that the limitation of the low temperature reaction is a concentration of adsorbed ammonia on the surface of the AP. Experimenters generally agree[11] that the initial reaction is

$$NH_4ClO_4 \rightleftarrows NH_3 + HClO_4$$

which is an endothermic process; hence the subsequent decomposition of perchloric acid accounts for the exotherm.

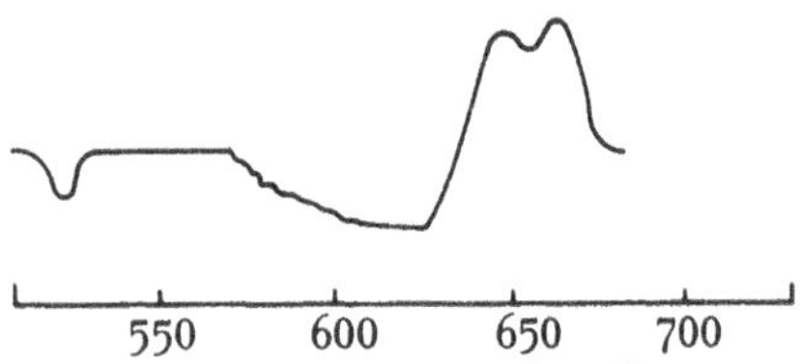

Figure 4. AP in a Polyphenyl Ether (Heating Rate 20 C/Minute)

The very high activation energy obtained for the high temperature decomposition (127 kcal/mole) from the DSC thermogram is not considered erroneous (Table II). TGA data give the more accepted value of 35 kcal/mole. It is pertinent here to note that the high temperature reaction involves several processes: the proton transfer reaction already cited; the decomposition of perchloric acid; and the reaction of ammonia and perchloric acid. The method of measurement determines what combination of these processes gives the effective activation energy. Although it has not as yet been rationalized which apparent activation energy will be most representative of solid propellant combustion, it appears that values will be somewhat higher than 35 kcal/mole.

TABLE II

ACTIVATION ENERGIES BY METHOD OF ROGERS AND MORRIS

Ammonium Perchlorate (Commercial)		
$\varphi^{(a)}$	$E_1^{(b)}$	$E_2^{(b)}$
5	28	--
10	30	--
20	--	127

Ammonium Perchlorate (Purified)		
	E	Remarks
20	118	Only One Exotherm
20	116	

(a) Heating rate in deg C/minute
(b) Activation energy in kcal/mole
E_1 - first exotherm;
E_2 - second exotherm

Propellant

Thermograms of propellant using small pieces (≤ 3 mg) obtained on the DSC look similar to those obtained by DTA apparatus. The special technique described in the experimental procedure section of this paper givesquite different results. This difference can be seen by comparing thermograms B and C in Figure 5. These thermograms were made from samples of the same propellant; C, from a small square piece, B, from a thin disc. The propellant contained a bimodal oxidizer comprising about 30% 15-micron AP and the balance 200-micron AP. From the figure, it is obvious that burning occurs over a broader temperature range for the disc sample, which has a geometry that allows physical effects of structure to enter into the kinetics of the heterogeneous combustion. Thermogram A in Figure 5 was made from a propellant containing only 15-micron AP. Here, the presence of the extremely high exotherm at a lower temperature is most pertinent to the construction of a combustion model. The study of this phenomenon is continuing.

Discussion

At present one of the most controversial issues in solid propellant combustion is whether control of steady-state burning exists in the gas phase or in the solid phase. If one considers only the thermal energy, it is easy to prove that the solid phase reactions contribute little. As a result of the current investigation it was concluded that thermal energy is not a good criterion to use. If we draw an analogy to liquid propellants, we must admit that control resides in the valves that release the oxidizer and fuel. Solid propellants are probably much the same, i.e., the combustion that established itself in the gas phase is dependent on what the solid phase provides in the form of oxidizer, fuel and diluent gas. It is well known in solid propellant technology that in some instances large changes, as much as 500 degrees (K), in flame temperature alter the burning rate of the solid phase very little. Conversely, small changes in the solid phase temperature and structure (most particularly oxidizer particle size) provide drastic differences in burning rate. With these and other considerations in mind, we have proceeded to relate solid phase reactions as determined by DSC to steady-state combustion of composite solid propellants.

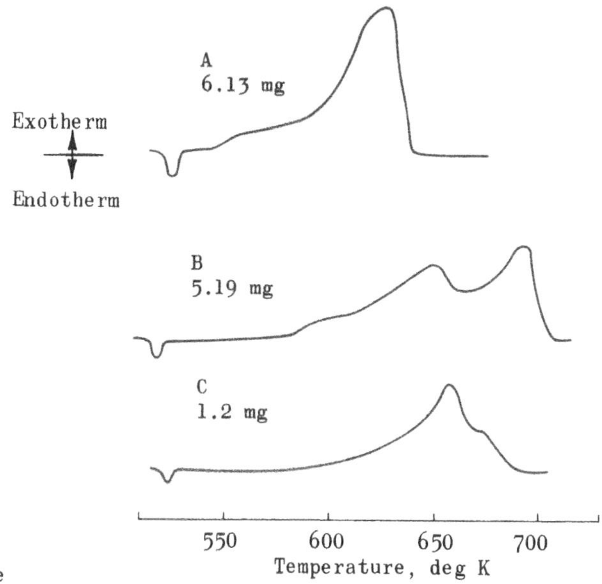

Figure 5. Propellant Thermograms (Heating Rate 20 C/ Minute)

The combustion model suggested is based on the data obtained with the DSC in combination with the results of others in the field using different techniques. It is certainly true that if the surface temperature of propellant undergoing steady-state burning is in the generally agreed on range of 725--1000 K, the thermal gradient existing, however steep, provides a temperature of 600 K at some depth within the solid phase. At this temperature both AP and binder start decomposing. This provides us with a liquid binder that boils with decomposition product gases and with a decomposition of the fine oxidizer that provides perchloric acid and ammonia. There then exists at some depth in the solid phase a mixing of fuel and oxidizer, which gives heat that increases the depth of subsurface reaction. Perspective in this model can be obtained by considering Figure 6. Mixing and boiling would be in the interstices between the large (200-micron) AP particles. The fine AP (10-micron) would be primarily responsible for the oxidizer gases or liquids in the interstices because of the large surface area of the fine AP. It is easy to see in this microscopic perspective that if only 200-micron AP were present the model would change drastically. Oxidizer gases and liquids could come only from the interfaces of the fuel and the large AP, allowing a diffusional control to heat production in the interstices. This type of reasoning can quite logically explain the effect of fine AP on burning rate.

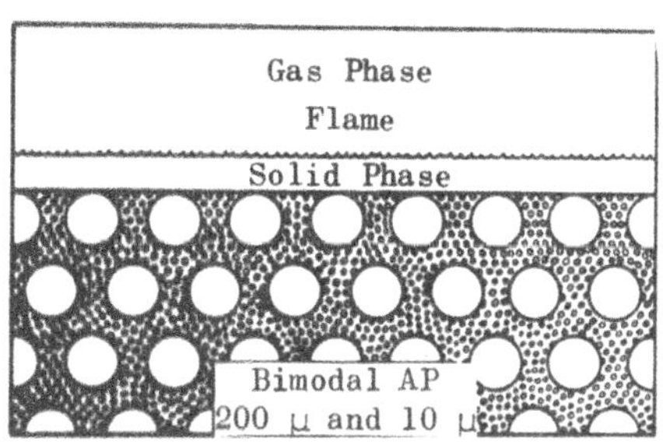

Figure 6. Scale Drawing of Propellant with 86% Total Solids

A study of ammonium nitrate propellant conducted in our laboratories provides an example of an application of this theory. While investigating new propellants for a turbine starter, the need for composites with a lower temperature sensitivity of burning rate and a higher burning rate with less catalyst was stressed. The theory of solid phase control of ballistic parameters was applied to this problem, using a newly developed technique designated the "Expanded Profile Concept." With this method a thermogram of the propellant of interest is run and the reactions observed are considered to represent subsurface reactions. A simple, laminar, one-dimensional model, such as that shown in Figure 7, is then constructed. Although this is a somewhat naive model for heterogeneous combustion, it provides a basis for rationalizing what type of compounds need to be added to the propellant to alter this profile. We noted from the model constructed in Figure 7 that the high temperature reaction (500 K, 227 C) occurs below the accepted surface temperature (580 K, 307 C) of AN propellants. It was then reasoned that altering this major reaction zone would most likely alter subsurface effects on temperature sensitivity of burning rate. Using this rather loose premise, compounds were selected that were either endothermic or exothermic in the temperature range of 235 C to 300 C. A list of compounds selected is shown in Table III. Propellants containing varying amounts of these compounds were then formulated. Thermograms were then run to ascertain the effect. Figure 8 shows the change noted in one instance. Ballistic testing of small propellant charges containing selected quantities of the additives was conducted. The results, shown in Table IV, must be considered far more than coincidental, lending direct support to the selection principle. We, therefore, concluded that the DSC can be used as a tool for the selection of additives for the ballistic variation of solid propellants using the "Expanded Profile Concept." This technique is admittedly very new; further, it requires considerable additional experimental work for verification and improvement.

The investigation discussed above is continuing with emphasis on ammonium perchlorate composite propellants. Studies are being made by DSC and TGA. Work with the DSC is being extended to greater-than-atmospheric pressures.

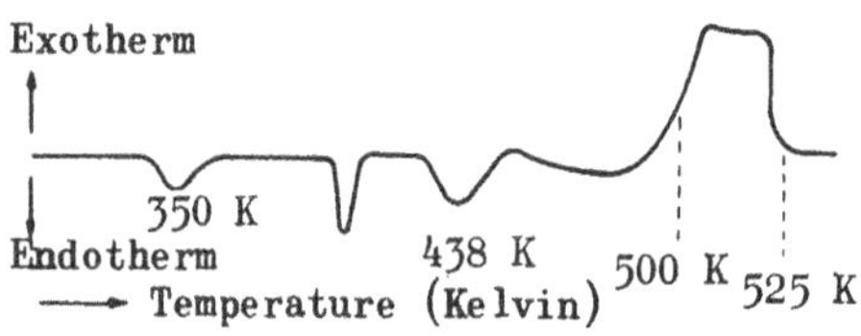

THERMOGRAM OF AMMONIUM NITRATE PROPELLANT

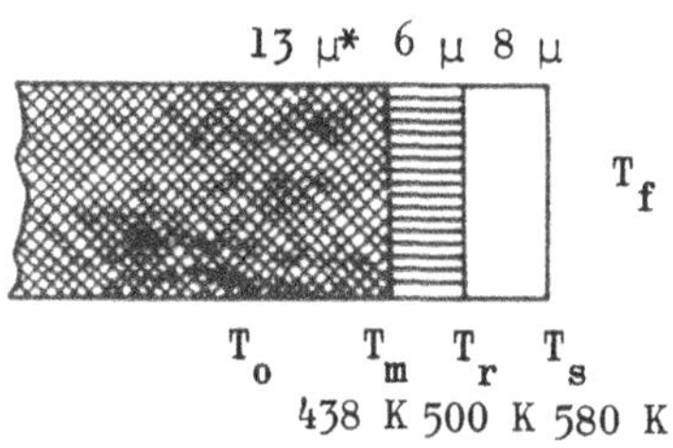

Legend:

T_f - flame temperature

T_s - average surface temperature

T_r - temperature at which fuel and oxidizer begin to decompose or react

T_m - melting point of AN

T_o - original temperature of propellant

*Distance in μ calculated without considering reaction exotherm; thus they represent a minimum

Figure 7. Expanded Profile Concept

TABLE III

ADDITIVES SELECTED BY EXPANDED PROFILE CONCEPT

Compound	Melting Point
A	239
B	240 d (exothermic)
C	251-3
D	236-238
E	344-346
F	293-295
G	246-247
H	273

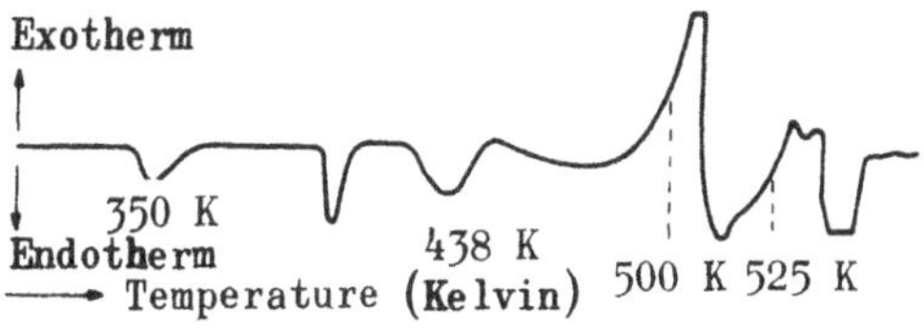

Figure 8. Effect of Compound A in Ammonium Nitrate Propellant

TABLE IV

RESULT OF EVALUATING PROPELLANT CONTAINING SELECTED ADDITIVES

Compound	Results
A	Zero π_p from 77 to 170 F
B	Increased burning rate Much lower π_p from -65 to 170 F
C	Lower or zero π_p indicated
D	Much lower π_p 77 to 170 F
E	Increased π_p
F	Very little effect
G	Very little effect
H	Much lower or zero π_p indicated
π_p = Temperature sensitivity of burning rate	

We are indebted to the National Aeronautics and Space Administration, Marshall Space Flight Center, for support of the ammonium nitrate work presented in this paper, which was done under Contract NAS-9-18.

References:

(1) E. S. Watson, M. J. O'Neill, J. Justin and N. Brenner, Anal. Chem. 36, 1233 (1964).

(2) M. J. O'Neill, Anal. Chem. 36, 1238 (1964).

(3) H. E. Kissinger, Anal. Chem. 29, 1702 (1957).

(4) H. J. Borchardt and F. Daniels, J. Amer. Chem. Soc. 79, 41 (1957).

(5) R. N. Rogers and E. D. Morris, Jr., Anal. Chem. 38, 412 (1966).

(6) Takeo Ozawa, Bul. Chem. Soc. Japan. 38, 1881 (1965)

(7) B. Carroll and E. P. Manche, J. Appl. Polymer Sci. 9, 1895 (1965)

(8) J. D. Hightower and E. W. Price, ICRPG 2nd Combustion Conf. CPIA Publication No. 105. Vol I, 421 (1966).

(9) P. W. M. Jacobs and A. Russell-Jones, A.I.A.A. Journal 5, 829 (1967).

(10) P. W. M. Jacobs and A. Russell-Jones, "Eleventh Symposium (International) on Combustion," The Combustion Institute, Pittsburgh, Pennsylvania, 1967, p 457.

ANALYSIS OF HIGHLY EXOTHERMIC REACTIONS BY DSC

by
A. A. Duswalt
Research Center
Hercules Incorporated, Wilmington, Delaware

In the handling of highly energetic compounds, it is important from a safety standpoint to be knowledgeable about their thermal instabilities. It is sometimes possible to gain information about the sensitivity of a compound by simple inspection of its DTA (or DSC) exothermic decomposition curve. One can, for example, compare the "starting" and maximum peak temperatures of two similar materials, for one of which the stability is known under the conditions in question. This is an admittedly crude test but could be adequate for detecting gross differences, changes during storage, effects of solvent, etc. A more quantitative but still simple test would be to heat a portion of a sample isothermally for a period of time at an appropriate temperature, and then to decompose it, and compare the size of its decomposition exotherm with that of a similar but untreated sample portion. The difference between the two exotherms is a measure of the rate of decomposition of the sample at the isothermal heating temperature. We have used this technique successfully on nitrocellulose formulations as a substitute for the German Test (1).

Occasionally insight into the nature of an exothermic reaction is obtainable from the shape of the decomposition peak. An example of this is the exotherm, shown in Figure 1 for diazo dinitrophenol. One can say that the reaction occurs in at least two steps. The abrupt, steep slope of the first step would normally suggest a zero order or autocatalytic reaction with a high activation energy. In this case it is due to melting of the compound to form a liquid which is more thermally unstable than the solid. A faster program rate will actually show a small melt endotherm just prior to the first decomposition (dotted curve in Figure 1). Although the total energy released by the first step reaction is relatively minor, its suddenness and proximity to the second, larger exotherm, will cause a detonation once a temperature of about 155 to 157°C. is reached.

Critical temperature ranges are not normally easy to define, however, as is illustrated by a decomposition curve for methyl parathion in Figure 2. Two exotherms are observed, a minor one peaking at approximately 210°C. and a major one at about 290°C. The exotherms appear to be completely separated under the conditions of decomposition, i.e., small sample size and slow, 5°C. per minute, heating rate. A chemist concerned with the safe handling of such a material would need to know under what conditions the major reaction could cause a detonation and also, if it could be triggered by the first reaction. Unfortunately a simple inspection of the decomposition curve will not supply this information. A more extensive analysis of the data from the curve is required and is based upon the following fundamental considerations. If the temperature of an unstable material is allowed to reach the point where the decomposition produces slightly more heat than can escape through the sample to the environment, the material will self heat. If the material has a high heat of decomposition, the temperature and reaction rate will increase automatically to the point of explosion. The critical factors determining whether this increase will occur and at what rate are: the amount of heat

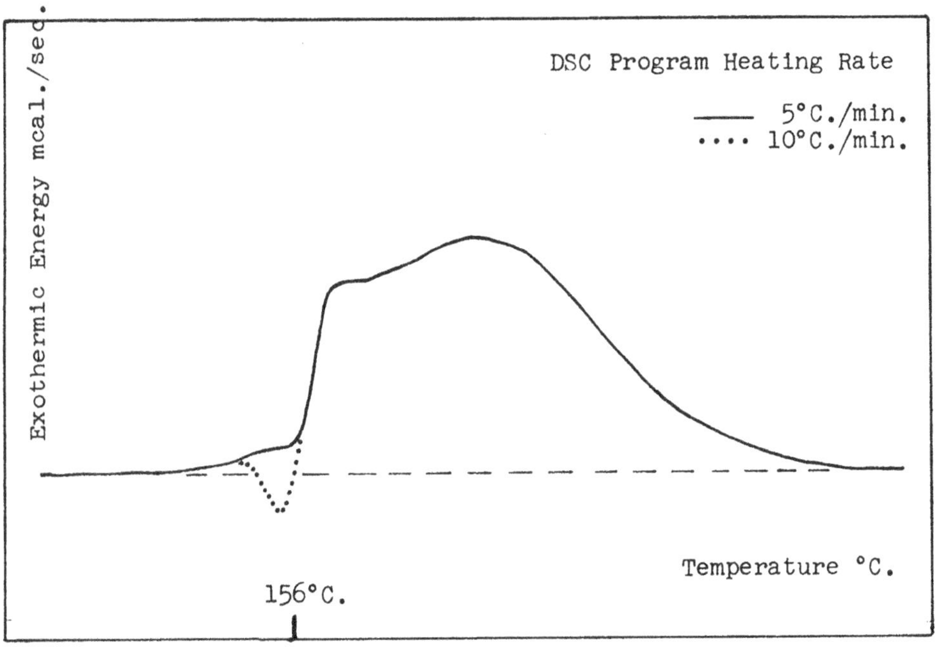

Figure 1. Diazodinitrophenol Decomposition Thermogram

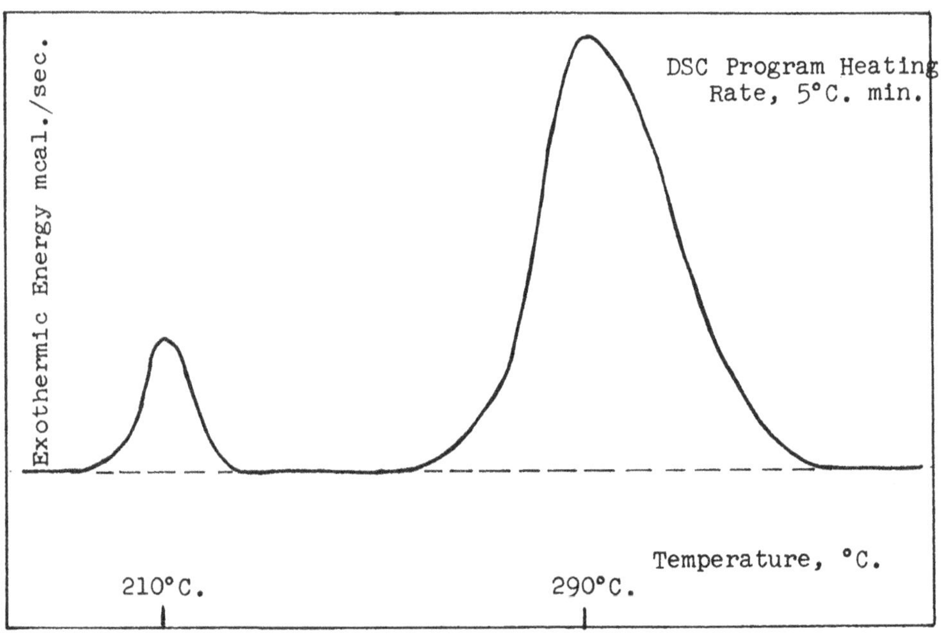

Figure 2. Methyl Parathion Decomposition Thermogram

generated by the reaction at a temperature T, the amount of heat that can escape from the reaction center, the heat capacity, which determines the temperature rise per calorie generated, and the kinetics, which determine the increase in rate of reaction per degree of temperature rise. These factors are included in a model proposed by Frank-Kamenetskii (2) to describe the behavior of unstable materials at a given ambient temperature as follows:

$$\partial = \left[\frac{\Delta H_d\ A\ Co\ E\ r_o}{K\ R\ T_o^2}\right] \left[e^{-E/RT_o}\right]$$

H_d = Heat of Decomposition
A = Frequence Factor
Co = Initial Concentration or Density
E = Activation Energy
r_o = Effective Bulk Radius (Semi-Thickness)
T_o = Temperature of Environment
K = Bulk Thermal Conductivity

∂ is a dimensionless parameter whose critical value depends upon the spatial configuration of the sample. For example, should ∂ exceed values of 0.88, 2.00 or 3.32 for the respective slab, cylinder or spherical configurations, a runaway condition for the reaction is predicted. An induction period, τ, for this condition may be calculated from the relation

$$\tau = \frac{C_p\ d\ RT_o^2}{\Delta H_d\ EAC_o\ e^{-E/RT_o}},$$

where C_p is the heat capacity and d is the bulk density.

In the specific case for methyl parathion, if ΔH_d, E, A, and K can be measured or calculated for the major exotherm, then the critical temperature for a runaway reaction can be determined for given physical conditions of sample bulk size, shape and concentration. If similar considerations show that the smaller, initial reaction can also approach a runaway condition at some lower temperature, then, from this heat of decomposition and the heat capacity of the sample, one can calculate whether the initial decomposition step can raise the sample temperature sufficiently to exceed the critical value for the major reaction. If it does, then the maximum safe temperature would be below the critical temperature of the smaller reaction.

The Perkin-Elmer Differential Scanning Calorimeter has been used to determine heats of decomposition, heat capacities and reaction kinetics. The methods we have used to determine kinetic data are described below.

Isothermal Method. This method measures directly the extent of reaction as a function of time and temperature. A number of samples are held in the DSC at a set temperature but each for a different period of time. The samples are then temperature programmed through their decomposition range and their reaction exotherms recorded. The decrease in the exothermic peak area per unit sample weight as a function of previous isothermal exposure time is a measure of the reaction rate at that set temperature. From this relation a specific reaction rate constant k can be calculated. For example, a first order decomposition would yield a straight line whose slope equals 2.3 k, for a plot of -log c/C_o vs. time. If the procedure is repeated for different temperatures, the slope of a plot of

2.3 R log k vs. 1/T yields the activation energy, E. From k at a given temperature and E the frequency factor, A, can be calculated. This method is the only one described which allows us to observe slow decomposition at relatively low temperature.

Direct Heat of Reaction Method. This has been described previously by M. Uricheck (3). The specific rate constant is measured directly in terms of the rate of heat output in calories per second, of the unreacted fraction, at a given temperature. As with the isothermal method, specific rate constants are calculated for various temperatures and a plot of log k vs. 1/T gives the activation energy and subsequently the frequency factor. In general, the method is rapid and with a well shaped curve, capable of good precision. For relatively uncomplicated, first order reaction of nonvolatile compounds, we prefer this method.

Heat of Reaction Ratio Method. This method, proposed by Rogers and Morris (4) requires only measurement of the initial slope of the decomposition exotherm. A ratio of log displacemetns from the baseline at two temperatures, T_1 and T_2, is assumed to equal that of the reaction rate constants for the same temperature range. From the equation $\log k_1/K_2 = E/2.3R\left(\frac{1}{T_1} - \frac{1}{T_2}\right)$ the activation energy can be estimated. This method is especially valuable for studies involving relatively volatile samples. Such samples must be run in sealed containers which, unfortunately, usually burst before the decomposition is complete. The Rogers and Morris method can be applied where only the initial portion of a decomposition can be recorded. The method used in this way is especially vulnerable to baseline error and, in effect, assumes zero order kinetics.

Varied Program Rate Method. The last method to be described is the most generally applicable of the four for kinetic studies in the DSC. It is an adaptation of a thermal gravimetric technique described by Takeo Ozawa (5). By observing the temperature at which a constant degree of conversion occurs, as a function of heating rates, an activation energy is obtained whose calculation is independent of reaction order effects. Situations where activation energy varies can be detected and to some extent resolved by this method. For simpler decompositions, reaction order and pre-exponential factors can be obtained graphically. The basis of the method for the activation energy is the derived equation:

$$\frac{d \log \beta}{d\,(1/T)} \cong (0.457/R)\,E$$

β = programmed heating rate.

Determination of the reaction order and frequency factor involves the fitting of a calculated curve for the reaction exotherm against a series of theoretical curves determined for each possible reaction order. The frequency factor is determined from the degree of lateral shift between the calculated experimental and theoretical curves. The following table compares results obtained by the four methods on a sample of a sulfonyl azide dispersed on Cab-O-Sil.

Isothermal		Uricheck		Rogers & Morris	Variable Heating Rate	
E*	Log A**	E	Log A	E	E	Log A
38.4	23.5	38.6	23.7	38.9	40.0	22.1
		39.3	23.0	38.5	38.9	21.8
				38.6		

*Kcal./mol.
**Minutes^{-1}

Experimental Considerations. We have observed at least two incidents where the effects of sample bulk have affected results. In one case a nitrocellulose formulation in solution was allowed to dry in a DSC aluminum sample pan and formed essentially a doughnut shaped ring around the bottom of the pan. The "doughnut" reacted considerably faster than a flat film of similar weight (6 mg.) when both were held isothermally at the same temperature. In a second case a 4 microliter sample of 80% methyl parathion solution decomposed more rapidly and at a considerably lower temperature than more dilute samples temperature-programmed at the same rate. From these instances we conclude that it is possible for highly energetic samples to self heat even at these small sample sizes. To avoid this effect as much as possible we have used three techniques to dissipate reaction heat; (1) use flat films where possible, (2) dilute sample with inert solvent, (3) disperse sample on an inert support such as Cab-O-Sil. The use of low heating rates also minimizes heat transfer problems. Concentrated samples become less prone to self heating and diluted or dispersed samples follow the programmed temperature more closely.

Since decomposition curves tend to be broad, baseline positioning becomes more liable to error especially if there is curvature. A technique we have used is to lower the temperature after a decomposition and then let the instrument draw a baseline on a second run. This is satisfactory provided the decomposition has not left deposits on the sample holder that change its surface radiation characteristics. For the most precise results the baseline should be adjusted by blackening the appropriate pan cover until the baseline curvature is negligible over the range of interest.

We have found in general that obtaining good results is less dependent on the precision of the instrument than on operator technique.

References

(1) Military Standards, Propellants, Solid, Mil-Std. 286 B, July 1966, Method 404.1.2.

(2) Frank-Kamenetskii, D. A., "Diffusion and Heat Exchange in Chemical Kinetics", Princeton Univ. Press, 1955.

(3) M. Uricheck, Perkin-Elmer Instrument News, 17, No. 2 (1966).

(4) Rogers and Morris, "Analytical Chemistry", 38, 412 (1966).

(5) Takeo Ozawa, Bull. Chem. Soc. Japan, 38, 1881 (1965).

Index

GPSR Compliance

The European Union's (EU) General Product Safety Regulation (GPSR) is a set of rules that requires consumer products to be safe and our obligations to ensure this.

If you have any concerns about our products, you can contact us on ProductSafety@springernature.com

In case Publisher is established outside the EU, the EU authorized representative is:

Springer Nature Customer Service Center GmbH
Europaplatz 3
69115 Heidelberg, Germany

The manufacturer's authorised representative in the EU is Springer
Nature Customer Service Centre GmbH, Europaplatz 3, 69115 Heidelberg,
Germany. If you have any concerns regarding our products, please
contact ProductSafety@springernature.com

Printed and bound by CPI Group (UK) Ltd, Croydon, CR0 4YY
29/07/2026
02180528-0014